Inosine Monophosphate Dehydrogenase

ACS SYMPOSIUM SERIES **839**

Inosine Monophosphate Dehydrogenase

A Major Therapeutic Target

Krzysztof W. Pankiewicz, Editor
Pharmasset, Inc.

Barry M. Goldstein, Editor
University of Rochester Medical Center

American Chemical Society, Washington, DC

Library of Congress Cataloging-in-Publication Data

Inosine monophosphate dehydrogenase : a major therapeutic target / Krzysztof W. Pankiewicz, Barry M. Goldstein, editors.

    p. cm.—(ACS symposium series ; 839)

  Includes bibliographical references and index.

   ISBN 0–8412–3780–8

    1. IMP dehydrogenase—Congresses. 2. IMP dehydrogenase—Inhibitors—Congresses.

  I. Pankiewicz, Krzysztof W., 1942- II. Goldstein, Barry M., 1952-.III. Series.

QP603.15 I566   2003
615′.7—dc21   2002035645

The paper used in this publication meets the minimum requirements of American National Standard for Information Sciences—Permanence of Paper for Printed Library Materials, ANSI Z39.48–1984.

Distributed by Oxford University Press

The cover art shows the crystal structure of human type II IMP dehydrogenase complexed with cofactor NAD and substrate analogue 6-Cl IMP (*see* Chapter 7).

PRINTED IN THE UNITED STATES OF AMERICA

Foreword

The ACS Symposium Series was first published in 1974 to provide a mechanism for publishing symposia quickly in book form. The purpose of the series is to publish timely, comprehensive books developed from ACS sponsored symposia based on current scientific research. Occasionally, books are developed from symposia sponsored by other organizations when the topic is of keen interest to the chemistry audience.

Before agreeing to publish a book, the proposed table of contents is reviewed for appropriate and comprehensive coverage and for interest to the audience. Some papers may be excluded to better focus the book; others may be added to provide comprehensiveness. When appropriate, overview or introductory chapters are added. Drafts of chapters are peer-reviewed prior to final acceptance or rejection, and manuscripts are prepared in camera-ready format.

As a rule, only original research papers and original review papers are included in the volumes. Verbatim reproductions of previously published papers are not accepted.

ACS Books Department

Contents

Structure and Mechanism

Inhibitor Design and Clinical Applications

Indexes

Preface

"IMP Dehydrogenase, an enzyme linked with proliferation and malignancy" is the title of a classic 1975 paper in *Nature* (*256*, 331–333) by Jackson and Weber. In the 1950s, pioneering work by Carter, Cohen, Lieberman, Lagerkvist and others had identified the enzymatically catalyzed conversion of inosine 5′ monophosphate (IMP) to xanthosine 5′ mononophosphate (XMP). The subsequent two decades saw the enzyme ultimately identified as inosine monophosphate dehydrogenase (IMPDH, E.C. 1.2.1.14) recognized as having a central role in both purine nucleotide metabolism and neoplasia. Since that time, a remarkable growth in interest has occurred in IMPDH, with more than 600 manuscripts now dealing with the characterization, mechanism, and biological functions of IMPDH, its role as a target for anticancer, antiviral, antibacterial, and immunosuppressive therapy; and its inhibition by a variety of chemotherapeutic agents. The goal of this volume is to provide an overview of this rapidly expanding field, as well as an introduction to some of the recent directions in basic and applied IMPDH research.

IMPDH occupies a key position in purine nucleotide metabolism. As noted in virtually every manuscript dealing with the enzyme, IMPDH catalyzes the rate limiting step in the de novo synthesis of the guanylyl nucleotides: the conversion of IMP to XMP. Inhibition of the enzyme reduces the cellular pool of guanylyl nucleotides, resulting in arrest of cell proliferation, and affecting G-protein-mediated signal transduction. IMPDH inhibition down-regulates the *ras* oncogene, induces apoptosis, and stimulates cell differentiation. A functional relationship between the IMPDH gene and the p53 tumor suppressor gene has been established. A decade ago, IMPDH research was given additional impetus by the discovery that the human enzyme exists as two isoforms, type I and type II. The type I isoform is expressed constitutively in normal cells, whereas type II expression is up regulated in neoplastic cells and activated lymphocytes. Thus, type II IMPDH has become a major target for the development of anticancer and immunosuppressive drugs. A mouse model with selective deficiency of the IMPDH type II gene has been established, and crystal structures of the human type I and type II enzymes are available (see cover), as are those from hamster, *Streptococcus pyogenes, Borrelia burgdorferi,* and *Tritrichomonas foetus.*

These findings have been applied in the development and clinical application of antiviral, antitumor, and immunosuppressant agents. Three inhibitors of IMPDH are currently on the market. Ribavirin (Virazole, Rebetol, ICN Pharmaceuticals), in combination with interferon-α has been approved for treatment of hepatitis C virus infections. Mycophenolic mofetil (Cell-Cept, Roche) is now routinely used as an immunosuppressant in the United States, and mizoribine (Bredinin, Asahi Kasei) has found a similar application in Japan. The antitumor agent tiazofurin has shown considerable success in the treatment of patients with end-stage acute nonlymphocytic leukemia or in myeloblastic crisis of chronic myeloid leukemia. This agent continues in phase II studies, where monitoring of tumor cell biochemical markers helps define dosage and identify resistance. Major pharmaceutical companies continue vigorous programs of IMPDH-targeted drug design, including Vertex Pharmaceuticals, Bristol-Myers Squibb, ICN Pharmaceuticals and Pharmasset, all of whom have new IMPDH inhibitors in preclinical studies. IMPDH enzymes from microbial pathogens are also beginning to be targeted, including those from *Haemophilus influenzae, Mycobacterium tuberculosis, Pneumocystis carinii,* and *Borrelia burgdorferi.*

Despite unquestionable progress in IMPDH research, as of 2000, no attempts had been made to integrate results from the medicinal chemists, biochemists, biologists, clinicians, and members of the pharmaceutical industry interested in the role of this enzyme as a drug target. To fill this gap, we organized the first national symposium on this topic, titled *Inosine Monophosphate Dehydrogenase (IMPDH): Perspectives on a Major Therapeutic Target,* held August 20–21, 2000 at the American Chemical Society (ACS) Annual Meeting in Washington D.C. The symposium covered the biology, structure, and mechanism of action of IMPDH, and the design and clinical applications of IMPDH inhibitors. Members of most major academic and industrial laboratories involved in IMPDH research participated in the symposium and contributed updated chapters to this book.

Organization of the book begins with the basic molecular biology and genetics of IMPDH. Here, the relationships between the control of purine metabolism and cell proliferation and differentiation are reviewed. The roles of the two IMPDH isoforms are outlined, and the effects of a selective deficiency of the IMPDH type II gene on both cell growth and lymphocyte stimulation in a mouse model is described. Lastly, the cloning and expressing of IMPDH from a potential microbial target, *Pneumocystis carinii,* is reported.

In the Structure and Mechanism section, crystal structures of mammalian and microbial IMPDH's are reviewed, differences in the mechanisms between these enzymes are outlined, and the requirements of IMPDH for cation activation are described. These chapters provide a foundation for the final section: Inhibitor Design and Clinical Applications. Here, antitumor nucleosides such as tiazofurin, its analogues, and benzamide riboside are described. In addition, preformed cofactor analogues in the form of metabolically stable bis(phosphonate) derivatives, including mycophenolic adenine dinucleotide (MAD) analogues, are discussed. In addition to mechanism-based approaches to IMPDH drug design, the screening of inhibitors generated by combinatorial chemistry applicable to any class of organism is also presented. Finally, the use of mizoribine as an immunosuppressant following renal transplantation is described as one example of the now routine use of IMPDH inhibitors in the clinic, and their limitations.

Significant progress has been made in understanding the roles of IMPDH in the control of cell proliferation and differentiation, and as a therapeutic target. New anticancer, antiviral, antibacterial, and immunosuppressive drugs based on IMPDH inhibition are on the horizon. The editors hope that this volume will serve as an introduction for newcomers to this exciting field, as well as a resource for experienced investigators.

Acknowledgments

The editors greatly appreciate the excellent contributions made by the authors of the chapters of this book. The majority of authors served as speakers at the first IMPDH Symposium held during the 220th ACS National Meeting, and their names and affiliations are provided under each chapter title. We also thank the additional participants in the Symposium, including D. L. Hollenbaugh (Bristol-Myers Squibb), J. A. Thomson (Vertex Pharmaceuticals), S. A. Raybuck (Vertex Pharmaceuticals), and D. G. Wright (Boston University Medical Center).

This meeting was sponsored by the ACS Division of Carbohydrate Chemistry, and we thank Z. J. Witczak, the Chairman of the Division, for his help and assistance. Also, we are grateful for the financial support of the pharmaceutical companies that showed interest in IMPDH research. Among these are Banyu Merck, Vertex Pharmaceuticals, ICN Pharmaceuticals, Bristol-Myers Squibb Company, and Pharmasset, Inc. We also appreciate the help of Dr. Steven E. Patterson of Pharmasset, and Kelly Dennis and Stacy VanDerWall in acquisitions and Margaret Brown in editing/production of the ACS Books Department for their extraordinary patience during the production of this volume.

Lastly, we remain indebted to our colleagues for their participation in the Symposium, for the preparation and reviewing of manuscripts, and for their invaluable comments.

Krzysztof W. Pankiewicz
Pharmasset, Inc.
1860 Montreal Road
Tucker, GA 30084
Phone: 678–395–0027
Fax: 678–395–0030
(Email) kpankiewicz@pharmasset.com

Barry M. Goldstein
Department of Biochemistry and Biophysics
University of Rochester Medical Center
601 Elmwood Avenue
Rochester, NY 14642
Phone: 716–275–5095
Fax: 716–275–6007
(Email) barry.goldstein@rochester.edu

Inosine Monophosphate Dehydrogenase

Chapter 1

Inosine Monophosphate Dehydrogenase and Its Inhibitors: An Overview

Krzysztof W. Pankiewicz[1] and Barry M. Goldstein[2]

[1]Pharmasset, Inc., 1860 Montreal Road, Tucker, GA 30084
(email: kpankiewicz@pharmasset.com)
[2]Department of Biochemistry and Biophysics, University of Rochester
Medical Center, 901 Elmwood Avenue, Rochester, NY 14642
(email: barry_goldstein@urmc.rochester.edu)

Introduction

Inosine 5′-monophosphate dehydrogenase (IMPDH, E.C.1.1.1.205), the NAD-dependent enzyme that controls *de novo* synthesis of purine nucleotides, catalyzes the oxidation of inosine 5′-monophosphate (IMP) to xanthosine 5′-monophosphate (XMP), which is then converted to guanosine 5′-monophosphate (GMP) by GMP synthase (Fig. 1). IMP also serves as a substrate for the biosynthesis of adenosine 5′-monophosphate (AMP). An adequate pool of purine nucleotides is essential for cell proliferation, cell signaling, and as an energy source. Consequently, inhibition of IMPDH causes a variety of biological responses, and it is not surprising that this enzyme has emerged as a major target for antiviral, antileukemic and immunosuppressive therapies.

The inhibition of IMPDH is accompanied by a reduction in guanine nucleotide pools. This results in an interruption of DNA and RNA synthesis *(1)*, a decline in intracellular signaling *(2-6)*, and down-regulation of c-myc and Ki-ras oncogenes *in vitro* *(7-9)* and in leukemic cells of patients treated with inhibitor *(9)*. IMPDH inhibition results in apoptosis in both neoplastic cell lines and activated T lymphocytes *(7, 10)* and induces differentiation in myeloid and erythroid cell lines, as well as chronic granulocytic leukemic cells after blast transformation *(11)*. The links between the enzymatic regulation of guanine nucleotide production and neoplastic transformation are only beginning to be understood *(12)* and are discussed further in the chapter by Shirley. The central role of IMPDH in this process has long been advocated by Weber, and is reviewed in chapter 2.

This potent combination of both antiproliferative and differentiation-inducing effects has made IMPDH inhibitors attractive candidates as antileukemic agents. Inhibition of IMPDH also leads to suppression of both T and B lymphocyte proliferation, as growth and differentiation of human lymphocytes are highly dependent upon the IMPDH-catalyzed *de novo* pathway for purine nucleotide synthesis *(10, 13, 14)*. Thus IMPDH inhibitors are also employed as immunosuppressive agents. Inhibition of IMPDH can also lead to antiviral activity, although the origins of this effect are less clear. While there is an increased demand for purine nucleotides for RNA and DNA synthesis in virus-infected cells, not all IMPDH inhibitors have antiviral properties. Ribavirin, the IMPDH inhibitor that is most commonly used as an antiviral agent, may induce error-prone replication via incorporation of ribavirin triphosphate. However even this mechanism appears to be dependent on adequate guanine nucleotide levels *(15)*.

The role of IMPDH as a chemotherapeutic target was further advanced by the discovery, in 1990 by Natsumeda *et al*, that the enzyme exists as two isoforms, labeled type I and type II *(16)* (see the chapter by Natsumeda). The isoforms are of identical size and share 84% sequence identity. However, the type I "housekeeping" isoform is constitutively expressed in both normal and neoplastic cells *(17, 18)*, while type II expression is preferentially up-regulated in human neoplastic cell lines *(16, 19)* and in leukemic cells from patients with chronic granulocytic, lymphocytic and acute myeloid leukemias *(17)*. Conversely, type II expression is down regulated in HL-60 cells induced to differentiate *(18)*, or in patient leukemic cells treated *in vitro (17)*. The disproportionate increase in type II IMPDH activity in neoplastic cells has made this isoform a key target for specific antileukemic chemotherapy *(9, 14)*.

The type II isoform is also a target for immune suppression. The role of this isoform in immunosuppression has been elegantly elucidated by a series of mice knock-out models described in the chapter by Gu *et al*. While the type II enzyme is the major isoform in normal human T lymphocytes, these cells appear to induce both type I and type II enzymes when stimulated by mitogen *(20, 21)*. However, mice deficient in IMPDH type II and the salvage pathway enzyme hypoxanthine-guanine phosphoribosyl-transferase (HPRT) have decreased lymphocyte response to antigen. Thus, development of an isoform-specific immunosuppressant with an improved therapeutic ratio remains an actively sought goal (below). Further, IMPDH from bacterial, parasitic and mammalian sources differs significantly. These differences may also be exploited in the development of specific antibacterial or antiparasitic agents. The therapeutic potential of IMPDH has been recently reviewed by Saunders and Rybuck *(22)* and inhibitors of the enzyme have been described by Pankiewicz *(23, 24)*.

Figure 1

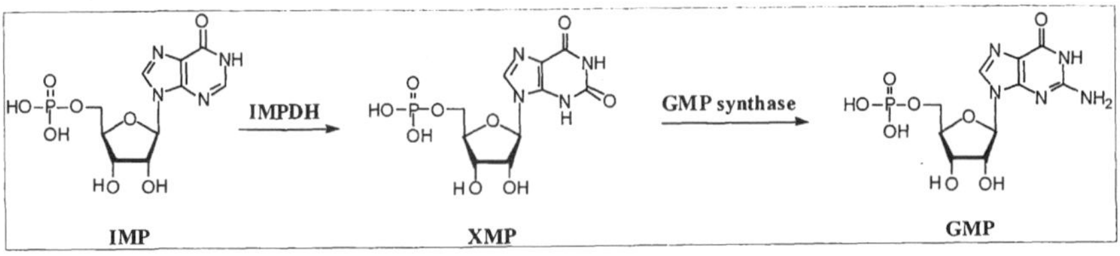

IMP   XMP   GMP

Mechanism of catalysis

The chemical mechanism of IMPDH catalysis has been studied in detail *(25)* (see the chapters by Markham and by Hedstrom and Digits). The IMPDH reaction involves an active site cysteine (Cys-331 in the human IMPDH numbering scheme). Nucleophilic addition of the cysteine thiol at the C2 of the IMP base forms a tetrahedral E-IMP adduct (Fig. 2). Subsequently, the C2-hydride is transferred to the *pro-S* face of NAD to give (after NADH release) an intermediate thioimidate covalently bound to the enzyme, E-XMP*.

Figure 2

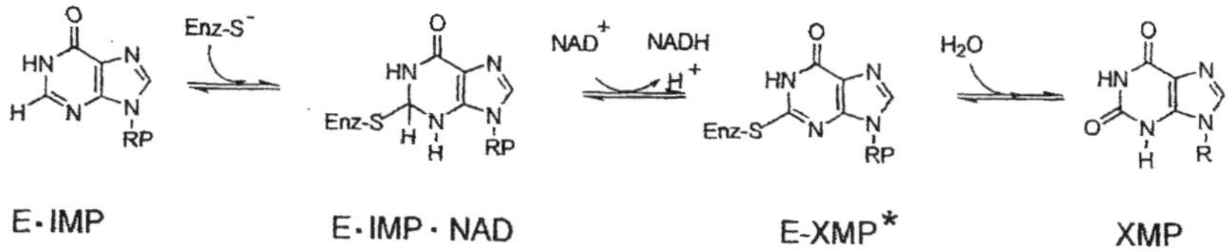

E·IMP   E·IMP·NAD   E-XMP*   XMP

Finally, water attacks at C2 to regenerate free Cys-331, and the product XMP is released. The final steps of the reaction are rate limiting for the human enzyme. Addition of substrate and cofactor have traditionally been thought to follow an ordered bi-bi mechanism, but more recent findings suggest that addition of substrate and cofactor is random *(26, 27)*.

The catalytic Cys-331 residue can also form a covalent adduct with some IMP analogues designed as mechanism-based inhibitors. Among these, monophosphates of 6-chloropurine riboside and EICAR (below) are irreversible inactivators of IMPDH. Interestingly, some analogues of NAD in which adenine is replaced by xanthine or guanine can also be used (although less efficiently) as cofactors by IMPDH *(28)*. No divalent metal cofactors are associated with IMPDH, although the enzyme is activated by K^+ and other monovalent cations. Despite identification of a K+ binding site in hamster IMPDH, the mechanism of monovalent cation activation in IMPDH remains unknown. This is discussed further in the chapter by Markham.

Crystal structures of IMPDH

Within the past several years, a wealth of crystallographic data has become available for IMPDH from a variety of sources. These structures are reviewed in the chapter by Goldstein *et al.* However, it is worth mentioning here some of the major findings from these studies. Structures have been solved for the hamster and human type I and type II enzymes, *(29, 30)* as well as for the enzymes from the pathogens

Streptococcus pyogenes (31), *Borrelia burgdorferi (32)* and *Tritrichomonas foetus (33, 34)*. Binary and ternary complexes have been obtained with a number of ligands, including the substrate IMP, substrate analogues 6-Cl IMP and ribavirin 5'-monophosphate, the cofactor NAD and NAD analogues SAD, β-TAD and C2-MAD, and the chemotherapeutic agent mycophenolic acid. A number of general principles have emerged.

All IMPDH structures show a tetrameric organization. With the exception of the *B. burgdorferi* enzyme, each 55kDa monomer is composed of two domains. A larger catalytic domain (394 residues in the human enzyme) consists of an eight-stranded α/β barrel. As is usually the case in α/β barrel enzymes, the active site is located on the C-terminal face of the barrel. A potassium binding site has been identified in the hamster and *T. foetus* structures adjacent to the active site at the monomer-monomer interface *(30, 34)*. In IMPDH, a smaller flanking domain (120 residues in the human enzyme) of as yet unknown function lies adjacent to the catalytic domain. This domain is not required for catalytic activity, and has as its closest homologue cystathionine beta synthase.

The catalytic mechanism outlined above is reflected in the geometry of both substrate and cofactor binding. The IMPDH active site consists of a long cleft in the barrel face located close to the monomer-monomer interface. The cleft has IMP and NAD-binding subregions. Binding of IMP places the inosine base close to the catalytic residue Cys 331. This residue forms part of a mobile loop that covers a portion of the IMP binding subsite. This geometry allows formation of the covalent adduct between the Cys 331 thiol group and the C2 carbon on the inosine base. This adduct has now been directly observed in several complexes *(30, 35)*.

The NAD cofactor binds in the continuation of the active site cleft adjacent to the IMP subsite. In this position, the NAD nicotinamide ring stacks with its B-face against the inosine base, insuring efficient stereochemical hydride transfer between the two moieties. The nicotinamide phosphate and adenosine moieties lie in the remainder of the active site cleft. In the human and hamster enzymes, the adenosine moiety forms interactions with residues from the adjacent monomer *(29, 30)* .

Both the substrate and cofactor are anchored in the active site by a series of polar interactions involving conserved residues on the walls of the active site cleft and residues on the active site loop. In addition, a highly mobile 50-residue flap covers portions of both the IMP and NAD sites, and also forms interactions with the substrate and cofactor. This flap may be involved in stabilizing the IMP-NAD complex. Unfortunately, the segment of the flap adjacent to the cofactor is disordered to varying degrees in all structures to date.

Inhibitors of IMPDH can generally be divided into two categories: those that bind at the substrate site, and those that bind at the cofactor site. Examples of the former include the 5'-monophosphate anabolites of ribavirin and mizoribine. Examples of cofactor-site binders include mycophenolic acid and TAD, the dinucleotide anabolite of the antitumor drug tiazofurin. These and other agents are discussed below and in chapters 11-15.

The available IMPDH structural data has provided useful information about overall molecular architecture, enzyme mechanism and ligand and inhibitor binding. Nevertheless, structure-based design of an isoform-specific agent remains hampered by the 84% identity between the type I and type II enzymes. Of the 514 residues in human IMPDH, 84 differ between the two isoforms. Of these, ~25% are conservative substitutions, and only a handful of the remaining substitutions are found in the active site itself. Interestingly, several of these occur in the active site flap, and three of the four residues that interact with the cofactor adenine moiety also differ between the type I and II isoforms. Strategies for exploiting these interactions are discussed in the chapter by Pankiewicz *et al.*

Inhibitors of human IMPDH in clinic

Three inhibitors of IMPDH (Fig. 3) are currently in use in the clinic: 1) ribavirin (Virazole, Rebetol) is used as a broad spectrum antiviral drug, and has recently been approved in combination with interferon for treatment of hepatitis C, 2) mizoribine (Bredinin), in use as an immunosuppressant in Japan, and 3) mycophenolic mofetil (CellCept, a prodrug form of mycophenolic acid) is in use as an immunosuppressant in the United States. None of these agents shows significant selectivity against IMPDH type II.

Figure 3

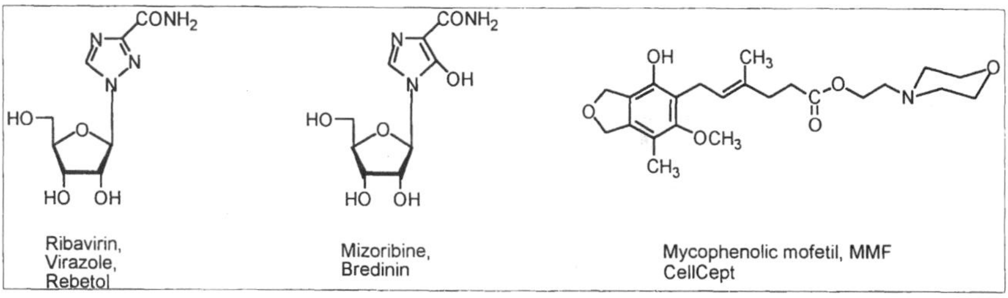

Ribavirin, Virazole, Rebetol

Mizoribine, Bredinin

Mycophenolic mofetil, MMF CellCept

In vivo, ribavirin is converted into its 5′-monophosphate (RMP) and binds at the substrate site of the enzyme. The crystal structure of a deletion mutant of human IMPDH type II with bound RMP has been published *(35)*, and, as expected, RMP closely mimics IMP binding. Given the similarity between ribavirin and IMP, it is perhaps not surprising that ribavirin displays broad biological activity and has several modes of action. It directly inhibits replication of many DNA and RNA viruses and can also act as an immunomodulator by augmentation of type 1 cytokine expression (IL-2), gamma interferon (IFN-γ), and tumor necrosis factor alpha (TNF-α). Recently, ICN Pharmaceuticals reported the synthesis of ICN 17261, the L-enantiomer of ribavirin *(36)*. In contrast to ribavirin, ICN 17261 was found to be inactive against a panel of DNA and RNA viruses. Interestingly, the L-enantiomer showed preserved immunomodulatory activity with a significantly improved toxicology profile.

Mizoribine (see the chapter by Ishikawa) was discovered in 1971 in Japan, where it is used for prevention of rejection following renal transplantation, and for treatment of lupus nephritis, rheumatoid arthritis and nephrotic syndrome. After phosphorylation, mizoribine 5'-monophosphate (MZMP) inhibits GMP synthesis by inhibiting both IMPDH and GMP-synthase. MZMP is a potent inhibitor of both type I and II isoforms of IMPDH (K_i = 4 nM and 8 nM respectively) and is thought to act as a transition state analogue. The formal negative charge located on the 5-oxygen of the MZMP aglycone ring mimics the negative charge which develops on N3 of IMP during the reaction with Cys-331 (25).

Mycophenolic mofetil (MMF, CellCept) is a prodrug form of mycophenolic acid (MPA, Fig. 4), another potent inhibitor of IMPDH. An extensive review by Bentley titled "A One Hundred Year Odyssey from Antibiotic to Immunosuppressant" has recently been published (37).

Mycophenolate mofetil is approved in the United States as an immuno-suppressant in the treatment of acute rejection in renal transplants (38, 39). Acute rejection is the most common cause of kidney loss, with over half of transplant patients experiencing an episode of rejection in the first year (38). Three multicenter studies have shown a 50-60% decline in the incidence of rejection with MMF (38, 39).

In vivo, mycophenolate mofetil is metabolized to mycophenolic acid (MPA, Fig. 4). MPA remains the NAD site binding inhibitor with the highest affinity and specificity for type II IMPDH. MPA is an uncompetitive inhibitor with respect to NAD, binding with Ki's of 33 and 7 nM to the type I and II isoforms respectively (14) .

The immunosuppressant activity of MPA is linked directly to its ability to reduce both T and B lymphocyte proliferation via inhibition of IMPDH (14). However, the efficacy of MPA is limited by its rapid conversion to the glucuronide via uridine 55' -diphosphophoglucuronyl transferase(14, 40, 41). MPA is metabolized into the inactive MPA-7-O-glucuronide (Fig. 4) and as much as 90% of the drug circulates in this form. Therefore, high doses (2-3 g a day) of MMF are needed in order to maintain therapeutic level of MPA. In addition, glucuronidation of MPA is likely responsible for the fact that the agent lacks any anticancer activity. Glucuronidation is vigorous in many cancer cells (42). Thus, this inactivating metabolism may prevent MPA from achieving a sufficient therapeutic concentration in neoplastic cells.

Figure 4

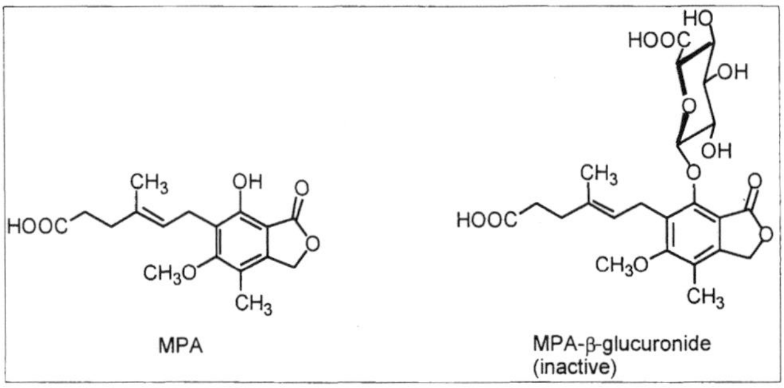

MPA

MPA-β-glucuronide
(inactive)

Glucuronidation of MPA also results in the drug's dose-limiting gastro-intestinal (GI) toxicity. The MPA glucuronide is removed from circulation by biliary excretion into the gut. Here, the compound is then deglucuronidated by bacterial enzymes, resulting in a high local concentration of MPA and subsequent GI toxicity *(43)*.

Inhibitors of human IMPDH under development

Clearly, new inhibitors of IMPDH with improved therapeutic profile are of interest, and several such compounds have been designed and prepared. For example, Vertex Pharmaceuticals undertook a *de novo* design program in which novel peptide mimics were constructed based on a hydrophobic core containing polar substituents. More than 150 compounds were synthesized and VX-497 (Fig. 5) was selected as the lead candidate. VX-497 binds at the nicotinamide end of the cofactor site, albeit in a novel manner (see the chapter by Goldstein *et al.*). Since it cannot be glucuronidated (due to lack of a phenolic group), its therapeutic potential should be improved. Indeed, it does show potent inhibition of IMPDH (K_i = 7-10 nM) and a broad spectrum of antiviral activity. VX-497 was tested in combination with INF-α in Phase II clinical trials against hepatitis. Unfortunately, no substantial improvement was found over interferon monotherapy *(44)*.

Figure 5

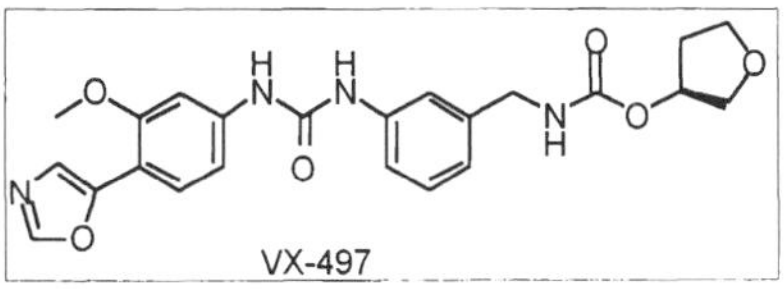

VX-497

Recently, Bristol-Myers Squibb disclosed a series of structurally related analogues of VX-497. Some of these compounds inhibit IMPDH in the nanomolar range. One of these analogues, BMS-337197 (Fig. 6), inhibited the antibody response to soluble antigen in BALB/c mice with a potency comparable to CellCept. These compounds were also equally potent in the induced arthritis model in rats *(45, 46)*.

Zeneca has developed a novel series of nonnucleoside inhibitors of IMPDH with immunosuppressive activity *(47)*. A high throughput screening of 80,000 compounds resulted in the discovery a single pyridazine derivative (**1**, Fig. 7) This agent

Figure 6

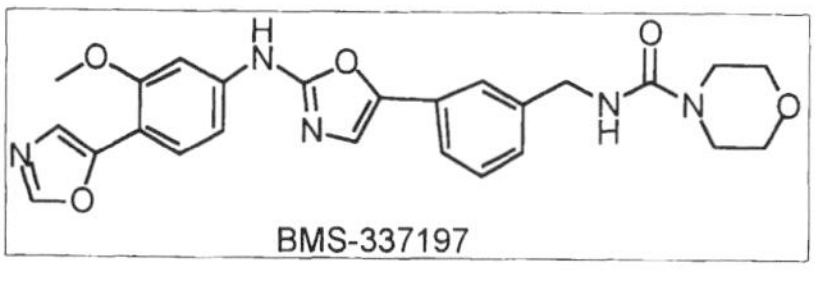

BMS-337197

functions via a mechanism similar to that of MPA, e.g. the compound traps the covalent intermediate E-XMP[*]. The lead compound was chemically modified to give other pyridazine derivatives, the most active being compound **2** ($K_i = 0.8$ μM). Compound **2** showed improved immunosuppressive activity over MPA in the inhibition of delayed-type hypersensitivity in mouse models.

A novel mechanism-based inhibitor of IMPDH, 5-ethynyl-1-β-D-ribofuranosylimidazole-4-carboxamide (EICAR, Fig. 8), shows potent cytostatic activity against a number of human solid tumors as well as a broad spectrum of antiviral activity *(48)*. This alkylating nucleoside is phosphorylated in cells, binds at the substrate site on IMPDH, and reacts irreversibly with the catalytic Cys-331. Recently, it has been discovered that the mode of action of this compound may be more complex than originally thought. Its broad biological activity is apparently due to the formation of several anabolites, including di- and triphosphates, as well as an NAD analogue (EAD, Fig. 8) *(48)*.

Tiazofurin (2-β-D-ribofuranosylthiazole-4-carboxamide, NSC 286193, Fig. 9) was discovered two decades ago and was found to produce potent and specific inhibition of IMPDH. Tiazofurin itself does not inhibit IMPDH, but undergoes a unique metabolic activation. In the cell, tiazofurin is phosphorylated by adenosine kinase and/or 5'-nucleotidase to the mononucleotide. This is then coupled with AMP by NMN adenylyltransferase, producing the active metabolite, tiazofurin adenine dinucleotide *(49, 50)* (TAD, Fig. 9). TAD is an analogue of NAD in which tiazofurin replaces the nicotinamide riboside moiety of the normal cofactor. TAD mimics NAD binding at the IMPDH active site, but the thiazole ring cannot participate in hydride transfer, resulting in potent inhibition IMPHD ($K_i = 0.1$ μM). TAD as a pyrophosphate is metabolically unstable, and it is degraded back to its components tiazofurin and adenosine by the

Figure 7

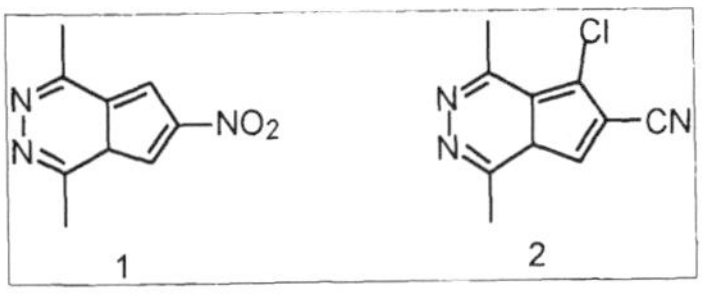

Figure 8

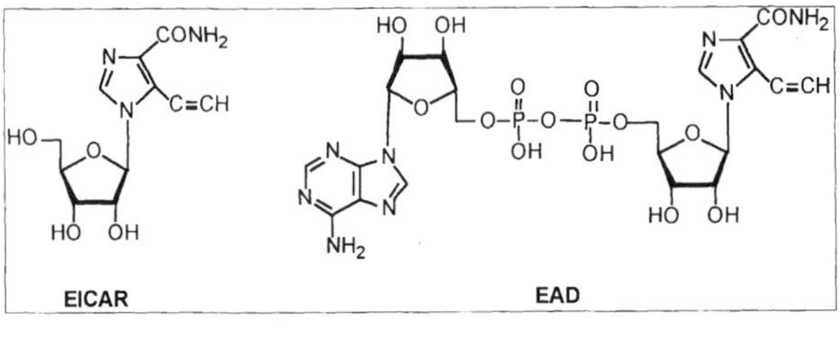

action of cellular phosphodiesterases and phosphatases. Resistance to tiazofurin is associated with impaired accumulation of TAD. It was found that TAD buildup was lower in resistant than sensitive tumors due to low activity of NMN adenylyltransferase in tumor cells. In addition, the activity of TADase, a specific phosphodiesterase which cleaves TAD, was higher in resistant cell lines *(50, 51)*.

It was reported by Tricot *et al.* *(52)* that tiazofurin induced complete hematologic remissions in patients with chronic myelogenous leukemia in blast crisis (CML-BC). Examination of sequential bone marrow biopsies did not show marrow

Figure 9

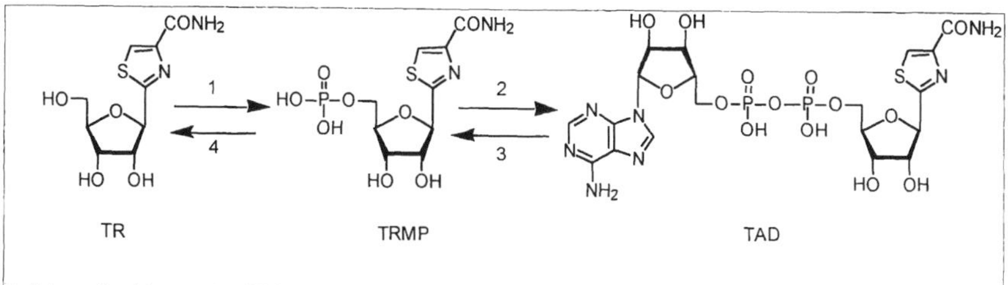

TR   TRMP   TAD

1. Adenosine kinase, 2. NMN - adenylyl transferase, 3. TADase, phosphodiesterase, 4. Phosphatases

hypoplasia, but revealed significant maturation of the myeloid elements in the marrow. The observation that the curative effect of tiazofurin was due to induction of cell differentiation in humans was confirmed by Wright *et al.* *(53)* in their Phase II clinical studies. The ability to induce erythroid cell differentiation is commonly studied *in vitro* in a human erythroleukemia K562 cell line derived from CML. Tiazofurin was the most effective inducer of erythroid differentiation in K562 cells among numerous compounds studied *(8)*. Recently, the Food and Drug Administration (FDA) has granted orphan-drug status to tiazofurin (Tiazole, ICN Pharmaceuticals) for treatment of CML-BC. Tiazole is now in Phase II/III clinical trials in several cancer centers in the US.

Figure 10

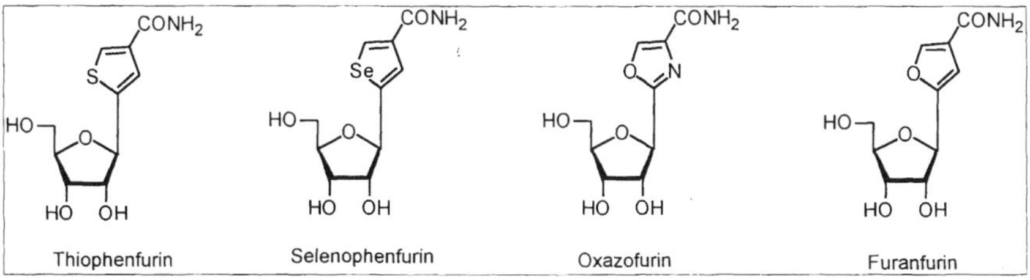

Thiophenfurin   Selenophenfurin   Oxazofurin   Furanfurin

Figure 11

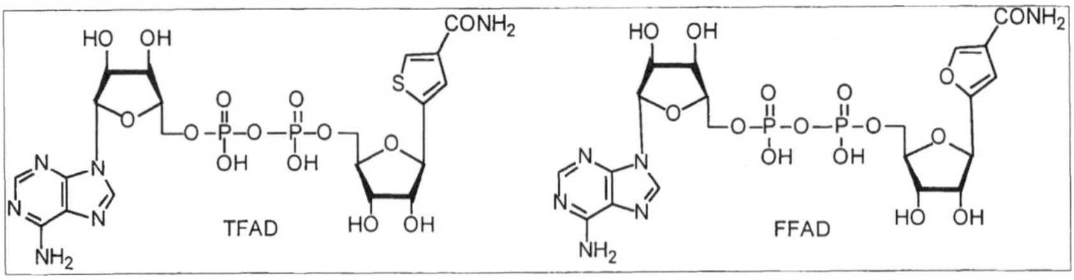

The above studies prompted the synthesis of a number of nicotinamide ribosideand tiazofurin mimics, in hopes that these agents would be similarly metabolized to the corresponding NAD analogues and act as IMPDH inhibitors. Thus, thiophenfurin, and selenophenfurin (Fig. 10) were found to be as potent as tiazofurin. In contrast, oxazofurin and furanfurin (Fig. 10) were inactive. Thiophenfurin was converted in cells into the corresponding dinucleotide analogue TFAD (Fig. 11) which was approximately as potent as TAD. Activity of selenophenfurin was similar. In contrast, furanfurin was not only poorly converted into its NAD analogue (FFAD), but FFAD was also a weak inhibitor of IMPDH *(54)*. Recently, synthesis of new derivatives in this series was reported by Franchetti and Grifantini, including imidazofurin and compounds containing 4-thio-D-ribofuranose (Fig. 12, see the chapter by Franchetti *et al.*).

Figure 12

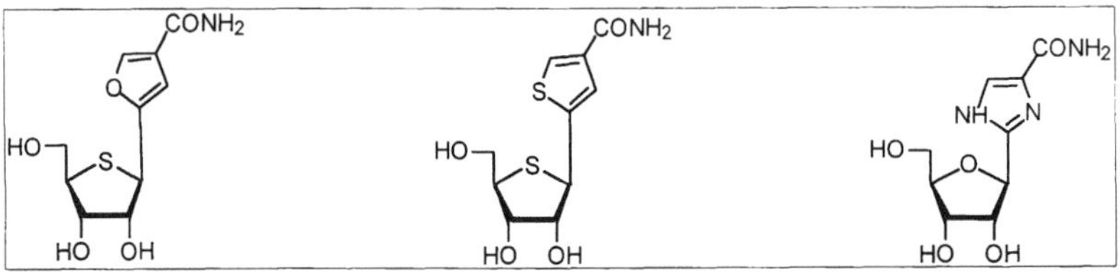

Benzamide riboside (BR, 3-β-D-ribofuranosylbenzamide, Fig. 13), a promising oncolytic agent discovered in the mid-1990's, was significantly more cytotoxic against human cancer cell lines than tiazofurin *(55)*. Tiazofurin and BR share a similar metabolic pathway. Like tiazofurin, BR is converted in the cell into an active metabolite, benzamide adenine dinucleotide (BAD), which also binds at the cofactor site of IMPDH. However, BR is converted into BAD more efficiently than tiazofurin is converted into TAD. In addition, BR induces apoptosis in human ovarian carcinoma N.1 cell lines, whereas tiazofurin does not. Thus, BR is an attractive candidate for anticancer drug development. The latest developments in BR research have been recently described *(56, 57)* (see also the chapter by Jayaram *et al.*).

Figure 13

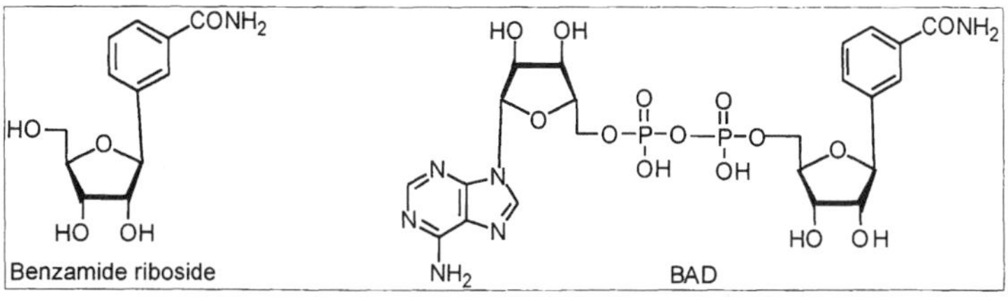

Interestingly, the closest mimics of nicotinamide riboside, pyridine C-nucleo-tides (Fig. 14), showed poor antitumor activity due to the poor conversion of these nucleosides to the corresponding NAD analogues in the target cells (58). These studies show that not all nicotinamide riboside analogues are metabolized into their NAD analogues. Therefore, synthetic NAD analogues, which do not require metabolic activation, are of primary interest.

Figure 14

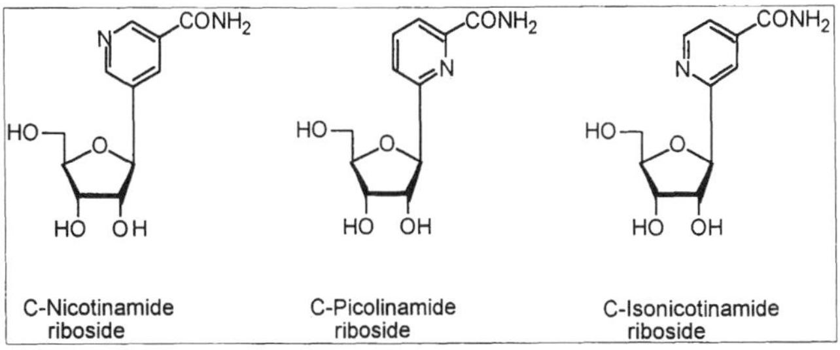

Methylenebis(phosphonate) and difluoromethylenebis(phosphonate) analogues of TAD and BAD (Fig. 15) are examples of metabolically stable NAD analogues that do not require activation. They differ from their parent pyrophosphates only by replacement of the pyrophosphate oxygen with a -CH$_2$- or -CF$_2$- group, and show as potent inhibition of IMPDH as the parent compounds TAD and BAD (see the chapter by Pankiewicz et al.). These compounds are capable of entering cells, are resistant to the action of phosphodiesterases, and are chemically stable in serum and cell. The latest developments in BR research have been recently described (56, 57) (see also the chapter by Jayaram et al.). extracts. Of particular significance is the observation that bis(phosphonate) analogues of TAD and BAD are active in tiazofurin resistant cell lines.

12

Figure 15

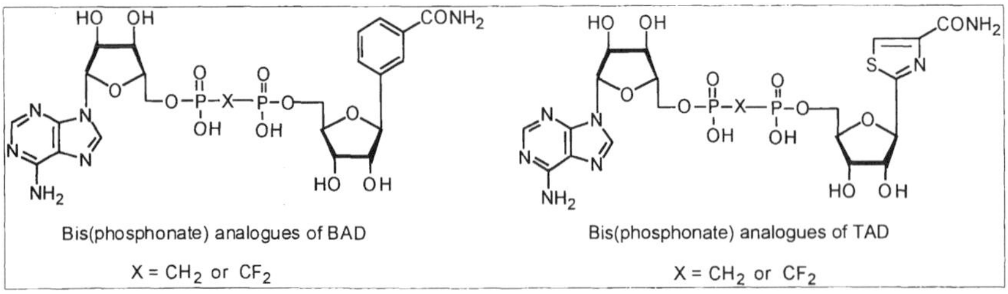

Bis(phosphonate) analogues of BAD
X = CH₂ or CF₂

Bis(phosphonate) analogues of TAD
X = CH₂ or CF₂

Bis(phosphonate) analogues of mycophenolic adenine dinucleotide (MAD) (Fig. 16) are under development at Pharmasset. These compounds are mimics of NAD, in which mycophenolic acid replaces the nicotinamide riboside moiety of NAD. A crystal structure of human type II IMPDH complexed with one of these compounds, C2-MAD (Fig 16) shows that the agent binds at the NAD site, mimicking both MPA and NAD binding (see the chapter by Goldstein *et al.*).

Bis(phosphonate) analogues of MAD were found to be potent and specific inhibitors of IMPDH. In contrast to mycophenolic acid these compounds are resistant to glucuronidation *(59)*. In addition, they trigger vigorous differentiation in K562 cells. Two compounds, C2-MAD and C4-MAD, were found to be an order of magnitude more potent inducers of K562 cell differentiation than tiazofurin itself *(60)*. As expected, MAD analogues are also active in tiazofurin resistant cell lines. They are now being evaluated as potential agents against human leukemia.

Figure 16

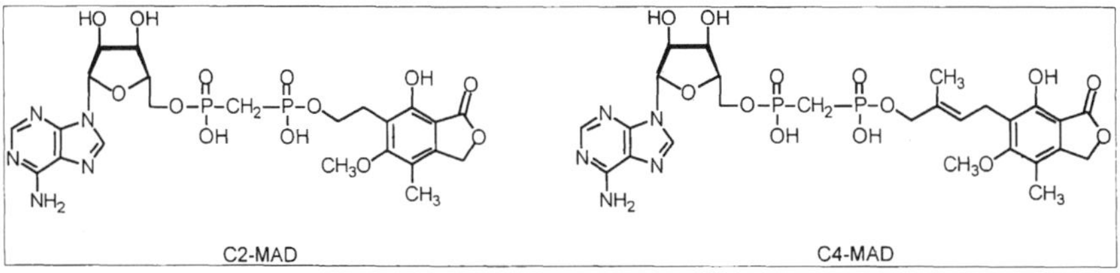

C2-MAD

C4-MAD

Towards specific inhibition of human IMPDH type II.

As noted above, a mouse model with selective deficiency of the IMPDH type II gene has been established in order to determine the effects of loss of IMPDH type II expression on both cell growth and lymphocyte stimulation *in vivo* (see the chapter by Gu *et al.*) . Homozygous loss of the IMPDH type II gene results in early embryonic lethality, demonstrating the importance of the type II isoform for rapidly proliferating and differentiating cells. IMPDH II heterozygous mice with combined deficiency in the salvage pathway enzyme HPRT demonstrate significant impairment in T-cell activation and function. These studies support earlier suggestions that selective inhibition of type II IMPDH may provide a significant therapeutic advantage, avoiding potential toxicity caused by inhibition of the type I isoform.

It was recently discovered *(61)* that three derivatives of 1,5-diazabicyclo-[3.1.0]hexane-2,4-dione **3, 4,** and **5** (Fig. 17) showed specific inhibition of the type II isoform. Surprisingly, these compounds bind at the IMP-binding site, which is conserved in the type I and type II isoform. The pattern of inhibition is competitive with respect to IMP, with K_i's in $5 - 44$ μM range. Computer modeling studies based on the crystal structure of Chinese hamster IMPDH with compound **4** docked in the active site revealed that the alpha helix adjacent to the active site may be responsible for the observed specificity *(61)*. This helix contains sequence differences between the type I and type II isoforms. The authors speculate that the methylphenyl group of compound **4** extends from the IMP site, making contact with the alpha helix, and is thus able to distinguish between the isoforms. Although the K_i values of these compounds are not very low, they provide an excellent lead for isoform-specific chemotherapy.

In contrast to the substrate-binding domain, the binding subsite for NAD is not conserved between the two isoforms of the human enzyme. The crystal structure of the type II isoform *(29)* shows that the adenine end of the cofactor interacts with residues that are not conserved between the isoforms. This suggests that modifications of the adenine moiety of cofactor-type inhibitors may be exploited in the design of isoform-specific agents. Further, the cofactor adenine binding region of IMPDH is also a source of species-specific differences. The importance of the NAD site in the development of species-specific inhibitors has recently been emphasized *(62)*.

Species-specific inhibition of IMPDH

Recently, IMPDH enzymes from other than human sources have emerged as attractive targets for the development of new drugs. Species differences exist in terms of substrate, cofactor, and ion binding, in addition to differences in the biochemistry of purine metabolism itself. Structural differences in the enzymes are reflected as differences in affinities and patterns of inhibition, suggesting that the development of species-specific chemotherapeutic agents is possible.

IMPDH from *Tritrichomonas foetus*, a protozoan that causes sterility and miscarriages in cows, is a target for the control of this parasitic infection. As described in

Figure 17

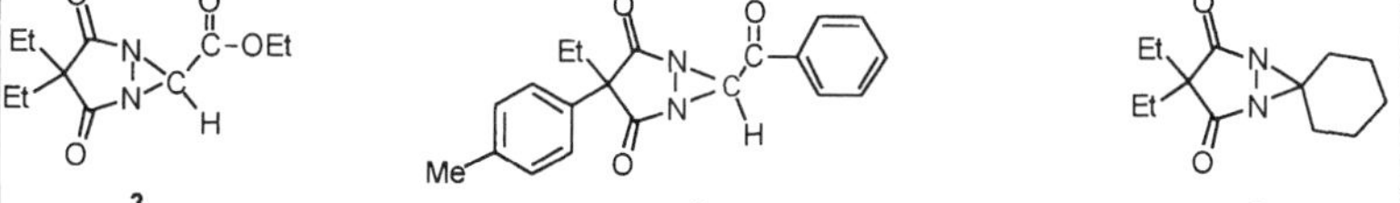

the chapter by Goldstein *et al.*, crystal structures of this enzyme have been solved in complex with a variety of agents. As noted above, most IMPDH enzymes crystallize as a tetramer of α/β barrels, binding the cofactor and substrate in a long active-site cleft covered by a flexible loop and flap. However, species differences do exist. For example, unlike IMPDH from mammalian sources, the *T. foetus* enzyme has a single arginine responsible for binding the substrate phosphate group in the active site. Residue differences also exist at the adenine end of the cofactor binding site. Further, the pattern and degree of inhibition by such agents as mycophenolic acid can differ dramatically, as described in the chapter by Hedstrom and Digits. All of these unique features may be exploited in the identification of inhibitors specific for the enzyme.

IMPDH from *Pneumocystis carinii*, an opportunistic pathogen in AIDS patients, has been cloned, expressed, and is now available for screening potential drugs. Unlike mammalian systems, salvage pathways for purine synthesis are weak or nonexistent in *P. carinii*, making the organism particularly susceptible to IMPDH inhibitors. Further, different patterns of inhibition are observed with other agents, suggesting that selective drug design may be possible. These observations are discussed in the chapter by Ye *et al.*

A number of microbial IMPDH enzymes, including those from *Streptococcus pyogenes, Haemophilus influenzae,* and *Mycobacterium tuberculosis* have been expressed and characterized by Collart and co-workers, and an X-ray structure of the enzyme from *S. pyogenes* in complex with IMP has been solved *(31)*. Bacterial and human IMPDH share only 20-30% amino acid sequences, and their kinetic and biochemical characteristics differ from mammalian enzymes. Residues in the active site, and particularly in the cofactor-binding domain, show a different pattern of sequence conservation in bacteria and eukaryotes, indicating the possibility of chemotherapeutic intervention. The emergence of bacteria resistant to known antibiotics underlies the necessity of targeting not only the enzymes exclusively encoded by pathogens, but also those enzymes with biochemical characteristics that differ between pathogens and humans.

Despite the recent progress in IMPDH crystallography, the number of IMPDH structures remains well below that of the more than 50 coding sequences from eukaryotic, bacterial and archael organisms that are now available. In the chapter by Collart and Huberman , a high throughput microbial system is described for screening large numbers of inhibitors. Methods such as these will be particularly valuable in identifying species-specific agents, as it is likely that clinically useful inhibitors will result from the combined application of structure-based design with combinatorial methods.

Future Directions

Significant progress has been made in understanding the mechanism, structure, and role of IMPDH as both a key enzyme in purine metabolism, and as a drug target. However, interesting questions remain. At the most basic structure-function level, the role of the large, conserved flanking domain has yet to be defined. The complex links between the regulation of IMPDH activity, guanine nucleotide metabolism, and the biological events that control cell differentiation are only beginning to be understood.

On the chemotherapeutic front, three drugs based on inhibition of IMPDH are well established in the market: the antiviral agent Ribavirin, and the immunosuppresants Mizoribine, and CellCept. However, there remains much room for improvement in the development of second generation drugs.

First, despite the fact that IMPDH is strongly linked with neoplasia, no anticancer drugs based on inhibition of the enzyme have achieved routine clinical use, although tiazofurin remains a leading candidate. Further, like all agents, these drugs are limited by toxicity and metabolism. Targeting the inducible type II isoform of the human enzyme has the potential of yielding agents with improved activity and reduced toxicity, but this remains to be tested. Thus, new lead compounds that show specificity towards IMPDH type II will likely continue to be a focus of activity. New compounds that are resistant to the inactivating metabolism found in neoplastic and immune cells will also have therapeutic potential. Lastly, recent advances in our understanding of the structures and mechanisms of IMPDH enzymes from non-human sources holds promise for the development of new antibacterial, antiviral, and antiparasitic agents.

The coming decade promises to be an exciting one at all levels of our understanding of this key enzyme; from the most basic biological processes to therapeutic applications in the clinic.

References.

1. Jayaram, H.N.; Dion, R.L.; Glazer, R.I.; Johns, D.G.; Robins, R.K.; Srivastava, P.C.; Cooney, D.A., *Biochem. Pharmacol.* **1982**, *31*, 2371-2380.

2. Manzoli, L.; Billi, A.M.; Gilmour, R.S.; Martelli, A.M.; Matteucci, A.; Rubbini, S.; Weber, G.; Cocco, L., *Cancer Res.* **1995**, *55*, 2978-2980.

3. Mandanas, R.A.; Leibowitz, D.S.; Gharehbaghi, K.; Tauchi, T.; Burgess, G.S.; Miyazawa, K.; Jayaram, H.N.; Boswell, H.S., *Blood* **1993**, *82*, 1838-1847.

4. Kharbanda, S.M.; Sherman, M.L.; Kufe, D.W., *Blood* **1990**, *75*, 583-588.

5. Parandoosh, Z.; Robins, R.K.; Belei, M.; Rubalcava, B., *Biochem. Biophys. Res. Commun.* **1989**, *164*, 869-874.

6. Parandoosh, Z.; Rubalcava, B.; Matsumoto, S.S.; Jolley, W.B.; Robins, R.K., *Life Sci.* **1990**, *46*, 315-320.

7. Vitale, M.; Zamai, L.; Falcieri, E.; Zauli, G.; Gobbi, P.; Santi, S.; Cinti, C.; Weber, G., *Cytometry* **1997**, *30*, 61-66.

8. Olah, E.; Csokay, B.; Prajda, N.; Kote-Jarai, Z.; Yeh, Y.A.; Weber, G., *Anticancer Res.* **1996**, *16*, 2469-2477.

9. Weber, G.; Prajda, N.; Abonyi, M.; Look, K.Y.; Tricot, G., *Anticancer Res.* **1996**, *16*, 3313-3322.

10. Allison, A.C. and Eugui, E.M., *Immunopharmacology.* **2000**, *47*, 85-118.

11. Laliberte, J.; Yee, A.; Xiong, Y.; Mitchell, B.S., *Blood* **1998**, *91*, 2896-2904.

12. Yalowitz, J.A. and Jayaram, H.N., *Anticancer Research.* **2000**, *20*, 2329-2338.

13. Alison, A.C.; Hovi, T.; Watts, R.W.E.; Webster, A.D.B., *CIBA Foundstion Symposium* **1977**, *48*, 207-224.

14. Wu, J.C., *Perspectives in Drug Discovery and Design* **1994**, *2*, 185-204.

15. Lanford, R.E.; Chavez, D.; Guerra, B.; Lau, J.Y.; Hong, Z.; Brasky, K.M.; Beames, B., *Journal of Virology.* **2001**, *75*, 8074-8081.

16. Natsumeda, Y.; Ohno, S.; Kawasaki, H.; Konno, Y.; Weber, G.; Suzuki, K., *J. Biol. Chem.* **1990**, *265*, 5292-5295.

17. Nagai, M.; Natsumeda, Y.; Konno, Y.; Hoffman, R.; Irino, S.; Weber, G., *Cancer Res.* **1991**, *51*, 3886-3890.

18. Nagai, M.; Natsumeda, Y.; Weber, G., *Cancer Res.* **1992**, *52*, 258-261.

19. Konno, Y.; Natsumeda, Y.; Nagai, M.; Yamaji, Y.; Ohno, S.; Suzuki, K.; Weber, G., *J. Biol. Chem.* **1991**, *266*, 506-509.

20. Dayton, J.S.; Lindsten, T.; Thompson, C.B.; Mitchell, B.S., *J. Immunol.* **1994**, *152*, 984-991.

21. Gu, J.J.; Spychala, J.; Mitchell, B.S., *J. Biol. Chem.* **1997**, *272*, 4458-4466.

22. Saunders, J.O. and Raybuck, S.A., in *Annual Reports in Medicinal Chemistry.*, D.A. M, Editor. 2000, Academic Press: San Diego. p. 201-209.

23. Pankiewicz, K.W., *Expert Opin. Ther. Patents* **1999**, *9*, 55.

24. Pankiewicz, K.W., *Expert Opin. Ther. Patents* **2001**, *11*, 1161.

25. Hedstrom, L., *Current Medicinal Chemistry* **1999**, *6*, 545-560.

26. Xiang, B. and Markham, G.D., *Arch. Biochem. Biophys.* **1997**, *348*, 378-382.

27. Wang, W. and Hedstrom, L., *Biochemistry* **1997**, *36*, 8479-8483.

28. Carr, S.F.; Papp, E.; Wu, J.C.; Natsumeda, Y., *J. Biol. Chem.* **1993**, *268*, 27286-27290.

29. Colby, T.D.; Vanderveen, K.; Strickler, M.D.; Markham, G.D.; Goldstein, B.M., *Proc. Natl. Acad. Sci. U S A* **1999**, *96*, 3531-3536.

30. Sintchak, M.D.; Fleming, M.A.; Futer, O.; Raybuck, S.A.; Chambers, S.P.; Caron, P.R.; Murcko, M.A.; Wilson, K.P., *Cell* **1996**, *85*, 921-930.

31. Zhang, R.; Evans, G.; Rotella, F.J.; Westbrook, E.M.; Beno, D.; Huberman, E.; Joachimiak, A.; Collart, F.R., *Biochemistry* **1999**, *38*, 4691-4700.

32. McMillan, F.M.; Cahoon, M.; White, A.; Hedstrom, L.; Petsko, G.A.; Ringe, D., *Biochemistry* **2000**, *39*, 4533-4542.

33. Whitby, F.G.; Luecke, H.; Kuhn, P.; Somoza, J.R.; Huete-Perez, J.A.; Phillips, J.D.; Hill, C.P.; Fletterick, R.J.; Wang, C.C., *Biochemistry* **1997**, *36*, 10666-10674.

34. Gan, L.; Petsko, G.A.; Hedstrom, L., *Biochemistry* **2002**, in press.

35. Sintchak, M.D. and Nimmesgern, E., *Immunopharmacology.* **2000**, *47*, 163-184.

36. Tam, R.C.; Ramasamy, K.; Bard, J.; Pai, B.; Lim, C.; Averett, D.R., *Antimicrobial Agents & Chemotherapy.* **2000**, *44*, 1276-1283.

37. Bentley, R., *Chem. Rev.* **2000**, *100*, 3801.

38. Behrend, M., *Clinical Nephrology* **1996**, *45*, 336-341.

39. Shaw, L.M.; Sollinger, H.W.; Halloran, P.; Morris, R.E.; Yatscoff, R.W.; Ransom, J.; Tsina, I.; Keown, P.; Holt, D.W.; Lieberman, R.; Jaklitsch, A.; Potter, J., *Ther. Drug Monit.* **1995**, *17*, 690-699.

40. Franklin, T.J.; Jacobs, V.; Bruneau, P.; Ple, P., *Adv. Enzyme Regul.* **1995**, *35*, 91-100.

41. Franklin, T.J.; Jacobs, V.; Jones, G.; Ple, P.; Bruneau, P., *Cancer Res.* **1996**, *56*, 984-987.

42. Franklin, T.J.; Jacobs, V.N.; Jones, G.; Ple, P., *Drug Metabolism & Disposition* **1997**, *25*, 367-370.

43. Papageorgiu, C., *Mini-reviews, Med. Chem.* **2001**, *1*, 71.

44. Edelson, S., *BioCentury* **2001**, *5*, A6.

45. Dhar, T.G.; Shen, Z.; Guo, J.; Liu, C.; Watterson, S.H.; Gu, H.H.; Pitts, W.J.; Fleener, C.A.; Rouleau, K.A.; Sherbina, N.Z.; McIntyre, K.W.; Witmer, M.R.; Tredup, J.A.; Chen, B.C.; Zhao, R.; Bednarz, M.S.; Cheney, D.L.; MacMaster, J.F.; Miller, L.M.; Berry, K.K.; Harper, T.W.; Barrish, J.C.; Hollenbaugh, D.L.; Iwanowicz, E.J., *J. Med. Chem.*. **2002**, *45*, 2127-2130.

46. Dhar, T.G.; Guo, J.; Shen, Z.; Pitts, W.J.; Gu, H.H.; Chen, B.C.; Zhao, R.; Bednarz, M.S.; Iwanowicz, E.J., *Organic Letters.* **2002**, *4*, 2091-2093.

47. Franklin, T.J.; Morris, W.P.; Jacobs, V.N.; Culbert, E.J.; Heys, C.A.; Ward, W.H.; Cook, P.N.; Jung, F.; Ple, P., *Biochemical Pharmacology.* **1999**, *58*, 867-876.

48. Minakawa, N. and Matsuda, A., *Current Medicinal Chemistry* **1999**, *6*, 615-628.

49. Cooney, D.A.; Jayaram, H.N.; Glazer, R.I.; Kelley, J.A.; Marquez, V.E.; Gebeyehu, G.; Van Cott, A.C.; Zwelling, L.A.; Johns, D.G., *Adv. Enzyme Regul.* **1983**, *21*, 271-303.

50. Jayaram, H.N.; Zhen, W.; Gharehbaghi, K., *Cancer Res.* **1993**, *53*, 2344-2348.

51. Jayaram, H.N.; Pillwein, K.; Lui, M.S.; Faderan, M.A.; Weber, G., *Biochem. Pharmacol.* **1986**, *35*, 587-593.

52. Tricot, G. and Weber, G., *Anticancer Res.* **1996**, *16*, 3341-3347.

53. Wright, D.G.; Boosalis, M.S.; Waraska, K.; Oshry, L.J.; Weintraub, L.R.; Vosburgh, E., *Anticancer Res.* **1996**, *16*, 3349-3351.

54. Franchetti, P. and Grifantini, M., *Current Medicinal Chemistry.* **1999**, *6*, 599-614.

55. Jayaram, H.N.; Cooney, D.A.; Grusch, M., *Current Medicinal Chemistry* **1999**, *6*, 561-574.

56. Gharehbaghi, K.; Grunberger, W.; Jayaram, H.N., *Current Medicinal Chemistry.* **2002**, *9*, 743-748.

57. Jayaram, H.N.; Yalowitz, J.A.; Arguello, F.; Greene, J.F., Jr., *Current Medicinal Chemistry.* **2002**, *9*, 787-792.

58. Pankiewicz, K.W. and Goldstein, B.M., *The Chemistry of Nucleoside and Dinucleotide inhibitors of Inosine Monophosphate Dehydrogenase (IMPDH). NAD analogues 18.*, in *Recent advances in Nucleosides: Chemistry and Chemotherapy*, C.K. Chu, Editor. 2002, Elsevier Science. p. 71-90.

59. Lesiak, K.; Watanabe, K.A.; Majumdar, A.; Powell, J.; Seidman, M.; Vanderveen, K.; Goldstein, B.M.; Pankiewicz, K.W., *J Med Chem* **1998**, *41*, 618-622.

60. Pankiewicz, K.W.; Zatorski, A.; Watanabe, K.A., *Acta Biochimica Polonica* **1996**, *43*, 183-193.

61. Barnes, B.J.; Eakin, A.E.; Izydore, R.A.; Hall, I.H., *Biochemistry.* **2000**, *39*, 13641-13650.

62. Digits, J.A. and Hedstrom, L., *Biochemistry.* **2000**, *39*, 1771-1777.

Molecular Biology and Genetics

Chapter 2

IMP, DH, and GTP: Linkage with Neoplasia, Target of Chemotherapy, Regulation of *Ras*, Signal Transduction, and Apoptosis

George Weber

Laboratory for Experimental Oncology, Indiana University School of Medicine, Indianapolis, IN 46202–5119 (telephone: 317–274–7921; fax: 317–274–3939; email: gweber1@iupui.edu)

Introduction

Quest for pattern recognition

I followed two main lines of approach in my laboratory which led to the study of the behavior of IMP dehydrogenase activity.[1-3]

1. The molecular correlation concept and the key enzymes. I was concerned with the quest for the recognition of the pattern of behavior of gene expression in cancer cells. We anticipated that with the expression of neoplasia there would be enzymic correlates of transformation and progression. The analytic approach of the enzymic pattern of gene expression in cancer cells, "the molecular correlation concept," revealed the fact that not only the rate-limiting enzymes in carbohydrate, purine and pyrimidine metabolism but also a number of other enzymes were stringently linked with malignancy and progression.[1-3] These enzymes we termed the "key enzymes" because they were involved in the regulation of the metabolic behavior of the pathway.[3] Such a rate-limiting and key enzyme was IMP DH.[3] In GTP biosynthesis in neoplasia there was an increase not only in the activity of IMP DH but also in that of GMP synthase and GMP kinase.[3] We concluded that a reprogramming of gene expression was displayed in the cancer cells.

2. Metabolic regulation was multisite in a pathway. We have also been involved in the elucidation of the regulation of liver enzyme activity and amount in various metabolic pathways. Prevailing ideas on regulation had suggested that each pathway was controlled by one pacemaker enzyme, a single rate-limiting

enzyme in a sequence of metabolic steps. However, in our investigation in a series on the "Role of enzymes in homeostasis" we observed that when the metabolic pathway was stimulated not only the rate-limiting enzyme activity was increased but also that of several enzymes at key points of the metabolic pathway.[1,2] These were the "key enzymes." These studies led to the discovery that nutritional or hormonal regulation of the activity of a pathway was through multi-enzymic control of the metabolic sequence.

Results and Discussion

Enzymes involved in IMP utilization.

(Figure 1) In utilization of IMP for GTP biosynthesis, we elucidated that IMP DH is the rate-limiting enzyme; it also has the lowest activity of all the purine biosynthetic or degradative enzymes[3,4] (Table I). In utilization of IMP for *de novo* guanylate biosynthesis, the activities of IMP DH and GMP synthase increased in all the examined hepatomas and the rise correlated positively with the hepatoma proliferation rate.[3] In various hepatomas, the increase in activities of IMP DH and GMP synthase was transformation- and progression-linked, indicating the heightened capacity for the production of XMP, GMP, GDP and GTP in these tumors.[4-6] It is remarkable that the specific activity of IMP DH was markedly elevated in all animal tumors and samples from human neoplasms (Table II).

Biological significance of the increased ability to provide GDP and GTP

The increased capacity of the guanylate synthetic pathway provides GDP, the substrate for ribonucleotide reductase, which generates dGDP and with the subsequent action of a kinase produces dGTP, the immediate precursor of DNA. The hepatic specific activity of the reductase is ten-fold lower than that of IMP DH and it increased in a transformation- and progression-linked fashion in the hepatoma spectrum.[7] A transformation- and progression-linked increase in IMP DH and GMP synthase activities and the resulting elevation of GMP and dGTP concentrations indicate a close linkage with the core of the neoplastic program.[3,8]

22

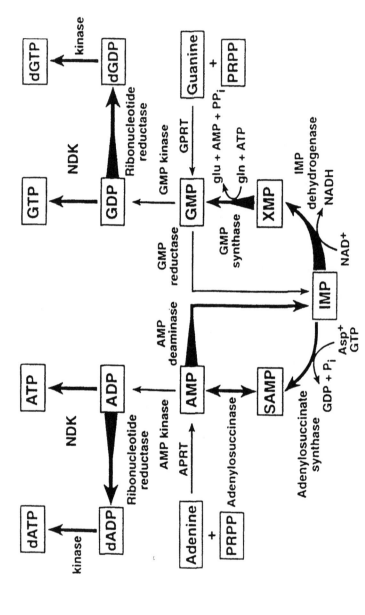

Figure 1. *Metabolic pathways of IMP utilization. Increased enzymic activities are indicated by heavy arrows.*

(Reproduced with permission from reference 9. Copyright 1992.)

Table I. Activities of Enzymes of the Anabolic and Catabolic Pathways of Purine Metabolism (From Ref. 3)

Enzymes	EC no.	Normal Liver nmol/hr/mg protein	Rapid hepatoma 3683F (% of liver)
Synthesis			
IMP Dehydrogenase	1.2.1.14	2	1350
GMP Synthase	6.3.4.1	33	548
Adenylosuccinate synthase	6.3.4.4	36	308
Amidophosphoribos-yltransferase	2.4.2.14	60	280
Adenylosuccinase	4.3.2.2	288	175
AMP deaminase	2.5.4.6	1,730	433
GMP Kinase	2.7.4.8	6,090	121
Catabolism			
Xanthine oxidase	1.2.3.2	100	10
Uricase	1.7.3.3	645	4
Inosine phosphorylase	2.4.2.1	28,800	19

*All hepatoma activities were significantly different from liver values ($p < 0.05$).

Table II. Increased IMP Dehydrogenase Activity in Cancer Cells

Samples	IMP dehydrogenase activity	
	nmol/hr/mg protein	% of control*
Rat		
Liver	2.1 ± 0.2	100
Hepatoma 3924A	27.3 ± 1.5	1,300
Human		
Normal leukocytes[+]	3.1 ± 0.5	100
CGL in blast crisis[+]	85.8 ± 23.6	2,768
K562 cells	44.6 ± 1.6	1,439
HL-60 cells	31.7 ± 2.2	1,023
Myeloma 8226 cells	44.5 ± 3.5	1,435
Normal breast HMEC cells	8.9 ± 0.2	100
MDA-MB-435 cells	29.5 ± 1.2	331
MDA-MB-435 tumors	25.0 ± 4.0	281
MCF-7 cells	26.0 ± 1.3	292
Normal ovary[+]	2.9 ± 0.9	100
Ovarian carcinoma[+]	19.6 ± 3.8	676
OVCAR-5 cells	61.2 ± 3.0	2,110

Means ± S.E. of 4 or more separate samples are given.

*All activities are significantly increased from relevant controls ($p < 0.05$).

[+]Clinical samples.

Biological functions of GTP

Whereas numerous studies focused on the functions and significance of ATP, the biological importance of GTP was perceived only recently.[9]

From the numerous biological functions of the metabolites of the guanylate pathway, the following may be emphasized. GTP, as mentioned above, is a precursor of RNA. GTP-dependent functions include 5'-"cap" formation in mRNA[10] and the production of small RNA species.[11] GTP is also involved in the promotion of polymerization of microtubules; this function has been exploited for the design of chemotherapy[12,13] (Figure2).

GTP is an important factor in protein biosynthesis and its role in RNA production was pointed out.[14] GTP is an activator of the rate-limiting enzyme of *de novo* CTP biosynthesis, CTP synthase,[3] and of adenylosuccinate synthase which is the first enzyme channeling IMP into ATP biosynthesis.[3] Therefore, a drop in GTP concentration should reduce the biosynthesis of CTP and ATP. Conversely, because GTP is an inhibitor of AMP deaminase,[8] a reduction in GTP pools would de-inhibit AMP deaminase activity with a subsequent drain on the AMP pools and a rise in IMP concentration. GTP is an activator of ribonucleotide reductase[7] and thus promotes the conversion of GDP into dGTP. Therefore, a pronounced decline in the GTP and GDP pools should reduce the synthesis of dGTP which, as the rate-limiting dNTP pool,[3] would result in an imbalance in the concentrations of the four dNTPs, limiting DNA biosynthesis (Figure 2).

A decrease in GTP concentration might curtail the assembly of complex lipids for which GDP-mannose is required[16] and might limit the synthesis of cyclic guanosine 3'-5'-monophosphate. GTP is an absolutely required cofactor of phospholipase C which is the third and final step in signal transduction leading to the production of the second messengers, DAG (diacylglycerol) and IP_3[17] (see below).

GTP is a required coenzyme of phosphoenolpyruvate carboxykinase, the second enzyme of hepatic gluconeogenesis. We showed that in starvation in the rat the liver activity of IMP DH is sustained in the face of a decrease of other enzyme activities and protein concentrations, with the exception of the enzymes of gluconeogenesis.[5] We also showed that, in starvation, liver GTP concentration is preferentially maintained.[5]

Design of enzyme-pattern-targeted chemotherapy for IMP DH.

Our approach for the design of chemotherapy directed against the unique enzyme pattern of neoplasms is based on four main considerations.[1-3] 1. The enzymic and metabolic imbalance provides selective reproductive advantages for cancer cells. 2. By identifying the biochemical alterations that characterize the commitment of cancer cells for replication, we can pinpoint biochemical targets for anti-cancer drug therapy, including the increased transformation- and progression-

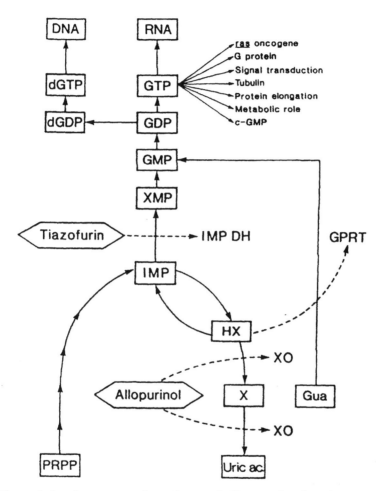

Figure 2. *Attacking points of tiazofurin and allopurinol and on de novo and salvage pathways of GTP biosynthesis.*

(Reproduced with permission from reference 24. Copyright 1989.)

linked activities of key enzymes and pools of nucleotides. 3. It is assumed that the selectivity of the drugs against cancer cells will be the higher the more marked is the biochemical difference from normal cells. The marked metabolic imbalance should reveal a heightened dependence on the enzyme or the metabolic process involved. 4. Because of the commitment to replication and utilization of precursors and the decreased responsiveness of cancer cells to physiological regulatory signals, the amplified and stringently linked enzymic and metabolic pattern in neoplasia should be more vulnerable to drug-induced perturbations than that of the normal tissues which have a wider range of repair, adaptability and recovery. This attack should provide higher drug selectivity.

The drug targeting approach that we proposed above identified IMP DH as a potentially sensitive target for chemotherapy.[6] IMP DH had low activity in normal resting tissues but was markedly increased in neoplasia, particularly in rapidly growing neoplasms. It was the rate-limiting enzyme of GTP biosynthesis and its inhibition should provide a selective decrease in the concentrations of guanylates, particularly GTP and GDP. These predictions also anticipated that the activity of IMP DH and the concentrations of the products, GTP, GDP and dGTP, might provide monitoring of the chemotherapeutic impact of a selective inhibitor of this enzyme.[3,6,14]

Our suggestion was picked up by Roland Robins, a foremost nucleoside chemist. Robins wrote, "The postulation of Weber that IMP dehydrogenase is a key enzyme in neoplasia and therefore a sensitive target for cancer chemotherapy has received considerable experimental support".[6] "It would appear that the search for potential inhibitors of IMP dehydrogenase will continue to be a fruitful area for the design and synthesis of potent antitumor agents".[18] One of the compounds Robins synthesized, tiazofurin, proved to be a potent inhibitor of IMP DH, particularly in sensitive cells where it was converted to the active metabolite, TAD (Figure 3). In NCI and other centers, it was shown that various experimental tumors were sensitive to tiazofurin action.[19,20]

Chemotherapeutic action of tiazofurin: preclinical studies in my laboratory.

Tiazofurin killed hepatoma cells in culture with an LD of 5 μM and markedly inhibited tumor growth in rats inoculated with hepatoma 3924A.[21] We demonstrated that a single *in vivo* injection of tiazofurin (200 mg/kg) in rats rapidly reduced in the hepatoma the activity of IMP DH followed by a marked decline in the pools of GDP and GTP and particularly in the concentration of dGTP. All levels returned to normal in one or two days after the injection; however, the dGTP pool remained depressed for 72 hr (Figure 4). These results revealed the chemotherapeutic impact of inhibition of IMP DH activity.[21] On the basis of these and other studies, we proposed a clinical evaluation of tiazofurin action in leukemia. This neoplastic disease was particularly suited for chemotherapeutic

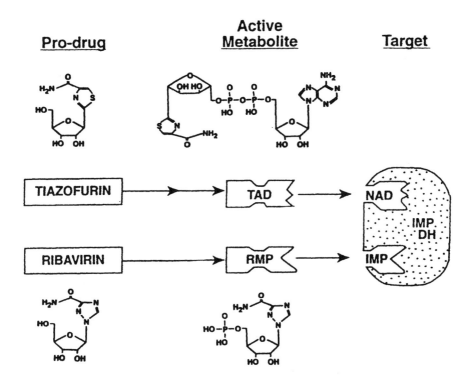

Figure 3. *Attachments of active metabolites of tiazofurin and ribavirin to the ligand sites of IMP dehydrogenase, TAD, tiazol-4-carboxamide adenine dinucleotide; RMP, ribavirin monophosphate.*
(Reproduced with permission from reference 26. Copyright 1996.)

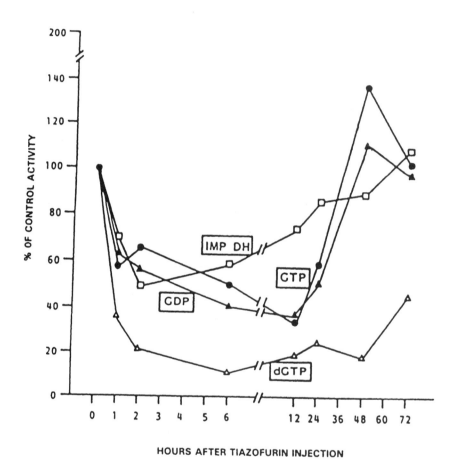

Figure 4. Metabolic impact of tiazofurin i.p. (150 mg/kg) injection in rat on
subcutaneously transplanted hepatoma 3924A. (21)

studies because it permitted sampling of blast cells during the course of treatment, allowing us to follow the activity of IMP DH and the concentration of GTP and other metabolites. The selectivity was provided by the fact that normal leukocytes converted very little of the tiazofurin to the active form, TAD, whereas a great deal of the infused drug was converted into TAD in the blast cells.[22]

Clinical impact of tiazofurin

The clinical studies outlined by Dr. Guido Tricot, who collaborated the most closely with us, showed that 1-hr infusion of tiazofurin yielded a pattern in the blast cells in the patients which was similar to that found in the rat (Figure 5). In the clinical situation there was a rapid decline in IMP DH activity, followed by the reduction in GTP concentration; subsequently, the blast cells were cleared from the periphery. Leukocyte counts remained in normal range. Treatment of a number of patients with CGL in BC resulted in a 77% response, including complete remissions.[23-27] These clinical and biochemical results were independently confirmed by Dr. D. Wright et al.[28]

Reprogramming of gene expression and isozymes

In our studies we examined the possibility that the cancer cell might express an isozyme that would show higher sensitivity to anti-cancer drugs. Our earlier studies on a number of enzymes revealed what we called an "isozyme shift" in hepatomas; the pattern of gene expression shifted from expressing liver type to muscle type enzymes.[1-3] We showed this for the glycolytic enzymes, glucokinase-hexokinase, 6-phosphofructokinase and pyruvate kinase and then for the nucleic acid enzymes, TdR kinase and AMP deaminase.[1-3] When we discovered the marked elevations in IMP DH activity in tumors we thought that, because it catalyzes the first committed step in the *de novo* biosynthesis of guanine nucleotides, isozymes might be present. Initial studies on crude and purified enzymes showed that in normal and cancer cells IMP DH had similar kinetic properties.[4,5] They also appeared immunologically similar but differed in the increased expression of the enzyme amount which was parallel with the increased activity. Subsequent studies, using molecular biology, revealed that this was not unexpected since the type I and II isoforms discovered in my laboratory shared an 85% sequence identity. Our studies indicated that type I is constitutively expressed, whereas type II expression is markedly up-regulated in cancer cells.[29-32] Because of the dependence of DNA biosynthesis, induced differentiation and signaling on IMP DH type II isozyme in tumor cells, this isozyme was targeted in chemotherapeutic drug design.[29-32,33] Both isoforms are homotetrameres consisting of four 56 kD monomers.

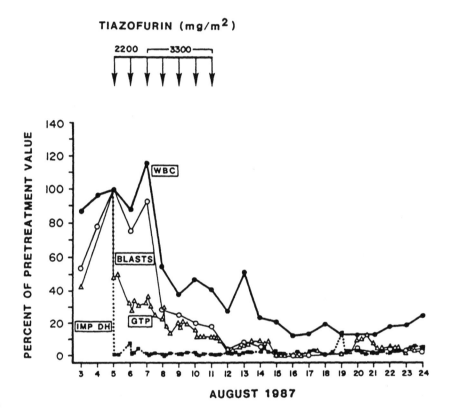

TIAZOFURIN (mg/m²)

Figure 5.Effects of i.v. tiazofurin treatment on white blood cell and blast cell counts and on activity of IMP dehydrogenase and concentration of GTP in leukemic cells. Values are given as percentages of the amount observed in the patient the day before treatment. Absolute values of pre-treatment samples were reported elsewhere (38).

(Reproduced with permission from reference 24. Copyright 1989.)

The studies of Nagai and Natsumeda in my laboratory showed that type II isozyme was markedly increased in various leukemic blast cells from patients and the elevation in the activity and amount of this isoform accounted for the increased total IMP DH activity in the leukemic blast cells of the patients.[31,32]

That the increased enzyme activity was due to an increased amount of the enzyme was demonstrated by enzyme kinetic and immunological evidence for hexokinase-glucokinase, 6-phosphofructokinase, pyruvate kinase, glucose-6-phosphate dehydrogenase, transaldolase, carbamoyl phosphate synthase II, CTP synthase, ribonucleotide reductase, amidophosphoribosyl-transferase, AMP deaminase, and IMP dehydrogenase.[1-3,8]

In most of these studies the tumor isozyme had higher affinity to its substrate and lost its regulatory responsiveness compared to the liver enzyme. This is in line with the clinical observation that in tumor progression the physiological responsiveness to regulatory molecules and to drugs is decreased or eventually lost.

In the isozyme shift the replacement of isozymes that respond to nutritional and hormonal regulation by non-responsive low K_m isozymes is characteristic of cancer cells and is linked with progression.[1-3]

Thus, the reprogramming of gene expression is reflected in both quantitative and qualitative alterations. "Reprogramming" is the appropriate term because the essential neoplasia-linked part of the genic program is displayed in an ordered, meaningful and integrated enzymic and metabolic imbalance.[3]

Inhibition of IMP DH activity and down-regulation of the expression of the *ras* oncogene.

Since GTP is a critical factor in the maturation and translocation into the membrane and expression of the *ras* oncogene, we tested with Dr. Olah the hypothesis that tiazofurin might reduce *ras* expression. In K-562 cells addition of tiazofurin did result in down-regulation of *ras* expression and in induced differentiation.[35] This *in vitro* observation on differentiation agreed well with the increased maturation of blast cells in the bone marrow of patients during tiazofurin treatment.[23-25] Further studies showed that tiazofurin down-regulated *ras* in various other cells, including the blast cells of patients treated with this drug.[36] (Figure 6)

The tiazofurin-induced differentiation of the blast cells in the patients was important because after completion of therapy the patients could be discharged without the need for antibiotic or other bone marrow protection treatment. This is in sharp contrast to treatment with all other commonly used drugs which destroy the bone marrow.[25]

Impact of tiazofurin on signal transduction activity in cancer cells

In signal transduction PI is converted through the action of three enzymes, PI-4 kinase, PIP kinase, and phospholipase C (PIP$_2$ phosphodiesterase), into the second

32

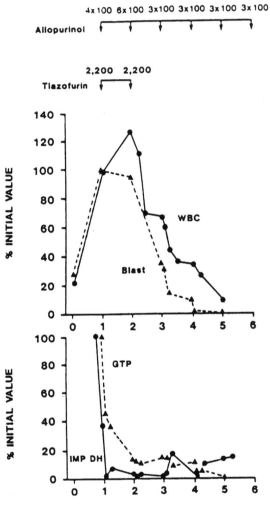

Figure 6. *Down-regulation of the ras and myc oncogene in the blast cells of a patient with chronic granulocytic leukemia in blast crisis treated with two infusions of tiazofurin.*

(Reproduced with permission from reference 38. Copyright 1991.)

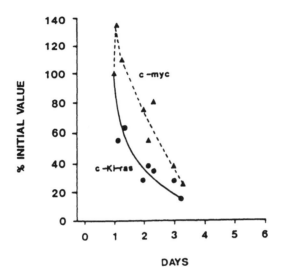

Figure 6. *Continued.*
(Reproduced with permission from reference 38. Copyright 1991.)

messengers, DAG and IP_3[17,39] (Figure 7). Phospholipase C activity is entirely dependent on GTP.

Tiazofurin has two separate impacts on the signal transduction process. 1. By limiting macromolecular biosynthesis, tiazofurin caused a decrease in activities of enzymes with short half lives. Our cycloheximide studies in the rat showed that PI and PIP kinases have the shortest half lives among ten examined enzymes: $t_{1/2} = 12$ min in rat bone marrow cells.[39] A single injection of tiazofurin in the rat caused a marked reduction in activities of PI and PIP kinases and in the concentration of IP_3 in the bone marrow cells.[40] A similar reduction of signal transduction activity was observed in the transplanted hepatoma. 2. In vivo PLC activity followed the reduction of GTP concentration which is necessary for the activity of this enzyme. Through these two mechanisms, tiazofurin was able to reduce signal transduction activity (IP_3 concentrations) in a dose- and time-dependent fashion.

The significance of GTP in chemotherapeutic action, induced differentiation, apoptosis, down-regulation of *ras* expression and signal transduction

Our studies showed that the focal point of the impact of tiazofurin, as an inhibitor of IMP DH, is the reduction in the concentration of GTP. We provided evidence that the restoration or the maintenance of GTP concentration through supplying guanine or guanosine abrogated all impacts of the drug even if tiazofurin was present.[35] Thus, the impact of the inhibiting IMP DH, the induced differentia-

Table III. Molecular and Clinical Impact of Tiazofurin

Drug action	Targets	Impact
1. Chemotherapy:	IMP DH GTP, dGTP	Target enzyme -- inhibited Target metabolites -- decreased
2. Non-targeted action:	Enzymes with rapid turnover*	Decreased enzyme amounts
3. Oncogene activity:	ras, myc	Down-regulated
4. Differentiation	Maturation	Up-regulated
5. Signal transduction:	PI and PIP kinase, IP_3	Down-regulated
6. Apoptosis:	Programmed cell death	Activated

*TdR kinase, dTMP synthase, GPRT, PI 4-kinase and PIP kinase.

tion, the apoptosis[36] and the down-regulation of *ras* expression and of signal transduction were all abrogated if the GTP concentration was maintained through salvage.[17] This was true also in the tiazofurin treatment of patients where we increased serum hypoxanthine levels through daily multi-doses of allopurinol to inhibit the salvage enzyme, GPRT. If we failed to increase serum hypoxanthine to approximately 60 μM, the therapeutic response was not adequate. Conversely, in highly sensitive cases where, along with inhibition of IMP DH activity we also reached sustained high serum hypoxanthine levels, the results were good, sometimes amazingly rapid.[38,42] Enzyme-pattern-targeted chemotherapy required inhibition of IMP DH activity in the cancer cells and of xanthine oxidase activity in the liver.[38,42] Inhibition of xanthine oxidase activity resulted in elevated serum hypoxanthine concentrations which inhibited the activity of the salvage enzyme GPRT (guanine phosphoribosyltransferase). See also Figure 2.

Synergistic action of tiazofurin with other compounds

Tiazofurin provided good responses in patients with CGL in BC. [23-25,27,28] There was pre-clinical evidence that tiazofurin was effective against hepatocellular carcinoma,[21] human ovarian carcinoma,[43-45] human breast carcinoma[46] and colon carcinoma cells.[47] Recent studies also indicated activity in human myeloma cells and in myeloma grown in SCID mice. The clinical effectiveness in leukemia and myeloma is currently explored in other Centers.

We sought to strengthen the impact of tiazofurin by combining it with other compounds and we developed the following guidelines.

1. Tiazofurin and the other agent should impact at different sites on the IMP DH molecule.

2. Tiazofurin and the other drug should attack in the same metabolic pathway at two targets where one requires GTP.

3. Tiazofurin should be used in combination with drugs which act at different phases of the cell cycle.

Some of the drugs we considered and their impacts on the cell cycle phases are shown in Figure 8.

1. **Synergistic impact on different sites of the IMP DH molecule.** Kinetic investigations in my laboratory by Yamada and Natsumeda on the purified enzyme showed that TAD had a better affinity to the NAD pocket of IMP DH than the endogenous coenzyme NAD itself.[34] Ribavirin had affinity to the IMP binding site.[34] (Figure 3) Since tiazofurin inhibited at the NAD site of the enzyme and ribavirin at the IMP site, we tested the combination of these compounds.[48] The two drugs were synergistically cytotoxic in hepatoma 3924A cells and synergis-tically enhanced inhibition of *de novo* purine and guanylate synthesis, studied with radioactive precursors. Selenazofurin and ribavirin also provided synergistic cytotoxicity because selenazofurin acts at the NAD site. Tiazofurin plus

36

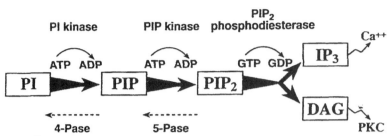

Figure 7. Enzymic steps of the signal transduction pathway in the biosynthesis of second messengers, diacylycerol and IP$_3$ (inositol 1,4,5,-trisphosphate). (Reproduced with permission from reference 17. Copyright 2000.)

ATTACKING POINTS OF DRUGS IN THE CELL CYCLE

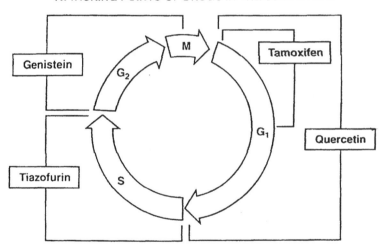

Figure 8. Impact of drugs at different phases of the cell cycle. (Reproduced with permission from reference 17. Copyright 2000.)

selenazofurin because they attack the same NAD site were additive, as expected.[48] We concluded that ribavirin might strengthen tiazofurin action against cancer cells and in turn tiazofurin might enhance the action of ribavirin in the treatment of hepatitis C infection.

2. **Synergistic interaction of tiazofurin with agents which attack the same pathway at two targets where one step requires GTP: the mitotic spindle formation.** The drugs act at different phases of the cell cycle. An example is our studies on tiazofurin and taxol which targeted at two sites in the microtubular synthetic process. Figure 9 shows the attacking point where tiazofurin, through depressing the GTP pool, inhibits guanylation; taxol attacks mitotic spindle production at a subsequent site different from that of the attachment of GTP.[13] Tiazofurin inhibits chiefly in G_1 and S phases; taxol blocks primarily in the M phase.

For taxol in human OVCAR-5, PANC-1, pancreatic and lung carcinoma H-125 cells and rat hepatoma 3924A cells the antiproliferative IC_{50}s were 0.05, 0.06, 0.03 and 0.04 μM, respectively; for tiazofurin IC_{50} values were 8.3, 2.3, 1.8 and 6.9 μM. Taxol concentrations for inhibiting cell proliferation were 38- to 173-fold lower than those for tiazofurin. Combination of tiazofurin and taxol should have implications in the clinical treatment of human solid tumors with particular relevance to ovarian, pancreatic, lung and hepatocellular carcinomas.[13]

In human MDA-MB-435 estrogen receptor negative breast cancer cells in culture the growth inhibition for tiazofurin and taxol yielded IC_{50}s of 12.5 and 0.016 μM, respectively, and in clonogenic assays 4.5 and 0.004 μM.[49] In these cells synergism was obtained in growth inhibitory and clonogenic assays only when tiazofurin was followed 12 hr later by taxol; thus, the sequencing of the drugs was critical in achieving synergism. In these cells incubation time with tiazofurin added after taxol should be long enough (12 hr) to allow cells to traverse the G_1, S and G_2 phases and enter the M phase where they can be killed by taxol. Since taxol alone provided clinical responses in breast carcinoma at about 60%, the pre-clinical protocols yielding synergism might be of value in the design of taxol-based clinical trials for breast cancer.[49]

3. **Synergistic impact of tiazofurin with other drugs attacking in the same pathway: signal transduction activity.** Our work showed that tiazofurin reduced the increased signal transduction pathway activity in cancer cells, as measured by the concentration of the end product, IP_3.[26] Tiazofurin inhibited chiefly in the S phase, quercetin at G_1 phase, genistein at the G_2 and early M phases and tamoxifen in the early G_1 phase interval (Figure 8). Therefore, we postulated that combination of tiazofurin with drugs that might inhibit or reduce the activity of one or other of the enzymes in IP_3 biosynthesis might be synergistic with tiazofurin.

38

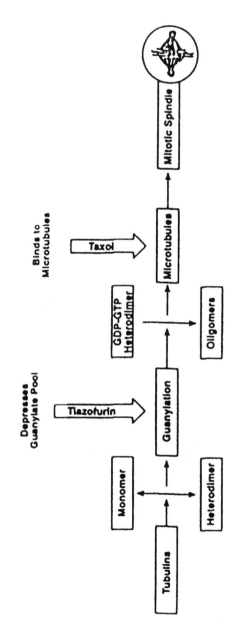

Targets of tiazofurin and taxol in the process of mitotic spindle formation.

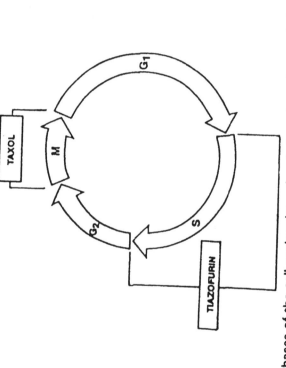

Phases of the cell cycle where tiazofurin and taxol exert their main effects.

Figure 9. Impact of tiazofurin and taxol in the pathway of mitotic spindle biosynthesis.

(Reproduced with permission from reference 13. Copyright 1993.)

Tiazofurin is synergistic with quercetin.

The outcome of these studies showed that quercetin, which inhibits primarily PI 4-kinase activity, was synergistic with tiazofurin in antiproliferative and cytotoxic action.[26,38,50] In human ovarian carcinoma OVCAR-5 cells in growth inhibition assay, the IC_{50}s for tiazofurin and quercetin were 13 and 66 µM; in clonogenic assays they were 6 and 15 µM, respectively. Because tiazofurin reduced GTP concentration in cells by 50% at 12 hr after administration[35] tiazofurin was followed 12 hr later by quercetin. In this sequential administration synergism was observed in both growth inhibition and clonogenic assays. The combination also yielded synergistic reduction of IP_3 concentration in the cells which may explain, at least in part, the synergistic action of tiazofurin and quercetin in OVCAR-5 cells. These results support the concept that drugs such as tiazofurin and quercetin attacking different biochemical targets in the same pathway and blocking in separate phases of the cell cycle can provide synergistic interaction. These experimental protocols yielding synergism may have implications in the clinical treatment of human ovarian carcinoma.[50]

Tiazofurin is synergistic with genistein.

Genistein, an inhibitor of PIP kinase, tyrosine kinase and topoisomerase-II, induces arrest in G_2 and/or early M phase in most carcinoma cells. Individually, tiazofurin and genistein reduce second messenger IP_3 concentration in ovarian carcinoma cells. Because genistein and tiazofurin attack different enzymic targets in the same signal transduction pathway and arrest the cell cycle at different phases, we tested the hypothesis that the two drugs might be synergistic. In growth inhibition assay for tiazofurin and genistein IC_{50}s were 26 and 18 µM, respectively, and in clonogenic assays LC_{50}s were 17 and 4 µM, respectively. Various combinations were examined. The best protocol took into consideration the above-mentioned fact that tiazofurin decreased GTP concentration in cells by 50% at 12 hr after administration.[35] Tiazofurin (20 µM) and genistein (20 µM) as single agents reduced cell counts to 60 and 50%, respectively. The predicted value, as a sum of the effect of the two drugs, would have been 30% of controls. However, genistein, added 12 hr after tiazofurin, decreased cell counts to 8%, showing synergistic action of the two drugs for growth inhibition. Similar results were observed in the clonogenic assays, which revealed synergistic cytotoxicity.[51] Because the cytotoxicity of tiazofurin is synergistically enhanced by genistein, lower concentrations of tiazofurin may be employed in treatment, thus decreasing potential side effects.

The utilization of tiazofurin, quercetin and genistein should be of interest for clinical trials of ovarian carcinomas and also for leukemias. These drugs are also of relevance because novel targets in signal transduction are provided for ovarian carcinoma cells (IMP dehydrogenase, PI kinase, PIP kinase, and through reduced

GTP concentrations, phospholipase C activity, and IP_3 concentration) (Figure 7). These biochemical targets may also be helpful in monitoring the impact of treatment and in early heralding impending relapse.

Tiazofurin and genistein, as single agents, can induce differentiation. Because of their different biochemical targets we tested the hypothesis that tiazofurin might be synergistic with genistein in inducing differentiation. In human leukemic K-562 cells in growth inhibition assay for tiazofurin and genistein IC_{50}s were 7 and 37 µM, respectively. The concentrations of tiazofurin and genistein that induce differentiation in 50% of the cells were 35 and 45 µM, respectively. Since tiazofurin decreased GTP concentration in cells by 50% at 12 hr after administration, genistein (10 to 30 µM) was added 12 hr after tiazofurin (5 to 15 µM). Synergistic action on inducing differentiation was observed from all tiazofurin and genistein combinations and in most combinations also on growth inhibition. The two drugs caused a 5.9-fold elevation in synergistically inducing differentiation. Similar action was also observed on inhibition of proliferation.[52]

Tiazofurin is synergistic with tamoxifen

We tested the hypothesis that tamoxifen and tiazofurin may be synergistic in MDA-MB-435 estrogen receptor-negative human breast carcinoma cells because the two compounds decrease IP_3 concentration through different mechanisms and inhibit the cell cycle in different phases: tiazofurin in S phase and tamoxifen in early G_1 phase. The breast carcinoma cells were selected because, compared to normal human parenchymal breast cells, they have increased IMP DH activity (3.3-fold), elevated activities of enzymes of signal transduction (15- to 95-fold) and increased concentrations of IP_3 (7- to 9-fold).[46]

In growth inhibition assay the IC_{50}s for tiazofurin and tamoxifen were 17 and 12 µM, respectively. In clonogenic assay the LC_{50}s for tiazofurin and tamoxifen were 4 and 0.7 µM, respectively. Therefore, in breast cancer cells tamoxifen is 1.4 times and 5.7-fold more effective than tiazofurin in growth inhibition and clonogenic assay, respectively. These IC_{50}s are achievable in the serum of patients treated with these drugs. Antiproliferative synergism was obtained when tamoxifen (1 to 10 µM) was added 12 hr after tiazofurin (5 to 15 µM). Synergistic cytotoxicity was also observed in clonogenic assay.[46]

Tiazofurin and tamoxifen each decreased IP_3 concentration in these cells and the combined results indicated borderline synergism.

The antiproliferative and cytotoxic synergism of tiazofurin and tamoxifen is a significant novel observation. This new combination should be of interest for clinical trial for estrogen receptor-negative breast carcinoma.[46]

42

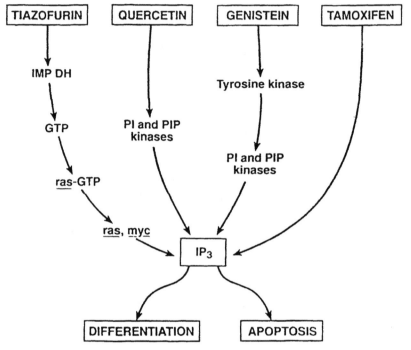

Figure 10. *The final common pathway in the actions of tiazofurin, tamoxifen, quercetin and genistein in the induction of apoptosis and differentiation. It is postulated that the biological impact depends on the reduction of IP$_3$ concentration by these drugs.*

Conclusions

The presented evidence indicates the important role of IMP DH in the strategic program of cancer cells. The results further document the significance of GTP in the impact of tiazofurin. The targets of tiazofurin include *ras* oncogene expression, cell maturation and apoptosis (see above Table III).

The results show that combinations of tiazofurin with quercetin, genistein, tamoxifen and ribavirin are capable of inhibiting proliferation and clonogenic activity as well as signal transduction activity. The synergistic combinations involve concentrations of the drugs that are in the range that is readily achievable in patients. The synergistic power of these compounds might be ready for translation to the clinic for various neoplastic diseases.

Figure 10 shows the final common pathway in the biochemical and biological impact of tiazofurin, ribavirin, tamoxifen, quercetin and genistein in the induction of apoptosis and in differentiation. It is postulated that the biological impact depends on the reduction of IP_3 concentration by these drugs.[17]

Acknowledgment

This investigation was supported by a Milan Panič Professorship and a grant from the Ladies Auxiliary of the Veterans of Foreign Wars, Indiana Division, to GW.

References

1. Weber, G. Enzymology of cancer cells, Part 1. *New Engl. J. Med.* **1977**, 296, 486-493.
2. Weber, G. Enzymology of cancer cells, Part 2. *New Engl. J. Med.* **1977**, 296, 541-551.
3. Weber, G. Biochemical strategy of cancer cells and the design of chemotherapy: G. H. A. Clowes Memorial Lecture. *Cancer Res.* **1983**, 43, 3466-3492.
4. Jackson, R. C.; Weber, G.; Morris, H. P. IMP dehydrogenase, an enzyme linked with proliferation and malignancy. *Nature (Lond)* **1975**, 256, 331-333.
5. Jackson, R. C.; Morris, H. P.; Weber, G. Partial purification, properties and regulation of inosine 5'-phosphate dehydrogenase in normal and malignant rat tissues. *Biochem. J.* **1977**, 166, 1-10.
6. Weber, G.; Prajda, N.; Jackson, R. C. Key enzymes of IMP metabolism: transformation- and proliferation-linked alterations in gene expression. *Advan. Enzyme Regul.* **1976**, 14, 3-24.
7. Takeda, E.; Weber, G. Role of ribonucleotide reductase in expression of the neoplastic program. *Life Sci.* **1981**, 8, 1007-1014.

8. Jackson, R. C.; Morris, H. P.; Weber, G. Enzymes of the purine ribonucleotide cycle in rat hepatomas and kidney tumors. *Cancer Res.* **1977,** 37, 3057-3065.

9. Weber, G.; Nakamura, H.; Natsumeda, Y.; Szekeres, T.; Nagai, M. Regulation of GTP biosynthesis. *Advan. Enzyme Regul.* **1992,** 32, 57-69.

10. Shatkin, A. J. Capping of eucaryotic mRNA's. *Cell* **1976,** 9, 645-653.

11. Zleve, G. W. Two groups of small stable RNA's. *Cell* **1981,** 25, 296-297.

12. Manfredi, J. J.; Horwitz, S. B. Taxol: an antimitotic agent with a new mechanism of action. *Pharmacol. Ther.* **1984,** 25, 83-125.

13. Taniki, T.; Prajda, N.; Monden, Y.; Weber, G. Synergistic action of taxol and tiazofurin in human ovarian, pancreatic and lung carcinoma cells. *Cancer Biochem. Biophys.* **1993,** 13, 295-302.

14. Cohen, M. B.; Maybaum, J.; Sadee, W. Guanine nucleotide depletion and toxicity in mouse T lymphoma (S-49) cells. *J. Biol. Chem.* **1981,** 256, 8713-8717.

15. Kizaki, H.; Williams, J. C.; Morris, H. P; Weber, G. Increased cytidine 5'-triphosphate synthetase activity in rat and human tumors. *Cancer Res.* **1980,** 40, 3921-3927.

16. Wyngaarden, J. B.; Kelley, W. N. Disorders of purine and pyrimidine metabolism. In: J.B. Stanbury, J.B. Wyngaarden, and D.S. Fredrickson (eds). *The Metabolic Basis of Inherited Disease* **1966,** 4, 821-822. New York: McGraw-Hill Book Co.

17. Weber, G.; Shen, F.; Prajda, N.; Abonyi, M. Down-regulation by drugs of the increased signal transduction in cancer cells. *Advan. Enzyme Regul.* **2000,** 40, 19-35.

18. Robins, R. K. Nucleoside and nucleotide inhibitors of inosine monophosphate (IMP) dehydrogenase as potential antitumor inhibitors. *Nucleosides and Nucleotides* **1982,** 1, 35-44.

19. Jayaram, H. N.; Dion, R. L.; Glazer, R. I.; Johns, D. G.; Robins, R. L.; Srivastave, P. C.; Cooney, D. A. Initial studies on the mechanism of action of a new oncolytic thiazole nucleoside, 2-β-D-ribofuranosylthiazole-4-carboxamide (NSC-286193). *Biochem. Pharmacol.* **1982,** 31, 2371-2380.

20. Jayaram, H. N.; Smith, A. L.; Glazer, R. I.; Johns, D. G.; Cooney, D. A. Studies on the mechanism of action of 2-β-D-ribofuranosylthiazole-4-carboxamide (NSC-286193) II. Relationship between dose-level and biochemical effects in P388 leukemia *in vivo*. *Biochem. Pharmacol.* **1982,** 31, 3839-3845.

21. Lui, M. S.; Faderan, M. A.; Liepnieks, J. J.; Natsumeda, Y.; Olah, E.; Jayaram, H. N.; Weber, G. Modulation of IMP dehydrogenase activity and guanylate metabolism by tiazofurin (2-β-D-ribofuranosylthiazole-4-carboxamide). *J. Biol. Chem.* **1984,** 259, 5078-5082.

22. Jayaram, H. N.; Pillwein, K.; Nichols, C. R.; Hoffman, R.; Weber, G. Selective sensitivity to tiazofurin of human leukemic cells. *Biochem. Pharmacol.* **1986,** 35, 2029-2-2032.

23. Tricot, G. J.; Jayaram, H. N.; Nichols, C. R.; Pennington, K.; Lapis, E.; Weber, G.; Hoffman, R. Hematological and biochemical action of tiazofurin in a case of refractory acute myeloid leukemia. *Cancer Res.* **1987**, 47, 4988-4991.

24. Tricot, G. J.; Jayaram, H. N.; Lapis, E.; Natsumeda, Y.; Yamada, Y.; Nichols, C. R.; Kneebone, P.; Heerema, N.; Weber, G.; Hoffman, R. Biochemically directed therapy of leukemia with tiazofurin, a selective blocker of inosine 5'-phosphate dehydrogenase activity. *Cancer Res.* **1989**, 49, 3696-3701.

25. Tricot, G.; Jayaram H. N.; Weber, G.; Hoffman, R. Tiazofurin: Biological effects and clinical uses. *Intl. J. Cell Cloning* **1990**, 8, 161-170.

26. Weber, G.; Prajda, N.; Abonyi, M.; Look, K.Y.; Tricot, G. Tiazofurin: Molecular and clinical action. *Anticancer Res.* **1996**, 16, 3313-3322.

27. Tricot, G.; Weber, G. Biochemically targeted therapy of refractory leukemia and myeloid blast crisis of chronic granulocytic leukemia with tiazofurin, a selective blocker of inosine 5'-phosphate dehydrogenase activity. *Anticancer Res.* **1996**, 16, 3341-3348.

28. Wright, D. G.; Boosalis, M. S.; Oshry, L.; Warasjam, K. Tiazofurin effects on IMP dehydrogenase activity and expression in the leukemic cells of patients with CML in blast crisis. *Anticancer Res.* **1996**, 16, 3349-3354.

29. Natsumeda, Y.; Ohno, S.; Kawasaki, H.; Konno, Y.; Weber, G.; Suzuki, K. Two distinct cDNAs for human IMP dehydrogenase. *J. Biol. Chem.* **1990**, 265, 5292-5295.

30. Konno, Y.; Natsumeda, Y.; Nagai, M.; Yamaji, Y.; Ohno, S.; Suzuki, K.; Weber, G. Expression of human IMP dehydrogenase types I and II in *Escherichia coli* and distribution in human normal lymphocytes and leukemic cell lines. *J. Biol. Chem.* **1991**, 266, 506-509.

31. Nagai, M.; Natsumeda, Y.; Konno, Y.; Hoffman, R.; Irino, S.; Weber, G. Selective up-regulation of type II inosine 5'-monophosphate dehydrogenase messenger RNA expression in human leukemias. *Cancer Res.* **1991**, 51, 3886-3890.

32. Nagai, M.; Natsumeda, Y.; Weber, G. Proliferation-linked regulation of type II IMP dehydrogenase gene in human normal lymphocytes and HL-60 leukemic cells. *Cancer Res.* **1992**, 52, 258-261.

33. Goldstein, B. M.; Colby, T. D. Conformational constraints in NAD analogs: implications for dehydrogenase binding and specificity. *Advan. Enzyme Regul.* **2000**, 40, 405-426.

34. Yamada, Y.; Natsumeda, Y.; Weber, G. Action of the active metabolites of tiazofurin and ribavirin on purified IMP dehydrogenase. *Biochemistry* **1988**, 27, 2193-2196.

35. Olah, E.; Natsumeda, Y.; Ikegami, T.; Kote, Z.; Horanyi, M.; Szelenyi, J.; Paulik, E.; Kremmer, T.; Hollan, S. R.; Sugar, J.; Weber, G. Induction of erythroid differentiation and modulation of gene expression by tiazofurin in K562 leukemia cells. *Proc. Natl. Acad. Sci. U.S.A.* **1988**, 85, 6533-6537.

36. Olah, E.; Csokay, B.; Prajda, N.; Kote-Jarai, Z.; Yeh, Y. A.; Weber, G. Molecular mechanisms in the antiproliferative action of taxol and tiazofurin. *Anticancer Res.* **1996**, 16, 2469-2478.
37. Vitale, M.; Zamai, L.; Falcieri, E.; Zauli, G.; Gobbi, P.; Santi, S.; Cinti, C.; Weber, G. IMP dehydrogenase inhibitor, tiazofurin, induces apoptosis in K562 human erythroleukemia cells. *Cytometry* **1977**, 30, 61-66.
38. Weber, G.; Nagai, M.; Natsumeda,Y.; Eble, J. N.; Jayaram, H. N.; Paulik, E.; Zhen, W.; Hoffman, R.; Tricot, G. Tiazofurin down-regulates expression of c-Ki-*ras* oncogene in a leukemic patient. *Cancer Commun.* **1991**, 3, 61-66.
39. Weber, G.; Shen, F.; Prajda, N.; Yeh, Y. A.; Yang, H.; Herenyiova, M.; Look, K. Y. Increased signal transduction activity and down-regulation in human cancer çells. *Anticancer Res.* **1996**, 16, 3271-3282.
40. Weber, G.; Prajda, N.; Yang, H.; Yeh, Y. A.; Shen, F.; Singhal, R. L.; Herenyiova, M.; Look, K. Y. Current issues in the regulation of signal transduction. *Advan. Enzyme Regul.* **1996**, 36, 33-55.
41. Weber, G.; Singhal, R. L.; Prajda, N.; Yeh, Y. A.; Look, K. Y.; Sledge, G. W., Jr. Regulation of signal transduction. *Advan. Enzyme Regul.* **1995**, 35, 1-21.
42. Weber, G.; Nagai, M.; Natsumeda, Y.; Ichikawa, S.; Nakamura, H.; Eble, J. N.; Jayaram, H. N.; Zhen, W.; Paulik, E.; Hoffman, R.; Tricot, G. Regulation of *de novo* and salvage pathways in chemotherapy. *Advan. Enzyme Regul.* **1991**, 31, 45-67.
43. Look, K. Y.; Sutton, G. P.; Natsumeda, Y.; Eble, J. N.; Stehman, F. B.; Ehrlich, C. E.; Olah, E.; Prajda, N.; Bosze, P.; Eckhardt, S.; Weber, G. Inhibition by tiazofurin of inosine 5'-phosphate dehydrogenase (IMP DH) activity in extracts of ovarian carcinomas. *Gynecol. Oncol.* **1992**, 47, 66-70.
44. Weber, G.; Shen, F.; Li, W.; Yang, H.; Look, K. Y.; Abonyi, M.; Prajda, N. Signal transduction and biochemical targeting of ovarian carcinoma. *Europ. J. Gynaec. Oncol.* **2000**, 21, 231-236.
45. Shen, F.; Herenyiova, M.; Weber, G. Synergistic down-regulation of signal transduction and cytotoxicity by tiazofurin and quercetin in human ovarian carcinoma cells. *Life Sci.* **1999**, 64, 1869-1876.
46. Shen, F.; Weber, G. Tamoxifen down-regulates signal transduction and is synergistic with tiazofurin in human breast carcinoma MDA-MB-435 cells. *Oncol. Res.* **1998**, 10, 325-331.
47. Zhen, W.; Jayaram, H. N.; Weber, G. Antitumor activity of tiazofurin in human colon carcinoma HT-29. *Cancer Invest.* **1992**, 10, 505-511.
48. Natsumeda, Y.; Yamada, Y.; Yamaji, Y.; Weber, G. Synergistic cytotoxic effect of tiazofurin and ribavirin in hepatoma cells. *Biochem. Biophys. Res. Comm.* **1988**, 153, 321-327.
49. Yeh, Y. A.; Olah, E.; Wendel, J. J.; Sledge, G. W., Jr.; Weber, G. Synergistic action of taxol with tiazofurin and methotrexate in human breast cancer cells: schedule-dependence. *Life Sci. Pharmacol. Letters* **1994**, 54, 431-435.

50. Shen, F.; Herenyiova, M.; Weber, G. Synergistic down-regulation of signal transduction and cytotoxicity by tiazofurin and quercetin in human ovarian carcinoma cells. *Life Sci.* **1999,** 64, 1869-1876.

51. Li, W.; Weber, G. Synergistic action of tiazofurin and genistein in human ovarian carcinoma cells. *Oncol. Res.* **1998,** 10, 117-122.

52. Li, W.; Weber, G. Synergistic action of tiazofurin and genistein on growth inhibition and differentiation of K-562 human leukemic cells. *Life Sci.* **1998,** 63, 1975-1981.

Chapter 3

Biological Significance of Guanylate Synthesis and IMP Dehydrogenase Isoforms

Yutaka Natsumeda

Clinical Development Institute, Banyu Pharmaceutical Company, Ltd.,
AIG Kabutocho Building, 5–1 Nihombashi Kabutocho, Chuo-ku,
Tokyo 103–0026, Japan

IMP dehydrogenase (IMPDH) is the rate-limiting enzyme for *de novo* guanine nucleotide synthesis and a potential target for anticancer and immunosuppressive chemotherapy. Human IMPDH was regarded as a single molecular species until the discovery in 1989 of two isoforms derived from different genes. The two isoforms showed striking differences in expression during neoplastic transformation and proliferation, lymphocytic activation and cell maturation. The type II IMPDH expression is stringently linked with immature characteristics and type I appears to be expressed constitutively. Selective inhibition of the inducible type II IMPDH may mitigate toxicity caused by inhibition of the type I isoform. Although mycophenolic acid has a slight selectivity to type II in its inhibition, the selectivity may not be sufficient. Further studies will be required to elucidate biological, biochemical and structural differences between type I and type II IMPDHs and to better understand the selective inhibition of these two isoforms in anticancer and immunosuppressive chemotherapy.

Introduction

IMP dehydrogenase (IMPDH) catalyzes the conversion of IMP to XMP at the IMP metabolic branch point and appears to be the rate-limiting enzyme for biosynthesis of guanine nucleotides which play important roles in many anabolic and regulatory processes (*1-6*). Weber *et al.* conducted extensive studies using rat transplantable hepatomas with different growth rates and demonstrated quantitative and qualitative changes in key purine metabolic enzymes in tumors as compared to those in normal rat liver (*7*). Among these enzymes, IMPDH has attracted great interest as a target for anticancer chemotherapy because a positive correlation was demonstrated between the rate of tumor proliferation and the activity of IMPDH. The specific activity of IMPDH, which is the least in rat normal liver among key purine metabolic enzymes, was increased the most in rat rapidly growing hepatoma (*7*). In fact, inhibition of IMPDH has been proved to have a potent antiproliferative effect against various tumor cells (*8-21*) and also lymphocytes (*22-24*). The mechanism of the antiproliferative effect has been attributed to the decrease in intracellular concentrations of guanine nucleotides, especially GTP and dGTP in the target cells (*8, 12-14, 22-25*). Sensitivities to the IMPDH inhibitors such as tiazofurin and ribavirin vary with cell types depending on the level of drug activation to their active metabolites and on requirement of *de novo* guanine nucleotide biosynthesis for proliferation in the cells.

An exciting discovery was made in 1989 and reported in 1990 (*26*) that there exist two distinct cDNAs (type I and type II) encoding human IMPDH with 84% amino acid sequence identity (*26, 27*). The type I IMPDH gene is located on chromosome 7 (*28*), and the type II gene is on chromosome 3 (*29*). The type II enzyme is regarded as an important chemotherapy target because type II mRNA expression is specifically up-regulated during neoplastic transformation and lymphocytic activation and down-regulated during cancer cell differentiation (*30-34*). On the other hand, type I IMPDH mRNA appears to be expressed constitutively in the various states of proliferation and differentiation (*31, 32*). The striking differences in regulation of the expression of IMPDH isoforms during transformation and cell proliferation may open a novel approach to isozyme-targeted chemotherapy.

Biological Significance of Guanine Nucleotide Synthesis in Proliferating Cells

Guanine nucleotides are required not only in nucleic acid biosynthesis but also in many anabolic processes such as synthesis of protein, phospholipids, adenine nucleotides and biopterines, protein glycosylation, cytoskeletal organization, and transmembrane signaling (*1-3*). Intracellular concentrations of guanine nucleotides are one order of magnitude lower than those of adenine nucleotides (*8, 12, 13, 24, 25*). At the IMP metabolic branch point, IMP is predominantly utilized to adenine nucleotide synthesis in the resting condition (*35*). When cells start growing, IMP utilization for guanine nucleotide synthesis is preferentially increased and that for adenine nucleotide synthesis inversely decreased (*35*). The link between preferential re-direction of metabolic switching at the IMP branch point toward guanylate synthesis and growth stimulation supports the potential significance of the guanylate pathway and IMPDH as targets of chemotherapy. In a strict sense, type II IMPDH should be the crucial target enzyme, because the expression of this isozyme is selectively and inherently linked with cell proliferation and immature characteristics.

Since guanine nucleotides are synthesized not only by the *de novo* pathway but also by the salvage pathway, it is relevant to elucidate relative contributions of both synthetic pathways for guanylate production in cells. Hypoxanthine-guanine phosphoribosyltransferase (HGPRT) catalyzes one step production of IMP and GMP from hypoxanthine and guanine, respectively, with the co-substrate 5-phosphoribosyl 1-pyrophosphate (PRPP) (*36*). Extensive studies comparing specific activities of the purine salvage enzyme, HGPRT, and *de novo* enzymes involved in IMP and GMP synthesis, amidophosphoribosyltransferase (amidoPRT) and IMPDH, showed that potential capacities of the salvage pathway were much higher than those of the *de novo* pathways in all the tissues and cells examined (*37-39*) except HGPRT-deficient mutants. Also the affinity to the common substrate PRPP of HGPRT was orders of magnitude higher than that of amidoPRT (*37*). This fact explains why the drug action associated with guanine nucleotide depletion induced by IMPDH inhibitors is circumvented by adding the salvage precursor guanine or guanosine (*12, 17, 21, 22, 24, 40*). The high capacity of the salvage pathway is biologically reasonable, because reutilization of a preformed purine ring can save three molecules of ATP to produce one molecule of purine nucleotide. Nevertheless, the salvage synthesis of guanine nucleotides seems to be limited *in vivo* in humans because the salvage precursor, guanine or guanosine, is not fully available in plasma (*41, 42*). Although hypoxanthine is available in plasma (*41, 42*) and is converted to IMP, guanine nucleotide synthesis could still be blocked in the presence of an inhibitor of IMPDH or GMP synthase.

Biological Significance of Depletion in Guanine Nucleotides

In rats carrying subcutaneously transplanted hepatoma 3924A solid tumors, a single intraperitoneal injection of tiazofurin, which is bio-activated to a potent IMPDH inhibitor, thiazole-4-carboxamide adenine dinucleotide (TAD), depleted GDP, GTP and dGTP pools in the tumor; and concurrently, IMP and PRPP pools expanded (*8*). The increase of PRPP concentration was attributed to the enhancement of the inosinate cycle consisting of 5'-nucleotidase, inosine phosphorylase, phosphoribomutase, PRPP synthase and HGPRT (*43*). In normal liver the effect of tiazofurin was less pronounced than in the hepatoma (*8*). The differential response in normal liver and hepatoma might be attributable to different metabolic turnover rates of guanylate production and its utilization in those tissues rather than the different metabolic rates producing the active metabolite TAD. In fact IMPDH was inhibited in normal liver to much lower activity than the remaining activity in the tumor after tiazofurin injection.

The decrease in GTP pool was closely correlated with the cytotoxic effect of tiazofurin in hepatoma 3924A cells (*8*). It has also been shown that depletion of the GTP pool caused by an IMPDH inhibitor results in a decrease in glycosylation of protein (*17, 44*) including adhesion molecules (*45, 46*), biopterin synthesis (*5*), RNA-primed DNA synthesis (*47*), IgE receptor-mediated degranulation (*48, 49*), secretion of serotonin (*50*), and cellular ras-GTP complex (*51*). It also results in an increase in type II IMPDH mRNA level (*52*) and purine salvage pathway capacity (*53*). Recently mycophenolic acid, an inhibitor of IMPDH, was demonstrated also to suppress cytokine-induced nitric oxide production in vascular endothelial cells (*54, 55*).

Treatment of immature proliferating cells with an inhibitor of IMPDH induces differentiation and maturation of the cells (*13, 14, 17, 18, 40, 56-59*). Differentiation induced by a known inducer such as retinoic acid,

12-O-tetradecanylphorbol-13-acetate (TPA) or dimethyl sulfoxide, which does not inhibit IMPDH activity, is associated with down regulation of type II IMPDH (31, 32). Thus, type II IMPDH is inherently linked with cell proliferation and should be a crucial target for anticancer and immunosuppressive chemotherapy.

Structural Similarities and Differences between IMPDH Isoforms

Human type II IMPDH has a similar primary structure to mouse and chinese hamster IMPDHs in which only 6 and 7 amino acids out of 514 are different, respectively (26, 27, 60). The difference between the human type I and type II IMPDH sequences is much more extensive: 84 out of 514 amino acids differ (26). Among these changes, 52 are conservative amino acid substitutions and 32 diverge with respect to their chemical properties. The consensus nucleotide-binding motif of β-α-β has been predicted in *Escherichia coli* IMPDH on the basis of steric and physicochemical properties of amino acids from pattern searches of protein-sequence databases (61). The domains are located from Asp-319 to Lys-349 in human IMPDHs and the alignments are well conserved in IMPDHs through evolution from bacteria to mammals (26, 27, 60, 62-65). The domain includes a cysteine which is the only sulfhydryl group conserved in various IMPDHs (64, 66) and appears to be involved in IMP-binding (66-69).

Comparison of Kinetic Properties of Human Type I and Type II IMPDHs

To elucidate differences in kinetic properties of human IMPDH isoforms, unmodified recombinant sequences of each isoform were overexpressed in an IMPDH-deficient strain of *E.coli* and purified to homogeneity (70). Both recombinant IMPDHs were tetramers which was in agreement with the subunit structure of the native mammalian IMPDH (70, 71). Recombinant IMPDH fusion proteins were a mixture of monomer, dimer, tetramer and various sizes of aggregated forms, even though the enzyme was purified to homogeneity as judged by sodium dodecyl sulfate-polyacrylamide gel electrophoresis.

In the studies using the recombinant enzymes, human type I and II IMPDHs both exhibited an ordered Bi Bi kinetic mechanism in which IMP binds to the free enzyme first, followed by XMP (70). Both IMPDH isoforms showed similar affinities for the substrates. The K_m values determined from the steady-state fit for type I and type II IMPDHs were in agreement with kinetic data previously reported for murine (71, 72) and human IMPDH (73), which presumably included a mixture of the two isoforms, and were 18 and 9 μM, respectively, for IMP, and 46 and 32 μM, respectively, for NAD (70). The isoforms had similar k_{cat} values of 1.5 and 1.3 turnovers/molecule of enzyme/second at 37°C for types I and II, respectively (70). Two alternative substrates, nicotinamide hypoxanthine dinucleotide and nicotinamide guanine dinucleotide were able to act as hydride acceptors, although their affinities were 20- to 150-fold lower than the natural substrate, NAD (70). In spite of the significant changes in K_m values for these alternative hydride acceptors, k_{cat} values remained unchanged from those obtained with NAD, suggesting that the rate-limiting step of the reaction does not involve binding of NAD to the enzyme (70). Inhibitory mechanisms by the products, XMP and NADH, were identical for human type I and type II IMPDHs. XMP inhibited in a competitive manner against IMP and K_i values of type I and II isoforms were 80 and 94 μM, respectively (70). NADH inhibited in an uncompetitive manner against IMP and with a mixed-type pattern against NAD and K_i values were 102 μM for type I and 90 μM for type II IMPDH, respectively (70).

Inhibition of IMPDH Isozymes by Mycophenolic Acid

Mycophenolate mofetil, a morpholinoethyl ester of mycophenolic acid, is clinically used as an immunosuppressive agent for the prevention of graft rejection after organ transplantation. The esterification to the prodrug improves the bioavailability of mycophenolic acid, a potent inhibitor of IMPDH (74). Pharmacokinetic studies showed that oral administration of mycophenolate mofetil provided plasma concentrations sufficient to show immunosuppressive effect with limiting toxicity, even when higher doses than the therapeutic dose are given to non-human primates for long periods (75).

The mechanism of inhibition by mycophenolic acid was the same for both type I and type II IMPDHs (70). The drug inhibited in an uncompetitive manner with respect to IMP and NAD, indicating that mycophenolic acid interacts with the enzyme after both substrates bind to it. Uncompetitive inhibition by this drug is of particular benefit due to the substrate accumulation, which potentiates inhibition. The results of kinetic studies suggest that mycophenolic acid binds to either the enzyme-IMP (XMP)-NAD (NADH) ternary complex or the enzyme-XMP complex (70, 76). The affinity of mycophenolic acid to type II IMPDH was 4.8-fold higher than that to the type I isoform (70). Further analyses and attempts are in progress to design more selective compounds to type II IMPDH (77).

Acknowledgments

The author is indebted to Dr. T. Ikegami for his dedication in the early stage of this work especially for developing the radioassay (78) and the affinity column chromatography for IMPDH purification (79). Special thanks go to Dr. S. Ohno for his guidance on cDNA cloning and discovery of human IMPDH isoforms. The author is also grateful to Dr. G. Weber, Dr. K. Suzuki and Dr. K. Tsushima for their continuous encouragement and discussions for this work and to Ms. Yasuko Asano for assistance in preparing this manuscript.

References

1. Weber, G.; Natsumeda, Y.; Pillwein, K. *Adv. Enzyme Regul.* **1986,** *24,* 45-65.
2. Pall, M. L. *Curr. Top Cell Regul.* **1985,** *25,* 1-19.
3. Bourne, H. R.; Sanders, D. A.; McCormick, F. *Nature* (Lond) **1990,** *348,* 125-131.
4. Metz, S. A.; Rabaglia, M. E.; Pintar, T. J. *J. Biol. Chem.* **1992,** *267,* 12517-12527.
5. Hatakeyama, K.; Harada, T.; Kagamiyama, H. *J. Biol. Chem.* **1992,** *267,* 20, 734.
6. Nicchitta, C. V.; Joseph, S. K.; Williamson, J. R. *Biochem. J.* **1987,** *248,* 741-747.
7. Weber, G. *Cancer Res.* **1983,** *43,* 3466-3492.
8. Lui, M. A.; Faderan, M. A.; Liepnieks, J. J.; Natsumeda, Y.; Olah, E.; Jayaram, H. N.; Weber, G. *J. Biol. Chem.* **1984,** *259,* 5078-5082.
9. Natsumeda, Y.; Yamada, Y.; Yamaji, Y.; Weber, G. *Biochem. Biophys. Res. Commun.* **1988,** *153,* 321-327.

10. Fukui, M.; Inaba, M.; Tsukagoshi, S.; Sakurai, Y. *Cancer Res.* **1982,** *42,* 1098-1102.
11. Smith, C. M.; Fontenelle, L. J.; Muzik, H.; Paterson, A. R. P.; Unger, H.; Brox, L. W.; Henderson, J. F. *Biochem. Pharmacol.* **1974,** *23,* 2727-2735.
12. Lee, H-J.; Pawlak, K.; Nguyen, B. T.; Robins, R. K.; Sadée, W. *Cancer Res.* **1985,** *45,* 5512-5520.
13. Yu, J.; Lemas, V.; Page, T.; Connor, J. D.; Yu, A. L. *Cancer Res.* **1989,** *49,* 5555-5560.
14. Kiguchi, K.; Collart, F. R.; Henning-Chubb, C.; Huberman, E. *Cell Growth & Different* **1990,** *1,* 259-270.
15. Uehara, Y.; Hasegawa, M.; Hori, M.; Umezawa, H. *Cancer Res.* **1985,** *45,* 5230-5234.
16. O'Dwyer, P. J.; Wagner, B. H.; Stewart, J. A.; Leyland-Jones, B. *Cancer Treat. Rep.* **1986,** *70,* 885-889.
17. Sokoloski, J. A.; Blair, O. C.; Sartorelli, A. C. *Cancer Res.* **1986,** *46,* 2314-2319.
18. Kiguchi, K.; Collart, F. R.; Henning-Chubb, C.; Huberman, E. *Exp. Cell Res.* **1990,** *187,* 47-53.
19. Sood, A.; Spielvogel, B. F.; Shaw, B. R.; Carlton, L. D.; Burnham, B. S.; Hall, E. S.; Hall, I. H. *Anticancer Res.* **1992,** *12,* 335-344.
20. Koyama, H.; Tsuji, M. *Biochem. Pharmacol.* **1983,** *32,* 3547-3553.
21. Stet, E. H.; De Abreu, R. A.; Bökkerink, J. P. M.; Lambooy, L. H. J.; Vogels-Mentink, T. M.; Keizer-Garritsen, J. J.; Trijbels, F. J. M. *Ann. Clin. Biochem.* **1994,** *31,* 174-180.
22. Eugui, E. M.; Almquist, S. J.; Muller, C. D.; Allison A. C. *Scand J. Immunol.* **1991,** *33,* 161-173.
23. Turka, L. A.; Dayton, J.; Sinclair, G.; Thompson, C. B.; Mitchell, B. S. *J. Clin. Invest.* **1991,** *87,* 940-948.
24. Dayton, J. S.; Turka, L. A.; Thompson, C. B.; Mitchell, B. S. *Mol. Pharmacol.* **1992,** *41,* 671-676.
25. Fukui, M.; Inaba, M.; Tsukagoshi, S.; Sakurai, Y. *Cancer Res.* **1986,** *46,* 43-46.
26. Natsumeda, Y.; Ohno, S.; Kawasaki, H.; Konno, Y.; Weber, G.; Suzuki, K. *J. Biol. Chem.* **1990,** *265,* 5292-5295.
27. Collart, F. R.; Huberman, E. *J. Biol. Chem.* **1988,** *263,* 15769-15772.
28. Gu, J. J.; Kaiser-Rogers, K.; Rao, K.; Mitchell, B. S. *Genomics* **1994,** *24,* 179-181.
29. Glesne, D.; Collart, F.; Varkony, T.; Drabkin, H.; Huberman, E. *Genomics* **1993,** *16,* 274-277.

30 Konno, Y.; Natsumeda, Y.; Nagai, M.; Yamaji, Y.; Ohno, S.; Suzuki, K.; Weber, G. *J. Biol. Chem.* **1991**, *266*, 506-509.

31. Nagai, M.; Natsumeda, Y.; Konno, Y.; Hoffman, R.; Irino, S.; Weber, G. *Cancer Res.* **1991**, *51*, 3886-3890.

32. Nagai, M.; Natsumeda, Y.; Weber, G. *Cancer Res.* **1992**, *52*, 258-261.

33. Collart, F. R.; Chubb, C. B.; Mirkin, B. L.; Huberman, E. *Cancer Res.* **1992**, *52*, 5826-5828.

34. Collart, F. R.; Huberman, E. *Blood* **1990**, *75*, 570-576.

35. Natsumeda, Y.; Ikegami, T.; Murayama, K.; Weber, G. *Cancer Res.* **1988**, *48*, 507-511.

36. Natsumeda, Y.; Yoshino, M.; Tsushima, K. *Biochim. Biophys. Acta* **1977**, *483*, 63-69.

37. Natsumeda, Y.; Prajda, N.; Donohue, J. P.; Glover, J. L.; Weber, G. *Cancer Res.* **1984**, *44*, 2475-2479.

38. Natsumeda, Y.; Lui, M. S.; Emrani, J.; Faderan, M. A.; Eble, J. N.; Glover, J. L.; Weber, G. *Cancer Res.* **1985**, *45*, 2556-2559.

39. Natsumeda, Y.; Ikegami, T.; Weber, G. *Adv. Exp. Med. Biol.* **1986**, *195B*, 371-376.

40. Olah, E.; Natsumeda, Y.; Ikegami, T.; Kote, Z.; Horanyi, M.; Szelenyi, J.; Paulik, E.; Kremmer, T.; Hollan, S. R.; Sugar, J. *Proc. Natl. Acad. Sci. USA* **1988**, *85*, 6533-6537.

41. Simmonds, R. J.; Harkness, R. A. *J. Chromatography* **1981**, *226*, 369-381.

42. Leyva, A.; Schornagel, J.; Pinedo, H. M. *Adv. Exp. Med. Biol.* **1980**, *122B*, 389-394.

43. Willis, R. C.; Kaufman, A. H.; Seegmiller, J. E. *J. Biol. Chem.* **1984**, *259*, 4157-4161.

44. Sokoloski, J.; Sartorelli, A. C. *Mol. Pharmacol.* **1985**, *28*, 567-573.

45. Allison, A. C.; Kowalski, W. J.; Muller, C. J.; Waters, R. V.; Eugui, E. M. *Transplant Proc.* **1993**, *25*, 67-70.

46. Allison, A. C.; Eugui, E. M. *Springer Semin. Immunopathol.* **1993**, *14*, 353-380.

47. Catapano, C. V.; Dayton, J. S.; Mitchell, B. S.; Fernandes, D. J. *Mol. Pharmacol.* **1995**, *45*, 948-955.

48. Wilson, B. S.; Deanin, G. G.; Standefer, J. C.; Vanderjagt, D.; Oliver, J. M. *J. Immunol.* **1989**, *143*, 259-265.

49. Mulkins, M. A.; Ng. M.; Lewis, R. A. *Cell Biol.* **1992**, *141*, 508-517.

50. Wilson, B. S.; Deanin, G. G.; Oliver, J. M. *Biochem. Biophys. Res. Commun.* **1991**, *174*, 1064-1069.

51. Hata, Y.; Natsumeda, Y.; Weber, G. *Oncol. Res.* **1993**, *5*, 161-164.

52. Glesne, D. A.; Collart, F. R.; Huberman, E. *Mol. Cell Biol.* **1991**, *11*, 5417-5425.
53. Natsumeda, Y.; Ikegami, T.; Olah, E.; Weber, G. *Cancer Res.* **1989**, *49*, 88-92.
54. Senda, M.; DeLustro, B.; Eugui, E.; Natsumeda, Y. *Transplant* **1995**, *60*, 1143-1148.
55. Allison, A. C.; Eugui, E. M. *Immunopharmacology* **2000**, *47*, 85-118.
56. Lucas, D. L.; Webster, H. K.; Wright, D. G. *J. Clin. Invest.* **1983**, *72*, 1889-1900.
57. Knight, R. D.; Mangum, J.; Lucas D. L.; Cooney, D. A.; Khan, E. C.; Wright, D. G. *Blood* **1987**, *69*, 634-639.
58. Downs, S. M.; Eppig, J. J. *Biol. Reprod.* **1987**, *36*, 431-437.
59. Eppig, J. J. *Biol. Reprod.* **1991**, *45*, 824-830.
60. Tiedeman, A. T.; Smith, J. M. *Gene* **1991**, *97*, 289-293.
61. Bork, P.; Grunwald, C. *Eur. J. Biochem.* **1990**, *191*, 347-358.
62. Wilson, K.; Collart, F. R.; Huberman, E.; Stringer, J. R.; Ullman, B. *J. Biol. Chem.* **1991**, *266*, 1665-1671.
63. Kanzaki, N.; Miyagawa, K. *Nucleic Acids Res.* **1990**, *18*, 6710.
64. Collart, F. R.; Osipiuk, J.; Trent, J.; Olsen, G. J.; Huberman, E. *Gene* **1996**, *174*, 209-216.
65. Zhang, R.; Evans, G.; Rotella, F.; Westbrook, E.; Huberman, E.; Joachimiak, A.; Collart, F. R. *Current Medicinal Chemistry* **1999**, *6*, 537-543.
66. Natsumeda, Y.; Carr, S. F. *Annal. New York Acad. Sci.* **1993**, *696*, 88-93.
67. Wu, J. C. *Perspect. Drug Discovery Design* **1994**, *2*, 185-204.
68. Antonino, L. C.; Straub, K.; Wu, J. C. *Biochemistry* **1994**, *33*, 1760-1765.
69. Huete-Pérez, J. A.; Wu, J. C.; Whitby, F. G.; Wang, C. C. *Biochemistry* **1995**, *34*, 13889-13894.
70. Carr, S. F.; Papp, E.; Wu, J. C.; Natsumeda, Y. *J. Biol. Chem.* **1993**, *268*, 27286-27290.
71. Yamada, Y.; Natsumeda, Y.; Weber, G. *Biochemistry* **1988**, *27*, 2193-2196.
72. Holmes, E. W.; Pehlke, D. M.; Kelley, W. N. *Biochim. Biophys. Acta* **1974**, *364*, 209-217.
73. Anderson, J. H.; Sartorelli, A. C. *J. Biol. Chem.* **1968**, *243*, 4762-4768.
74. Lee, W. A.; Gu, L.; Miksztal, A. R.; Chun, N.; Leung, K.; Nelson, P. H. *Pharm. Res.* **1990**, *7*, 161.

75. Morris, R. E.; Wang, J.; Blum, J. R.; Flavin, T.; Murphy, M. P.; Almquist, S. J.; Chu, N.; Tam, Y. L.; Kaloosian, M.; Allison, A. C.; Eugui, E. M. *Transplant Proc.* **1991**, *23*, 19-25.
76. Hedstrom, L.; Wang, C. C. *Biochemistry* **1990**, *29*, 849-854.
77. Goldstein, B. M.; Colby, T. D. *Current Medicinal Chemistry* **1999**, *6*, 519-536.
78. Ikegami, T.; Natsumeda, Y.; Weber, G. *Analyt. Biochem.* **1985**, *150*, 155-160.
79. Ikegami, T.; Natsumeda, Y.; Weber, G. *Life Sci.* **1987**, *40*, 2277-2282.

Chapter 4

An Emerging Cell Kinetics Regulation Network: Integrated Control of Nucleotide Metabolism and Cancer Gene Function

James L. Sherley

The Division of Bioengineering and Environmental Health and Center for
Environmental Health Science, Massachusetts Institute of Technology, 77
Massachusetts Avenue, Building 16, Room 755, Cambridge, MA 02139
(telephone: 617–258–8853; fax: 617–258–8648; email: jsherley@mit.edu)

Until recently, metabolic genes and cancer genes have been studied as independent determinants of tissue cell kinetics. Among metabolic genes, those encoding nucleotide biosynthesis enzymes, in particular, have been extensively studied as essential regulators of cell function. As determinants of rates of nucleotide production, nucleotide biosynthesis genes are critical control factors in nucleic acid synthesis and molecular regulation of processes as diverse as protein translation and cell morphogenesis. Although, in retrospect, integrated regulation of cancer gene and nucleotide biosynthesis gene function seems an obvious expectation, only in recent years has evidence for such connections emerged. The inosine monophosphate dehydrogenase (IMPDH) gene was the first gene encoding a nucleotide biosynthesis enzyme demonstrated to be directly involved in cell kinetics regulation by a cancer gene. Since the seminal report of a functional relationship between the IMPDH gene and the p53 tumor suppressor gene, several other cancer gene-nucleotide gene interactions have been reported. Here, reports defining each of these regulatory interactions are critically evaluated towards a synthesis that reveals a cellular control network that integrates the action of cancer genes and nucleotide biosynthesis genes to control normal tissue cell kinetics. A prominent feature of the network is the function of both cancer gene products and nucleotide biosynthesis control enzymes as signal nodes from which input network signals are transmitted in parallel to two or more other network components. Though parallel in the network structure, these bifurcated paths often have quite distinct character because of signaling to either cancer gene or nucleotide metabolism gene pathways.

INTRODUCTION

Compared to nucleotide metabolism research, the field of cancer gene research is relatively young. Mature investigations of the nucleotide biochemistry of normal cell and tissue physiology and of disease states like cancer were well underway in the first half of the 20th century.[1] In contrast, the concept that cancer is a disease caused by altered expression of specific cellular genes is a development of just the last 30 years.[2,3] Typically, genes given the "cancer gene" designation encode proteins involved in one of three general classes of cellular function: growth signal transduction, regulation of gene expression, or maintenance of genome integrity[3]. The cancer gene paradigm is a powerful precept that continues to unify our understanding of the biological basis for diverse human cancers. However, there is an unfortunate consequence of the rapidly acquired preeminence of the cancer gene paradigm in cancer research. It has cast a shadow over work on cellular genes that are also important determinants of carcinogenesis, but which do not conform to the routine definitions that have been established for cancer genes. [3]

Among genes that play essential roles in normal and neoplastic tissue cell states, but not previously listed among cancer genes, are genes that encode nucleotide biosynthesis enzymes. During the era of the cancer gene paradigm, investigations of the roles of nucleotide metabolism genes and cancer genes in normal cell function and cellular carcinogenesis have proceeded along largely independent paths. However, recent advances in nucleotide metabolism gene research now set the stage for a nexus of thought regarding integrated regulation of cancer gene functions and nucleotide biosynthesis. The purpose of this review is to highlight recent developments at this research interface and to critically evaluate them as evidence for a cell kinetics control network that tethers cancer gene function to the activity of rate-limiting enzymes for the biosynthesis of cellular nucleotides.

Cancer genes

In general, previously described cancer genes encode proteins that function either directly or indirectly as determinants of tissue cell kinetics.[4] Cell kinetics is the quantitative description of the rates of change in tissue cell number. Tissue cell kinetics manifest the collective effect of a variety of individual cellular processes including mitosis, quiescence, apoptosis, differentiation, signal transduction, macromolecular synthesis, energy production, and cellular metabolism. Although defects in many genes can lead to changes in cell kinetics, only a subset promotes the formation of tumors. Cancer genes can be defined as those genes in which mutation or altered expression increases the frequency of cancer cell development. Although cancer cell development has been described as a multi-step genetic process[5,6], individual cancer genes are most clearly defined by the fact that their mutation independently increases the frequency of carcinogenesis in otherwise normal cells. A cancer cell arises only after acquisition of a complete set of several different necessary genetic alterations[3]. A cancer gene is a gene for which altered function or expression provides one member of the complete set of required transforming events.

Cancer genes have been classified into two main groups, oncogenes and tumor suppressor genes.[7] Roughly 50 genes have been identified in each category.[4,7-9] In general, oncogenes encode proteins that are involved in processes that promote increased rates of tissue cell accumulation. These include genes for growth factors, growth factor receptors, cytoplasmic factors that form signal transduction cascades and networks, nuclear transcription factors that activate growth-promoting genes, and components of the cell cycle regulation machinery.[3,4,7-10] Expression of mutant oncogenes or aberrant expression of their wild-type form (*i.e.*, proto-oncogenes) typically leads to increased tissue cell number. This change in tissue cell kinetics may occur by growth activation of otherwise quiescent cells, acceleration of cell cycle progression, suppression of cell differentiation, or suppression of cell apoptosis.[3,4]

Tumor suppressor genes are also called antioncogenes, because their normal cellular functions serve to restrict tissue cell accumulation. Unlike for oncogenes, cancer-inducing mutations in tumor suppressor genes often result in complete loss of the function of the encoded protein (though there are notable exceptions). Tumor suppressor genes encode negative regulators of growth signal transduction, cell cycle checkpoint genes, nuclear transcription factors, and enzymes involved in DNA synthesis and repair.[3,4,7] Mutations in tumor suppressor genes either directly or indirectly affect cell kinetics rates.[4] Defects in some tumor suppressor genes can result in activation of quiescent tissue cells, deregulated cell cycle progression, and decreased apoptosis. Mutations in others serve to either increase the frequency of base mis-incorporation errors during DNA replication or decrease the efficiency of the repair of replication mistakes and mutation-prone DNA damage. The latter indirect mechanisms increase the frequency of subsequent mutations in cancer genes that directly control tissue cell kinetics.

Nucleotide metabolism genes and cancer

In general, metabolism genes are not thought of as cancer genes. This sentiment is due in large part to the fact that many metabolism genes are essential. In particular, metabolism genes that encode rate-limiting enzymes that function at control points in metabolic pathways are often essential for cell viability. Therefore mutations that alter the function of metabolism genes often result in cell death which precludes cancer cell development. So, although they are critical determinants of cell kinetics, many metabolism genes are only passive contributors to carcinogenesis primarily in the way of performing the same cellular activities that they provide in normal cells. However, noted exceptions to this general rule are nucleotide metabolism genes.

·Enzymes that control the biosynthesis of cellular nucleotides are well known for their increased expression in malignant cells. Early cancer biochemists suggested a very rational basis for this phenomenon. The active proliferation of cancer cells might be expected to dictate increased levels of biosynthetic enzymes to meet increased demands for cellular nucleotides to support active DNA replication and RNA synthesis. However, as it turned out, in many cases a somewhat trivial mechanism accounted for the commonly observed elevated expression of nucleotide biosynthetic enzymes. As a group, these enzymes show periodic elevations during S-phase of the normal cell cycle.[11-13] Their elevated activity in tumor cell populations often simply reflects the higher S fraction of these actively dividing cell populations as compared to quiescent or terminally-arrested, differentiated cells in normal tissues.[11]

Several other mechanisms may contribute to elevated expression of particular nucleotide synthetic enzymes beyond that related to S phase-induction. Gene amplification and activating point mutations have been described, but in most cases these occur after treatments with chemotherapy agents that target nucleotide metabolism.[14,15] Recent *in vitro* molecular studies show that several genes that encode nucleotide biosynthetic enzymes can be regulated by oncogenic transcription factors (discussed below).[16] Alterations in the expression of these oncogenes in tumor cells may account for elevated expression of their regulated targets. However, such a mechanism has not been formally proven to operate in cancer cells.

Much of past research on the role of nucleotide biosynthetic enzymes in cellular regulation and cancer has focused on the role of nucleotides as building blocks for cellular nucleic acids. Less attention has been given to nucleotide metabolism as a major component of cell regulation mechanisms.[17-20] In many ways, the more recently emerged field of molecular genetics led to erosion of emphasis on nucleotide biosynthesis control as an instrument of molecular regulation. Nucleotides are often only regarded as fuels for regulatory kinases or nucleotidases, whereas variation in their concentration is seldom acknowledged as another potential mechanism of cellular regulation. Whereas nucleotides like GTP and ATP are well known as essential substrates for signal transduction pathways, there have been few evaluations of whether and how their production rate and concentration effect signaling function.[19-25] Most of these evaluations were performed in cells treated with cytotoxic inhibitors of specific nucleotide biosynthesis enzymes.[19,21-25] It seems likely that observations from inhibitor studies will reflect regulatory mechanisms in normal cells, but to what extent remains uncertain.

In this review, a synthesis of recent articles is presented to reveal a rich entwining of cancer gene function and nucleotide biosynthesis. The reports considered provide experimental evidence for the action of nucleotides in cell kinetics regulation and cancer mechanisms. The function of several important cancer genes is shown to be tied to enzymes that control the production of cellular nucleotides. Recent studies link four rate-limiting enzymes for nucleotide biosynthesis, inosine monophosphate dehydrogenase (IMPDH; EC 1.1.1.205), CAD, thymidylate synthase (TS; EC 2.1.1.45), and ribonucleotide diphosphate reductase (RDP; EC 1.17.4.1-2), to the regulation and/or function of four well-studied cellular cancer genes, p53, c-Myc, MAPK, and Ras. Together these four enzymes regulate the biosynthesis of all four classes of cellular nucleotides. The four cancer genes are major effectors of several aspects of tissue cell kinetics, including growth signal transduction, cell cycle regulation, cell kinetics symmetry, and apoptosis induction. Figure 1 provides a diagram of key interactions in this emerging cancer gene-nucleotide gene control network. The following discussion, though not exhaustive, will consider six cancer gene-nucleotide gene connections in the network. These connections are associated with signaling inputs, outputs, and recurrent loops that form four separable regulatory modules (Figure 1: I, II, III, IV). These four modules dovetail to create a highly integrated regulatory network that allows cells to match nucleotide production to demands of tissue cell kinetics and to coordinate differential regulation of growth factor signal transduction and cellular response.

Module I: IMPDH-p53

IMPDH has the distinction of being the first nucleotide biosynthesis enzyme recognized to mediate the function of a cancer gene.[26] IMPDH is an essential enzyme that catalyzes the conversion of inosine-5'-monophosphate

(IMP) to xanthosine-5'-monophosphate (XMP). This is the rate-limiting step for the production of cellular guanine nucleotides.[24] Like adenine nucleotides, in addition to functioning in energy metabolism and nucleic acid biosynthesis, guanine nucleotides act as molecular regulators of many diverse cellular processes required for cell viability and growth (e.g., protein translation control and growth factor signal transduction).[17,18,27,28] During the past decade, a series of studies established that cell proliferation suppression by the p53 tumor suppressor protein is due to its ability to modulate guanine ribonucleotide biosynthesis by controlling the expression of IMPDH.[20,26,29,30]

In recent years, the p53 gene has become known best for its effects in cells that have sustained DNA damage due to any of a number of examined insults. Cells with defects in p53 gene function exhibit associated deficiencies in cell cycle checkpoint mechanisms and apoptotic responses. These two cellular mechanisms are thought to greatly limit the accumulation in tissues of damaged mutant cells that are precursors for cancer cells.[31-33] The requirement of normal p53 gene expression for checkpoint and apoptotic function reflects the p53 protein's role as a transcriptional regulator of genes that function in cell cycle regulation and apoptosis.[32-34] Many agents that cause DNA damage lead to increased levels of p53 protein that is activated for gene regulation.[35,36] The p53 protein is a complex multi-dimensional transcription factor. It interacts directly with promoters or indirectly via other transcription factors to effect either transcriptional activation or repression in a gene specific fashion.[32,33,36]

Despite the current heavy emphasis of much of p53 research on the apparent role of the protein as a DNA damage response factor, p53 was originally described as a tumor suppressor gene because of its effects in undamaged cells.[37-39] In the absence of any known cellular damage, loss of p53 gene function results in increased cell proliferation. Consistent with this cellular response, restoration of normal p53 function in p53-deficient cells or elevated expression of p53 in cells of any p53 genotype often results in cell growth suppression.[20,26,30,32,37] All of these cell growth suppression effects are now typically gathered together under the rubric of p53-dependent cell cycle checkpoint control.[31-33]

Recently, a more detailed picture of the cell kinetics basis for p53-dependent growth suppression in undamaged cells was developed. This picture is based on findings with cultured murine fibroblasts and epithelial cells genetically engineered to express p53 conditionally. By simple changes in culture conditions, these cells can be transitioned from expressing subnormal or undetectable levels of p53 to levels that closely approximate the p53 concentration in normal murine cells.[20,26,30] Under conditions of low p53 expression, the model cells exhibit typical exponential cell kinetics. When normal p53 levels are induced, the cells undergo a complex form of growth suppression. Although the overall growth rate of cell cultures decreases, it is not caused by a uniform increase in cell generation time. Instead some cells undergo a complete cell cycle arrest, whereas others continue to cycle at the rate observed uniformly for cells with reduced p53 expression.[29,40]

Subsequent studies revealed the cell dynamics responsible for p53-dependent growth suppression in undamaged cells. Physiological p53 expression in immortal murine cells restores a rudimentary form of asymmetric cell kinetics that are characteristic of somatic stem cells in mammalian tissues. Each dividing cell produces another dividing cell and a daughter cell that is the precursor of a non-dividing cell lineage. In model cell cultures, the non-dividing lineage is composed of either a single viable arrested daughter cell or a daughter cell that divides once to produce two arrested cells.[29,40,41] Somatic tissue stem cells *in vivo* divide by a very similar asymmetric cell kinetics program. However, *in vivo* the non-dividing lineage can contain hundreds to thousands of cells that mature along specific differentiation paths to produce the diverse array of functional cell types that constitute a given tissue. Although a high degree of proliferation occurs in this lineage, all of the mature differentiated cell types produced undergo terminal cell cycle arrest.[42] The maturation and differentiation of cells in the non-dividing cell lineage are governed by extracellular factors such as cell-cell interactions, cell-extracellular matrix interactions, cytokines, and growth factors. The absence of these inputs in cell culture is the likely cause of the limited size and differentiation reported for the non-dividing lineage of asymmetrically dividing model cells. However, cell-autonomous mechanisms for asymmetric cell kinetics appear intact. For this reason, the conditional p53 expression lines were used as *in vitro* model systems to investigate molecular and biochemical mechanisms that control asymmetric cell kinetics.

The remarkable cell kinetics properties of cultured cell lines that were reconstituted with physiological levels of p53 protein, of course, suggested that p53 might have a similar function in normal mammalian tissues. A corollary to this hypothesis was that the essential cancer-inducing effect of loss of p53 function by gene mutation was the disruption of asymmetric cell kinetics in somatic tissue stem cells.[29,40,41] Years earlier, it had been postulated that the conversion of asymmetric stem cells to exponential cell kinetics would be a highly carcinogenic event in mammalian tissues.[43] However, it was not until the discovery of the ability of p53 to induce asymmetric cell kinetics in cultured cells that any known gene had been implicated as a determinant of asymmetric cell kinetics. Many features of p53 function in normal development and carcinogenesis can be accounted for by this single function in somatic stem cell kinetics regulation.[29,41] Reconciling effects of p53 in DNA damage responses with a function in asymmetric stem cell kinetics regulation is more problematic, but presently perceived distinctions may resolve when more is known about the biology of somatic stem cells.

Only very recently has it become clear that a reduction in cellular IMPDH activity is an obligatory requirement for p53-dependent asymmetric kinetics. [20,40,44] The identification of IMPDH as a determinant of p53-dependent "growth suppression" pre-dates the understanding that p53 suppresses the growth of undamaged cells in culture by virtue of inducing asymmetric cell kinetics.[26] In early studies with conditional p53 expression lines, it was noted that purine nucleosides like xanthosine could prevent p53-dependent growth suppression.[26] Xanthosine, by virtue of salvage enzyme conversion to XMP, can promote the formation of guanine ribonucleotides in the absence of IMPDH function (see Figure 1, I). Subsequent biochemical tracer studies showed that p53 expression induced changes in the kinetic rates of the guanine nucleotide biosynthesis pathway that were consistent with inhibition at the IMPDH step.[26] Later molecular studies showed that that p53 expression led to a reduction in IMPDH gene expression.[20,30] Through bypass of a p53-induced reduction in IMPDH function, xanthosine permitted cells to maintain a high proliferative rate. After the asymmetric cell kinetics features of p53-dependent growth suppression were known, xanthosine was shown to act by preventing the p53-induced switch from exponential to asymmetric cell kinetics. Nucleosides that expand only guanine deoxyribonucleotide pools were unable to prevent asymmetric cell kinetics.[29] This property suggested that a reduction in guanine ribonucleotide pools was a critical requirement for asymmetric cell kinetics.

The described pharmacological complementation studies with xanthosine suggested the involvement of IMPDH as a rate-determining factor for p53-dependent growth regulation. To confirm this interpretation, isogenic conditional p53 expression lines were derived with a constitutively expressed IMPDH transgene. These "impd-transfectant" lines had a modestly increased level of IMPDH protein. Upon p53-induction in these lines, although endogenous IMPDH expression was reduced, the constitutive expression of exogenous IMPDH was sufficient to keep total cellular IMPDH at the level found in p53-null cells. Although wild-type p53 function was intact in impd-transfectants, their proliferative rate was not significantly altered by p53 induction.[20] Just recently, it was established that the basis for their resistance to p53-induced growth arrest is the inability to switch to asymmetric cell kinetics.[44] Thus, this molecular genetics approach confirms the interpretation of earlier cell kinetics experiments with xanthosine and identifies IMPDH as a second genetic determinant of asymmetric stem cell kinetics that is connected to the first, p53, by a regulatory link.

The described studies uncovered the IMPDH-p53 connection in the context of normal tissue cell kinetics control. Many aspects of this cancer gene-nucleotide gene network connection remain to be elucidated. For example, the molecular basis of IMPDH gene regulation by p53 is unknown. Since IMPDH mRNA is reduced in p53-induced cells, a transcriptional repression mechanism seems likely.[30] A more intriguing question is how does down-regulation of IMPDH initiate the asymmetric cell kinetics program. Early studies with specific inhibitors of IMPD revealed a curious feature of the enzyme in cellular regulation. Although DNA synthesis is affected greater than RNA synthesis, reductions in rGNP pools appear to be more important for the associated effects on cell growth than reductions in dGNP pools. Thus, changes in IMPDH activity are thought to affect S phase through effects on rGNP-binding regulatory factors as opposed to effects on the immediate precursors for DNA.[23,45,46] Certainly, inhibition of IMPDH can be shown to reduce the activation states of

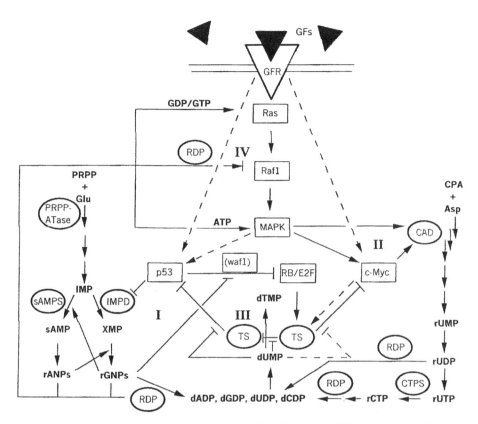

Figure 1. The emerging cancer gene-nucleotide gene cell kinetics control network. Incoming growth factor signals activate cancer gene products involved in cell kinetics control. Cancer gene products and other growth signal transduction factors function as signal nodes to split, transform, and transmit signal inputs to nucleotide biosynthesis control enzymes and cancer gene products. Five identified or potential signal nodes, p53, MAPK, c-Myc, RDP, and TS, are the essential integrating components of the network. They allow cells to regulate responses to incoming growth information by integrating external signal modulation with internal metabolic state. These signal nodes are key components of four network regulation modules (I-IV) that are based on the cancer gene-nucleotide gene connections described in the text. Closed triangles: GFs, growth factors. Open triangle: GFR, growth factor receptor. Rectangles: cancer gene products, signaling molecules, cell cycle regulators, transcription factors. Ovals: nucleotide biosynthesis control enzymes. Pointed arrows indicate substrate conversion or activation; blunted arrows indicate inhibition or repression; and dotted arrows identify poorly established pathways. Other abbreviations are CPA, carbamoyl phosphoric acid; Asp, aspartic acid; Glu, glutamine; rANPs, adenine ribonucleotides; and rGNPs, guanine ribonucleotides.

several GTP-binding signaling proteins.[21-23,25]   Since changes in the GTP/GDP ratio have also been shown to occur in cells treated with IMPDH inhibitors,[47,48] the potential for effects on Ras signaling has also been suggested. [20,22,23,26]

Recent studies with p53-inducible murine cells that contain stable constitutively expressed IMPDH transgenes implicate guanine ribonucleotides to play a role in the regulation of specific cyclin-dependent kinases that promote S phase progression. P53 has been shown to regulate the S-phase specific activity of cdk2/cyclin E cyclin kinase complexes by inducing the expression of the cyclin kinase inhibitor p21waf1.[49,50]   In p53-inducible murine cell lines, p21waf1 expression is induced by p53. Induction of p21waf1 is postulated to be a major mechanism of p53-dependent cell cycle arrest.   Inhibition of cdk2 containing complexes by p21waf1 prevents phosphorylation of one of their targets, the retinoblastoma (Rb) protein. Rb protein phosphorylation is needed to release E2F transcription factors that turn on genes whose products are required for S phase[16,51] (see also Module III section below).

During p53-dependent asymmetric cell kinetics, cells in the non-dividing lineage arrest in G1 and S phase of the cell cycle.[26,30]   In isogenic impd-transfectants, this arrest does not occur, but p21waf1 is still induced in response to p53 expression.[20,52]   Although elevated levels of p21waf1 are present, no inhibition of cdk2/cyclin E complexes is detected.[52]   These findings raise the possibility that guanine ribonucleotides modulate the ability of cdk2/cyclin E complexes to be inhibited by p21waf1.   This recurrent loop to p21waf1 is an important component of the IMPDH-p53 network module (Figure 1, I).

The p53 gene and protein receive a variety of both internal and external inputs that are related to cell health and growth.[36,53]   Within the proposed network, p53 appears to function as a signal node that splits incoming signals into two distinct components.[54]   One component activates a genetic pathway that functions to limit cell cycle progression. The other component dampens the production of purine nucleotides, which are unique in their role as fuels for growth factor signal transduction. Although, to date, effects of p53 expression on adenine nucleotide biosynthesis have not been detected, the known counter-balanced co-substrate regulation of the guanine and adenine branches of the purine nucleotide biosynthesis pathway[28,55] predicts that modest effects may occur. Clearly the most significant impact is on guanine ribonucleotides. Co-reduction of guanine ribonucleotides may cool down the rate of transduction of incoming growth activation signals when internal events mandate growth cessation via p53 activation. Such coordinate differential regulation could be critical for effective transient de-coupling of the cell cycle from growth factor stimulation. In this regard, it is noteworthy that p53-induced asymmetric non-dividing cells are quiescent in the presence of the full complement of serum growth factors.[26]

The postulated regulation of p21waf1 by guanine ribonucleotides provides a mechanism to sustain or amplify a cell cycle arrest that was initially signaled via p53 activation. On the other hand, cellular inputs that increase guanine ribonucleotide pools independent of IMPDH can use the p21waf1 link in the IMPDH-p53 module to counter effects of p53 activation (*e.g.*, salvage pathway activity). This regulatory dichotomy is fitting for a network module implicated in asymmetric stem cell kinetics. The molecular basis for asymmetric cell kinetics in both *in vitro* and *ex vivo* cell culture models is currently under investigation.[56] The IMPDH-p53 regulatory connection and its associated network module are postulated to be key elements of the cellular program for asymmetric cell kinetics.

Module II: CAD-c-Myc and CAD-MAPK

The rate limiting step for pyrimidine nucleotide biosynthesis in mammalian cells is catalyzed by the carbamoyl phosphate synthetase domain (CPS II; EC 6.3.5.5) of the multifunctional enzyme CAD. In addition to CPS II, CAD contains domains with aspartate transcarbamylase (ATCase; EC 2.1.3.2) and dihydro-orotase (EC 3.5.2.3) enzymatic activity. These enzymes catalyze the first three steps of pyrimidine nucleotide biosynthesis. The CPS II step is allosterically regulated by UTP, the end product of the pathway, and phosphoribosyl pyrophosphate (PRPP) a later substrate in the pathway.[57,58] Two different gene products, the c-Myc proto-oncogene and mitogen-activated protein kinases (MAPK), have recently been shown to regulate CAD gene expression and CPS II enzyme activity, respectively.

The c-Myc gene is among the most extensively studied cellular proto-oncogenes. In many human tumors the c-Myc gene is either amplified or activated by chromosome translocations, leading to high levels of expression of the c-Myc protein.[59] Whereas c-Myc is not an essential cellular gene, its over-expression can promote the immortalization of primary cells, and in cooperation with other mutated cancer genes it can transform primary cells in culture.[60]

C-Myc functions to induce the entry of quiescent cells into the cell cycle and to promote continued cell cycle progression.[61] It is induced as an immediate early gene after serum stimulation of resting cells[62] and is maintained at low basal levels in succeeding cell cycles.[63,64] The mechanisms of activation of c-Myc transcription in response to serum growth factors are not well defined, and there may be several (right, dotted arrow from "GFR" in Figure 1). At least one pathway involves the action of MAPKs activated by the Ras-signaling pathway.[65]

Despite the murkiness surrounding the basis of c-Myc induction, there is clear evidence that it is rate-determining for cell cycle entry and progression.

Genetically-engineered forced expression of c-Myc can induce serum-independent progression of quiescent cells into the cell cycle, whereas reduced expression or absence of c-Myc in cycling cells causes prolonged G1 and G2 periods.[61,66] In addition to its roles in cell growth activation and cell cycle progression, there is a large body of evidence that indicates that under some conditions c-Myc functions to induce apoptosis.[67]

The c-Myc protein binds to a sequence-related protein called Max to form a heterodimeric transcription factor.[59,68] Since its identification as transcription factor component, much of c-Myc gene research has focused on the challenge of identifying relevant gene targets whose altered regulation in c-Myc-transformed cells contributes to carcinogenesis. A number of candidate genes was derived from the collective efforts of many laboratories using several different molecular genetics techniques to isolate genes expressed specifically or preferentially in c-Myc expressing cells. These approaches lead to many false positives due differences in gene expression related to specific growth properties of c-Myc expressing cells.[69]

One approach applied to the identification of authentic c-Myc-activated target genes greatly reduced the number of candidates. Mateyak and others[66] used a gene knockout strategy to derive a c-Myc-null rat fibroblast cell line.[69,70] In addition to abrogated expression of the targeted c-Myc protein, the expression of two related proteins, N-Myc and L-Myc, was undetectable in the cell line. Consistent with c-Myc's role in promoting cell cycle progression, c-Myc-null cells exhibited a 2- to 3-fold increase in doubling time associated with delayed progression through G1 and G2 phases of the cell cycle. In serum stimulation studies with quiescent cells, c-Myc-null cells showed a significantly decreased rate of S-phase entry.

Out of 10 evaluated putative c-Myc-activated genes, only one, CAD, showed reduced expression in c-Myc-null cells.[71] Although CAD protein levels were not determined, CAD mRNA levels were reduced 3-fold in c-Myc-null cells. Subsequently, the human CAD promoter has been shown to contain two adjacent consensus c-Myc-specific DNA binding sites. At least one of these sites confers c-Myc-dependent gene expression.[72] These studies suggested that CAD transcriptional activation by c-Myc is an important mechanism that controls cell cycle progression. One caveat to this explanation is that the change in CAD expression might be a response to the slower growth of c-Myc-null cells.

The question of whether the slow cell cycle of c-Myc-null cells were due to reduced CAD activity was evaluated by culturing cells in medium supplemented with thymidine and uridine. Given effective uptake and functional salvage pathways, these nucleosides are predicted to complement defects in CAD function. In the reported studies, addition of thymidine and uridine did not increase the growth rate of c-Myc-null cells. However, their rates of uptake and conversion into pyrimidine nucleotides were not independently assessed.

Inefficiency in either of these processes would limit pharmacological rescue of a defect in cellular CAD function. Whereas the study showed that the c-Myc-null growth phenotype was prevented by introduction of a wild-type c-Myc transgene, the effect of introduction of a constitutively expressed CAD gene was not evaluated. In IMPDH-p53 regulation studies, the gene transfer complementation strategy was superior to the pharmacological approach for preventing growth effects of nucleotide biosynthesis gene regulation.[20]

MAPKs are the terminal phosphorylation engines of Ras-signaling pathways. Recently, they were shown to directly regulate CAD activity by a route that is completely independent of their involvement in inducing c-Myc gene expression. Graves and others[73] showed that *in vitro* the MAPK Erk2 phosphorylates the CPS II domain of CAD on a threonine residue in a conserved MAPK site. Erk2 phosphorylation of purified CAD alters the enzyme kinetics properties of CPS II in two ways. The enzyme activity becomes 3-fold to 4-fold more responsive to allosteric activation by PRPP, and 90% resistant to feedback inhibition by the pathway end product UTP. *In vivo*, these enzyme kinetics changes would be expected to lead to an increased rate of pyrimidine nucleotide production.

Analyses with cell lysates strongly support the conclusion of Graves *et al.*[73] that *in vivo* MAPK phosphorylation of CAD is an important determinant of pyrimidine nucleotide production in response to growth factor stimulation. After stimulation of serum-starved quiescent rat liver epithelial cells with epidermal growth factor (EGF) to activate MAPKs, CPS II activity in cell lysates became resistant to inhibition by UTP. Moreover, CPS II activity in EGF-stimulated cell lysates was 3-fold more responsive to PRPP-activation than the activity in quiescent cell lysates. A similar effect on PRPP-responsiveness was noted in cells stimulated with platelet-derived growth factor-BB, indicating that growth factor stimulation effects on CPS II enzyme kinetics were not limited to EGF. Development of both UTP resistance and increased PRPP responsiveness was associated with increased MAPK activity; and within 45 minutes of EGF addition, cellular UTP pools increased about 2-fold. EGF-stimulation was reported to lead to CAD phosphorylation *in vivo* at the same conserved threonine residue that was phosphorylated in the purified enzyme by Erk2. Finally, all of the effects of growth factor stimulation on CPS II enzyme kinetics were abolished by mutation of the conserved threonine residue to alanine.

In addition to Erk2, phosphorylation of CAD by purified preparations of cyclic AMP-dependent protein kinase (PKA) has been reported.[73,74] These experiments were performed with CAD purified from transformed hamster kidney cells that over-produce the protein.[74] Although a significant level of CAD phosphorylation occurred in these studies in the absence of added PKA, several enzyme kinetic and molecular criteria were used to distinguish this activity from PKA phosphorylation. PKA phosphorylated CAD on two serine residues in the CPS II enzyme domain and activated the CPS II enzymatic activity. Potential phosphorylation sites for other cellular protein kinases have also been identified in the CAD sequence.[58] Unlike phosphorylation of CAD by MAPK, the physiological significance of the observed *in vitro* phosphorylation of CAD by PKA has not been determined.

The identification of CPS II as a MAPK-regulated target defines a mechanism for integrated control of cell cycle activation and pyrimidine nucleotide production *in vivo*. Like p53's proposed role in integrated control of purine nucleotide production and asymmetric cell kinetics (discussed above; Figure 1, I), MAPK also acts as a network node to distribute incoming growth factor information to both gene transcription (including c-Myc) and pyrimidine nucleotide production. Activation of these two different processes from a common upstream signaling molecule provides a simple mechanism for their coordinate regulation.

An analogous direct signaling pathway from MAPK to IMPDH and other purine nucleotide biosynthesis control enzymes has not been reported.[58] Phosphorylation of p53 by MAPK has been reported, but the significance of this modification has not been established (see Figure1; dotted arrow from "MAPK").[53] This apparent asymmetry in the nucleotide regulation network prompts the idea that a yet to be discovered path exists for more direct growth factor signaling to purine nucleotide synthesis control point enzymes. Although regulatory connections of this type have not been reported for MAPK, recent studies suggest that both IMPDH and 5-phosphoribosyl pyrophosphate amidotransferase (PRPP-ATase; EC 2.4.2.14) may be regulated directly by kinases in other growth signaling pathways. PRPP-ATase catalyzes the first step of de novo purine biosynthesis.[75] The protein contains several potential cyclic AMP-dependent phosphorylation sites[76,77] and is phosphorylated *in vitro*.[78] More recently it has been shown that the type II human IMPDH protein can bind specifically to the pleckstrin homology domain of PKB/Akt, a serine/threonine kinase that participates in phosphoinositide-mediated growth signal transduction. Preliminary studies suggest that *in vitro* phosphorylation of IMPDH II by PKB/Akt may activate the enzyme.[79] The physiological significance of PRPP-ATase and IMPDH II phosphorylation must be evaluated before these potential connections can be fully factored into the present regulatory network of cancer genes and nucleotide metabolism genes.

Another interesting feature of the MAPK-nucleotide biosynthesis connections is that whereas purine nucleotide regulation is mediated through a repressing tumor suppressor gene, p53, pyrimidine nucleotides are regulated via an activating oncogene, c-Myc (compare I to II in Figure 1). In the case of pyrimidine nucleotides, which are also regulated by MAPK-dependent CAD phosphorylation, the c-Myc-CAD path may play an even greater role in allowing activation of pyrimidine biosynthesis in response to growth stimuli that activate c-Myc via MAPK-independent signal transduction.

Activation of c-Myc and inactivation of p53 appear to be equivalent cellular events in neoplastic transformation of mammalian cells. Either can cooperate with an activated Ras oncogene to transform primary cells.[8,60,80] This relationship does not mean that an increase in the production rate of either pyrimidine nucleotides or purine nucleotides suffices for the development cancer cells. Well-described, inter-nucleotide pathway allosteric regulation mechanisms function to maintain a tight balance of nucleotide pools.[28] A stable genetic change that raises the basal rate of purine nucleotide biosynthesis is predicted to trigger a corresponding increase in the basal rate of pyrimidine nucleotide production (and vice versa).

Supporting the hypothesis that elevation of nucleotide pools is a critical step in cancer cell formation, the respective rate-limiting nucleotide synthesis enzymes regulated by c-Myc and p53 show elevated activity in tumors.[11,44,81-83] Surprisingly, the molecular basis for elevated IMPDH and CAD activity in tumors has not been determined. A likely mechanism would seem to be genetic alterations in p53 and c-Myc gene function, which are both common events during tumorigenesis. These remarkable biological relationships are unlikely to be fortuitous. It is more likely that they are the result of the evolution of specific mechanisms to integrate control of cellular processes that must function in sync for robust tissue cell kinetics.

Module III: TS-c-Myc and TS-p53

Recent findings in studies of the molecular regulation of TS (EC 2.1.1.45) further underscore the importance of integrated cellular control of cancer gene function and nucleotide metabolism gene function. Once again, the p53 tumor suppressor gene and the c-Myc proto-oncogene are the cancer genes involved in an integrated control system to coordinate nucleotide production and cell division.[15] TS catalyzes the rate-limiting step for the production of thymine nucleotides, which are required uniquely for DNA synthesis. In a complex with dihydrofolate reductase (DHFR; EC 1.5.1.3), which generates the other substrate for the reaction, 5-10-methylenetetrahydrofolate (THF), TS converts dUMP to TMP. Because of its associated function in DNA-specific nucleotide metabolism and its dependency on folate metabolism, TS has been extensively targeted with nucleotide analogues and folate antagonists for cancer chemotherapy.[15,84]

The expression of TS is periodic during the cell cycle, being elevated in S phase. Both TS and DHFR are transcriptional activation targets of the E2F transcription factor. Activation of TS expression in S-phase is associated with release of an active form of E2F from complexes with the retinoblastoma (Rb) tumor suppressor protein. The p53 gene is an indirect regulator of this transcription factor complex (discussed in Module I section above). P53 activates the transcription of the cyclin-dependent kinase inhibitor p21wafl, which inhibits cyclin-dependent kinases that phosphorylate Rb in G1, leading to the release of active E2F.[4,16] Thus, as indicated in Figure 1 (Module III), p53 is ultimately a negative modulator of TS activity.

A more direct connection between p53 and TS, and c-Myc as well, was discovered by Chu and others in the course of their studies of TS molecular regulation.[85-90] Using rabbit reticulocyte *in vitro* translation systems, these investigators demonstrated that recombinant human TS protein could specifically inhibit the translation of its own mRNA. The enzyme substrates THF and dUMP or inhibitory nucleotide analogues prevented translational inhibition by recombinant TS protein. This effect of substrates suggested that, if the same auto-regulation occurred *in vivo*, it might serve to reduce TS production under conditions of low substrate concentration. Under conditions of TS-dependent translation inhibition, gel shift assays were used to detect a stable specific complex between recombinant TS protein and mRNA.[85] In subsequent work, Chu *et al.* showed that TS protein binds with nanomolar affinity to two independent regions of TS mRNA. One region includes the translation start site, and the other lies within the protein coding region.[86]

An immunoprecipitation-reverse transcription-polymerase chain reaction (IP-RT-PCR) method was employed to investigate whether analogous TS-ribonucleoprotein (RNP) complexes occurred in cells. For this method TS protein was precipitated from cell lysates with TS specific polyclonal antibodies. Cellular mRNA sequences associated with the precipitated immune complexes were detected by RT-PCR with gene specific primers. In addition to TS mRNA sequences, specifically associated sequences from several other cellular mRNAs were detected with this method, among them c-Myc and p53.[87-90] Like TS mRNA translation, both human c-Myc and p53 mRNA *in vitro* translation was inhibited by recombinant human TS protein.[87,89] Stable specific complexes between recombinant TS protein and either c-Myc or p53 mRNAs were also demonstrated.[87,89,90] Formation of TS-p53 mRNA complexes was also prevented by dUMP and the substrate analogue 5-flouro-dUMP. The effects of these compounds on TS-c-Myc complexes were not reported. Although each of these two cellular mRNAs competes with TS mRNA for binding to TS, no consensus binding site has been identified.[87,89] The inability to identify a specific binding site may indicate that the basis for recognition is a common feature of these mRNAs that is not strictly determined by nucleotide sequence.

An evaluation of the physiologic relevance of the *in vitro* interaction of TS and c-Myc mRNA has not been reported. The described IP-RT-PCR experiments suggest that TS-c-Myc mRNA-containing RNPs may exist in intact cells. The authors of these studies suggest that under conditions of elevated c-Myc expression (e.g., after growth factor stimulation), c-Myc mRNA might disrupt TS translational auto-regulation by competing for binding to TS. This would effectively induce cellular TS activity and thymidine nucleotide production for ensuing DNA replication in S phase. As DNA synthesis proceeds, the increase in TS protein would then dampen c-Myc protein expression and eventually restore the basal state of TS production.[87] In this model, there is an implicit assumption that TS bound to c-Myc mRNA is still active in TMP production. The fact that the substrates and substrate analogues interfere with TS binding to its own mRNA makes this assumption questionable. Neither the effect of TS substrates on the formation of c-Myc mRNA-TS complexes nor the effect of mRNA binding on TS enzyme kinetics was reported.[87]

The above model is a first proposal for the regulation of the expression of one gene by the *mRNA* of another through an indirect translational auto-regulation mechanism. This regulation model predicts that when TS levels rise, c-Myc levels will fall; and when c-Myc levels rise, TS levels should increase. Few studies exist that permit an evaluation of the existence of such relationships in cells. The authors of the hypothesis developed it to account for the sequence of TS and c-Myc induction in growth factor stimulated cells.[87] Tumors and tumor-derived cell lines would provide another basis for such an investigation. In cycling cells, TS expression is periodically elevated in S phase,[16,84,91] but c-Myc protein is maintained at a low and relatively constant level throughout the cell cycle.[63,64] Therefore, if the proposed mechanism operates in cells, it may be restricted to conditions of growth induction and cell cycle entry.

Chu and others performed a number of studies to provide evidence that the interaction between p53 mRNA and TS protein plays a role in p53 gene regulation in intact cells.[89,90] Their basic strategy was to evaluate wild-type p53 protein levels in cells that expressed elevated levels of TS protein. Cells with higher than normal levels of TS were derived either by selection for cells resistant to the TS inhibitor 5-fluoro-uracil or by stable transfection with TS-inducible gene constructs.[89] In these studies, no p53 protein was detected in cells that expressed TS at levels 15-fold to 40-fold higher than normal. Failure to detect p53 protein expression occurred with normal levels of p53 mRNA and greatly increased detection of TS-p53mRNA complexes by IP-RT-PCR. In cells derived with TS under the control of a tetracycline-inducible promoter, reducing TS expression allowed detection of p53 protein. Consistent with the idea that the absence of p53 was due to TS-induced inhibition of the p53 translation, the association of p53 mRNA with polysomes was shown to be greatly reduced in cells with elevated TS expression.

In a subsequent study[90], the effect of TS on p53 expression was evaluated by more direct experiments and under conditions of less extreme differences in TS protein concentration. For these studies, the TS-mutant cell line HCT-C, derived from the human colon cancer line HCT-8, was used. A missense mutation at amino acid 216 in TS in HCT-C cells renders the enzyme inactive. Although HCT-C and HCT-8 cells express the same level of wild-type p53 mRNA, HTC-C express significantly more p53 protein. Restoration of wild-type TS expression in HTC-C cells by stable transfection of a human TS cDNA expression plasmid, led to a 5-fold reduction in p53 protein levels. Direct measurements of p53 protein synthetic rates and half-life in HCT-C cells and their TS-transfected derivatives indicated that the observed differences in p53 protein levels were explained best by differences in the rate of p53 protein synthesis. Thus, it was concluded that TS protein regulates the translation rate of p53 mRNA *in vivo*. As in earlier studies with higher levels of TS protein expression, TS-p53 mRNA complexes were only detected in cells that expressed active TS protein.

Despite detection of TS-p53 mRNA complexes in several different cellular contexts, it is still of some concern that they were only detected in cells over-expressing TS to some extent. Even though the HCT-C studies do suggest that reductions in active TS protein also effect p53 translation, no complexes were reported for the parental HCT-8 cells.[90] Thus, the physiological relevance of these observations continues to be a point of concern. Despite significant differences in wild-type p53 protein expression associated with elevated TS expression, the doubling time of the affected cells was quite similar.[89,90] Given that the evaluated cell lines were either tumor-derived or specifically selected for rapid proliferation, other changes in them may have obscured p53-induced growth effects.

In DNA damage studies with gamma-irradiation, cell lines with elevated TS expression did fail to show the usual increase in G1/S ratio that is characteristic of a p53-dependent checkpoint arrest.[89,90] However, care must be taken in concluding from these experiments that decreased p53 protein is the responsible factor. With only two pairs of independently derived cell lines having been evaluated so far[89,90], the possibility that other differences, besides p53 expression level, are responsible must be considered. A similar evaluation using the tetracycline-inducible TS cell line[89] would have been very informative in this regard, but none was reported. The inducible cell line provides analyses of cellular effects at different TS-p53 expression levels without the variation in cell background that is inherent to comparisons of two independent cell lines that differ in expression levels. A complete evaluation of the observed effects of TS expression on cell cycle checkpoint responses will require studies in which TS expression is varied in p53-null cells. Thereby, the possibility of independent effects of TS on the cell cycle checkpoint function can also be addressed.

Notwithstanding the above criticisms, the discovery of a TS-p53 control link reveals a profoundly important integrating network junction (Figure 1,

Module I-III). Until now, the balanced production of purine and pyrimidine nucleotide pools was thought to be accomplished solely by allosteric regulation of the biosynthetic enzymes of one nucleotide class of by nucleotide metabolites of the other class.[28,92] This explanation alone has never been completely satisfactory. In the network shown in Figure 1, the TS auto-regulation module (Module III) is a central more complex signal node that provides integrated regulation of the production of purine and pyrimidine nucleotides.

This point can be illustrated by considering changes in nucleotide production in the S phase of cycling cells. During the S phase of normal cycling cells, when TS activity and dTMP production increase, TS protein is predicted to reduce p53 protein expression. The result of this down-regulation will be increased IMPD expression. Whereas dTMP production rate will increase dramatically, adenine and guanine ribonucleotide production rates will increase only modestly to support the new demand for nucleotides. This mechanism allows the cell to match purine nucleotide production to dTMP production, although the required change in respective production rates is quite different. In fact, TS down-regulation of p53 will initiate an amplification loop through Rb/E2F that will promote even higher dTMP production rates as S phase gets under way. When p53 levels are reduced, more free E2F is predicted to be available to activate TS gene transcription. As suggested for c-Myc regulation, at later times when TS levels are sufficiently high, the negative auto-regulation mechanism of TS can function to quench the explosion of S phase nucleotide synthesis that TS itself initiated.

Of course, the same initiating event, TS elevation, is predicted to suppress the production of other pyrimidine nucleotides due to inhibition of c-Myc translation followed by reduced CAD gene expression (Figure 1, Module II). The fact that CAD is independently activated by MAPKs, which are constitutively active in cycling cells, provides an alternate path for activation of cytidine and uridine nucleotide production. The fact that p53 represses IMPD, whereas c-Myc activates CAD, may account for the evolution of this alternate asymmetric CAD-MAPK activation path. The wiring of Module II in Figure 1 can account for why, like purine nucleotide pools[93,94], cytidine[95] and uridine nucleotide pools may show smaller fluctuations during the cell cycle as compared to dTNP pools.[92]

It is noteworthy, that like c-Myc, both p53 and IMPD show only modest fluctuations during the cell cycle.[93,96] This property suggests that either effects of TS on these proteins in normal cells are subtle or they occur on a time scale that made them obscure in previous studies. Of course, formally, the null hypothesis must also be considered. As depicted, network module III is predicted to be quite complex, finely tuned by evolution, and robust. Careful, precise, rapid measurements of the TS, p53, c-Myc, and CAD and their respective nucleotide metabolites will be necessary to confirm its operation in normal tissue cells and to define the full range of its regulatory functions.

Module IV: RDP-Ras

In the *Introduction* of this review, the observation was made that metabolism genes are generally not considered to be "cancer genes". Besides the explanation given that this situation reflects their function as essential cellular genes, another practical reason may also apply. Metabolic genes have rarely been evaluated in cancer gene assays. Given what is now known about cancer gene-nucleotide gene connections, it would be of interest to ask, for example, whether IMPD and CAD together can substitute for either mutant p53 or c-Myc in co-transformation of primary cells with an activated Ras oncogene. IMPD and CAD over-expression is predicted to have the same primary effects on nucleotide biosynthesis rates as loss of p53 function or c-Myc activation, respectively (see discussion above for modules I and II). If the main requirement for cell transformation met by p53 mutation or c-Myc expression is expansion of guanine and pyrimidine nucleotide pools, then ectopic IMPD and CAD expression might supply an equivalent transformation cooperation activity.

Given allosteric regulation of IMPD and CAD and the diverse cellular functions ascribed to p53 and c-Myc, which predate recognition of their roles in nucleotide regulation, the above suggestions may seem far-fetched. However, recent studies of the transforming properties of another important nucleotide metabolism gene make these ideas quite plausible. The establishment of a cancer gene link to RDP (EC 1.17.4.1-2), in fact, began with transformed foci experiments.[97,98]

RDP is the rate-determining enzyme for the production of all four cellular deoxyribonucleotides. The mammalian enzyme is a tetramer composed of two different homodimeric subunits. The R1 dimer contains the binding site for the four ribonucleoside diphosphate enzyme substrates, rUDP, rADP, rGDP, and rCDP. The R2 dimer contains a binuclear iron center that generates a stable radical at a nearby tyrosine residue. The tyrosyl radical is essential for RDP to catalyze the formation of the four deoxyribonucleotide diphosphates, dUDP, dADP, dGDP, and dCDP, from their respective ribonucleotide diphosphate precursors. The action of TS and nucleotide diphosphate kinases completes the formation of dTTP and dATP, dGTP, and dCTP, respectively, for DNA synthesis in S phase.[99]

The growth and cell cycle regulation of the RDP is bipartite with respect to its R1 and R2 subunits. In quiescent cells, neither component is detectable. However, when cells are induced to re-enter the cell cycle by growth factor stimulation, both subunits are induced. R1 and R2 proteins are also induced by treatment of quiescent cells with DNA damage-inducing agents. In cycling cells, the R1 subunit is produced in stoikiometric excess and does not vary significantly during the cell cycle. In contrast, the R2 subunit is highly regulated by several different molecular mechanisms. It approaches undetectable levels in G1 cells, rises in parallel with DNA synthesis in S phase, and then decreases again to low levels in G2/M.[99]

Fan and others[97,98] investigated whether the RDP expression could promote cell transformation. Initially, these investigators used retroviral infection to establish immortal 3T3 fibroblast cell lines that stably expressed a recombinant myc-tagged R2 protein. In actively dividing cells, the recombinant protein was expressed at levels similar to endogenous R2. However, the recombinant protein was expressed constitutively, without variation during the cell cycle. Its expression was associated with an approximate 2-fold increase in RDP activity and increased resistance to hydroxyurea, an R2 inhibitor. The growth and cell cycle properties of lines expressing recombinant R2 was indistinguishable from control 3T3 cell lines derived with an empty retroviral vector. On the other hand, they showed marked differences in their response to introduction of an activated Ras oncogene. Control immortal 3T3 fibroblasts could be transformed by an activated Ras oncogene, but at very low efficiency. In contrast, although the recombinant R2 protein had no apparent growth effect when expressed alone, its co-expression greatly increased the efficiency of cell transformation by activated Ras. Moreover, cells stably co-expressing activated Ras and recombinant R2 showed significantly increased tumorigenic and metastatic properties after implantation into mice.[97]

The molecular basis of R2-associated effects in Ras transformation is unclear. In culture, cells expressing recombinant R2 exhibited no significant changes in the cell cycle, possibly indicating that cell kinetics mechanisms are not involved. However, under conditions of growth factor limitation, cells co-expressing the recombinant protein and activated Ras exhibited increased cell accumulation rates. However, the assays used did not distinguish between effects on proliferation versus apoptosis.

Investigations of other Ras-signaling pathway components revealed R2-associated increases in Raf1 association with membranes and MAPK activation. Presently, it is unclear whether these changes are a direct result of R2 action and, if so, what is the nature of that action. Fan et al. suggest that "the R2 protein can participate in other critical cellular functions in addition to ribonucleotide reduction".[97] However, they provide little in the way of a proposed molecular function. One important caveat for their study is the question of whether similar transformation effects would occur in response to expression of untagged R2 protein. Although, even if the observed effects were unique to the myc-tagged R2, they still might open a window to another RDP function. An investigation of the effect of hydroxyurea on the ability of R2 to cooperate with activated Ras for enhanced cellular transformation could shed light on this question. Since lines co-transformed with RDP are resistant to hydroxyurea, it may be possible to assess effects of hydroxyurea on their transformed properties at doses that do not cause cell cycle effects.

Surprisingly, expression of a recombinant form of the R1 subunit of RDP had opposite effects in cell transformation studies. Analogous to the R2 studies, myc-tagged recombinant R1 protein was stably expressed in cells after retrovirus-mediated gene transfer. R1 expression was associated with a significant reduction in soft-agar colony formation efficiency of Ras-transformed and tumor-derived cells. Although the suppressive effect of R1 expression on primary tumor formation in mice was modest and inconsistent, it was associated with a marked and reproducible decrease in lung tumors after tail vein injection of cells.[98] These results suggested that recombinant R1 might induce cellular changes that prevent malignant progression. To demonstrate a role for endogenous R1 protein is malignancy mechanisms, Fan *et al.* used an anti-sense mRNA strategy to reduce R1 expression during transformation with activated Ras. Anti-sense induced reduction in endogenous R1 protein expression was associated with a dramatic increase in the efficiency of Ras-transformed colony formation in soft agar. The fact that only a modest reduction in R1 was required for these effects suggested that allelic changes in R1 gene dosage could play an important role in malignant progression *in vivo.*[98]

Like constitutive R2 expression, elevated R1 expression did not cause significant changes in cell proliferation rates under normal culture conditions. The authors of these studies suggested that the balance of R1 and R2 expression in cells might be an important determinant of cell transformation and malignancy. They proposed that the RDP subunits "have novel, but yet undefined, properties that effect signal pathways". However, no specific molecular mechanism(s) was advanced, and effects of R1 expression on the Ras/Raf1/MAPK signaling pathway were not reported.[98]

A role for such a counter-balanced regulation module (Figure 1, Module IV) in normal tissue cell function was not considered[98], but is highly intriguing. The stringency of cell transformation assays may provide conditions under which an otherwise imperceptible delicate regulation link can be detected. The utilization of an essential factor for DNA replication as a modulator of incoming stimuli for cell division constitutes yet another network node. Tissue cell kinetic information deposited at the RDP signal node can be channeled to both the DNA replication machinery and the flow of growth promoting signals.

The studies of Fan *et al.* suggest that RDP may be a conditional signal node as well, with the quality of its signal modulation varying as a function of properties of its DNA replication output.[97,98] For example, when RDP is actively engaged in the production of DNA precursors (*i.e.*, high R2/R1 ratio in S phase), it may positively influence signal transduction to insure completion of

S phase. In contrast, at times when the enzyme is relatively inactive (*i.e.*, low R2/R1 ratio in G1 phase) it may dampen incoming growth signals for better control of G1 phase length. These ideas are by no means exhaustive of the rich regulatory complexity that can be envisioned to exist in the RDP-Ras network module. Although the key to the discovery of this cancer gene-nucleotide gene interaction was cell transformation studies, future studies in non-transformed cells will be necessary to fully reveal its physiological nature and function.

V. Interrogating the cancer gene-nucleotide gene control network

The emergence of a regulatory network that integrates control of responses to external growth stimuli with control of nucleotide biosynthesis is an easily understood development in the evolution of mammalian cells. Nucleotides are the building blocks of nucleic acids, which store and transfer information for cell structure and function. They are also major sites of energy storage and catalysts for many essential energy transfer reactions.[27] In diverse tissue microenvironments, cells must relate a variety of fluctuating ambient signals to their internal energy stores and metabolic state in order to arrive at robust integrated cell kinetics outputs.[4]

The diagram in Figure 1 is likely to fall far short of a complete accounting of all the important cancer gene-nucleotide gene interactions that actually occur in mammalian tissue cells. Nucleotide salvage pathway enzymes, which were not considered here, constitute another complex array of metabolic pathways that are important for cell function.[14,28] In many tissues, salvage is the predominant pathway for nucleotide biosynthesis.[14] As was discussed for control enzymes in *de novo* nucleotide pathways, although there have been extensive studies of the regulation of some nucleotide salvage enzymes, their function in cell regulation beyond supplying nucleotides for nucleic acid synthesis has been largely neglected.

To date, four out of the seven mammalian control point enzymes in *de novo* nucleotide biosynthesis pathways have been directly linked to one or more cancer genes. As of this writing, the author is not aware of any reports of cancer gene associations for the remaining three enzymes, PRPP-ATase, CTP synthase (CTPS; EC 6.3.4.2) and adenylosuccinate synthase (sAMPS; EC 6.3.4.4). CTPS and sAMPS are the final control point enzymes in pathways for cytidine nucleotide and adenine nucleotide biosynthesis, respectively. The present absence of such links may predict future discoveries of new cancer gene-nucleotide gene connections. This seems highly likely for CTPS, which presently is depicted with no input in the network besides the concentration of its substrate rUTP. Given the role of cytidine nucleotides in lipid metabolism[28],

CTPS may receive input from cancer genes that are signal nodes with links to lipid energy metabolism. Similarly, the only inputs presently depicted for sAMPS are the concentrations of its two substrates IMP and GTP. However, this enzyme may not require additional inputs. Although sAMP catalyzes the final control step of adenine nucleotide biosynthesis, it is not the major determinant of ATP function. ATP is the largest cellular nucleotide pool. Its main cellular function, energy metabolism, is highly dependent on phosphate-charge state, which kinetically more that outweighs effects related to its rate of biosynthesis.[27,28]

The network diagrammed in Figure 1 was constructed by virtual assembly of four main regulatory modules (Figure 1, I-IV). Each module was discovered and characterized by one or two different laboratories that employed different experimental systems. An implicit assumption necessary for acceptance of this synthesis is that the cellular mechanisms independently defined in cells of different tissue origin, different species, different genotype, different culture conditions, and different cell kinetics do, in reality, co-exist and function in an integrated fashion in *individual* cells. Unlike regulation networks modeled from intensively studied bacterial and yeast strains, essentially all cell regulation network models for mammalian cells suffer from this potential Achilles' heal. Developing an accurate understanding of cellular regulation networks may be impossible until this situation is rectified.

Mathematical description of piecemeal data assemblies, followed by computer simulations, followed by prediction development, followed by experimental testing to validate and interrogate highly complex cellular control networks is an approach rich in academic pursuits, but it does not necessarily lead to meaningful and useful new insights to cell function. If important network connections based on assembled sets of experimental data are erroneous, a significant amount of effort and resources will be wasted in modeling efforts before problems are suspected, if indeed they ever are. It would advantageous, and potentially more rigorous, to organize a laboratory consortium to coordinate a concerted attack to elucidate cell control networks of particular interest, much in the same way that such groups have been created to tackle large technological problems in science like sequencing entire genomes.

A network elucidation consortium would have three main charges. The first charge would be to select several specific mammalian cell systems in which all elements of a selected control network could be systematically evaluated. The next charge would be to measure specified changes in concentration, activity, and interaction of all network components on the same time scale, in resting states, particular response states, and disease states. The final charge would be to develop a complete mathematical description of the network. Ideally this description should be modular, to permit concentrated evaluation of local features of the network, and should have sufficient power to predict dynamic cell behavior in response to specific external stimuli (*e.g.*, drugs, pathogens). Development of predictive integrated control network models will have major impacts in areas of biomedical research like molecular pathophysiology and pharmacology. Instead of performing, for example, simple in-series evaluations of disease-associated gene defects or drug structure-activity analysis based on single molecule effects, it will be possible to predict and, therefore, consider impacts on global cell function, which is, of course, the information that we really seek.

Recent developments in new technologies for genome-wide mRNA expression analysis have lead to a proliferation of studies with the goal of defining and modeling gene expression networks that control important cellular processes. For many years, biochemists, molecular geneticists, molecular biologists, and bioengineers have worked to define components of cellular regulation systems. These new global approaches hold great promise to confirm suspected regulatory connections and reveal previously unanticipated ones. However, it is already recognized that mRNA expression data alone will greatly under-represent the rich complexity of cellular control networks and systems integration. Proteomics, the discipline of global measurements of protein concentrations and interactions has been heralded as the solution to the shortfall of genomics. However, this view is shortsighted as well. In addition to proteomics, a full understanding of cellular regulation and control will require information coding for enzymatics. The cellular activity of many enzymes is governed by numerous factors other than their concentration. Global measures of the relationships among each level of gene information coding, including gene structure, macromolecule metabolism, macromolecule interaction, enzymatics, *and* small molecule metabolism, may bring the synthesis in understanding cellular life that we seek in its vast and presently incomprehensible complexity.

Acknowledgements

I am grateful to the editors for the opportunity to discuss these ideas in this forum. I thank J. Merok and other members of my research group for careful and expeditious review the manuscript. I am thankful to Karie Ng for procurement of citation materials and assistance in preparing the manuscript.

References

1. Haurowitz, F. *Progress In Biochemistry: A Report On Biochemical Problems And On Biochemical Research Since 1949*; Interscience Publishers, Inc.: New York, 1959; pp III-V.

2. Knudson, A. G. Mutation and Cancer: Statistical Study of Retinoblastoma. *Proc. Natl. Acad. Sci. USA* **1971**, *68*, 820-823.

3. Hanahan, D.; Weinberg, R. A. The Hallmarks of Cancer. *Cell* **2000**, *100*, 57-70.

4. Sherley, J. L. Tumor Suppressor Genes and Cell Kinetics in the Etiology of Malignant Mesothelioma. In *Asbestos Disease Control Sourcebook on Asbestos Diseases*, Vol. 21; Peters, G. A., Peters, B. J., Eds.; Matthew Bender & Co., Inc.: Charlottesville, 2000; pp 91-141.

5. Fearon, E. R.; Vogelstein, B. A Genetic Model for Colorectal Tumorigenesis. *Cell* **1990**, *61*, 759-767.

6. Herrero-Jimenez, P.; Thilly, G.; Southam, P. J.; Tomita-Mitchell, A.; Morgenthaler, S.; Furth, E. E.; Thilly, W. G. Mutation, Cell Kinetics, And Subpopulations At Risk For Colon Cancer In The United States. *Mutation Res.* **1998**, *400*, 553-578.

7. Bishop, J. M. Molecular Themes In Oncogenesis. *Cell* **1991**, *64*, 235-248.

8. Hunter, T. Cooperation Between Oncogenes. *Cell* **1991**, *64*, 249-270.

9. Hunter, T. Oncoprotein Networks. *Cell* **1997**, *88*, 333-346.

10. Keyomarsi, K.; Herliczek, T. W. The Role Of Cyclin E In Cell Proliferation, Development And Cancer. *Prog. Cell. Cycle. Res.* **1997**, *3*, 171-191.

11. Weber, G.; Olah, E.; Lui, M. S.; Kizaki, H.; Tzeng, D. Y.; Takeda, E. Biochemical Commitment to Replication in Cancer Cells. In *Advances in Enzyme Regulation*, Vol. 18; Weber, G., Ed.; Pergamon Press, Oxford: 1980; pp 3-26.

12. Hochhauser, S. J.; Stein, J. L.; Stein, G. S. Gene Expression and Cell Cycle Regulation. Int. Rev. Cytol. **1981**, *71*, 95-243.

13. Sherley, J. L.; Kelly, T. J. Regulation Of Human Thymidine Kinase During The Cell Cycle. *J. Biol. Chem.* **1988**, *263*, 8350-8358.

14. Murray, A. W.; Elliott, D. C.; Atkinson, M. R. Nucleotide Biosynthesis From Preformed Purines In Mammalian Cells: Regulatory Mechanisms And Biological Significance. *Progr. Nucl. Acid Res. Mol. Biol.* **1970**, *10*, 87-119.

15. Chu, E.; Allegra, C. J. The Role Of Thymidylate Synthase In Cellular Regulation. *Adv. Enzyme Regul.* **1996**, *36*, 143-163.

16. Bartek, J.; Bartkova, J.; Lukas, J. The Retinoblastoma Protein Pathway In Cell Cycle Control And Cancer. *Exp. Cell Res.* **1997**, *237*, 1-6.

17. Criss, W. E.; Pradhan, T. K. Regulation of Tumor Cell Metabolism by the Adenylate and Guanylate Energy Charges. In *Control Mechanisms in Cancer*; Criss, W. E., Ono, T., Sabine, J. R., Eds.; Raven Press, New York: 1976; pp 401-410.

18. Pall, M. L. GTP: A Central Regulator of Anabolism. *Curr. Topics Cell. Reg.* **1985**, *25*, 1-20.

19. Allison, A. C.; Kowalski, W. J.; Muller, C. D.; Eugui, E. M. Mechanisms Of Action Of Mycophenolic Acid. *Annals N. Y. Acad. Sci.* **1993**, *696*, 63-87.

20. Liu, Y.; Bohn, S. A.; Sherley, J. L. Inosine-5'-Monophosphate Dehydrogenase Is A Rate-Determining Factor For P53-Dependent Growth Regulation. *Mol. Biol. Cell* **1998**, *9*, 15-28.
21. Franklin, T. J; Twose, P. A. Reduction In β-Adrenergic Response Of Cultured Glioma Cells Following Depletion Of Intracellular GTP. *Eur. J. Biochem.* **1977**, *77*, 113-117.
22. Kharbanda, S.M.; Sherman, M. L.; Kufe, D.W. Effects Of Tiazofurin On Guanine Nucleotide Binding Regulatory Proteins In HL-60 Cells. *Blood* **1990**, *75*, 583-588.
23. Rizzo, M. T.; Tricot, G.; Hoffman, R.; Jayaram, H. N.; Weber, G.; Garcia, J. G. N.; English, D. Inosine Monophosphate Dehydrogenase Inhibitors. Probes For Investigations Of The Functions Of Guanine Nucleotide Binding Proteins In Intact Cells. *Cell. Signaling* **1990**, *2*, 509-519.
24. Weber, G.; Nakamura, H.; Natsumeda, Y.; Szekeres, T.; Nagai, M. Regulation Of GTP Biosynthesis. *Advan. Enzyme Regul.* **1992**, *32*, 57-69.
25. Metz, S. A.; Kowluru, A. Inosine Monophosphate Dehydrogenase: A Molecular Switch Integrating Pleitrophic GTP-Dependent b-Cell Function. *Proc. Assoc. Am. Physicians* **1999**, *111*, 335-346.
26. Sherley, J. L. Guanine Nucleotide Biosynthesis Is Regulated By The Cellular P53 Concentration. *J. Biol. Chem.* **1991**, *266*, 24815-24828.
27. Lehninger, A. L. *Bioenergetics*, 2nd ed.; W. A. Benjamin, Inc.: Philippines, 1971.
28. Lehninger, A.L. *Biochemistry*, 2nd ed.; Worth Publishers, Inc.: New York, 1975.
29. Sherley, J. L.; Stadler, P. B.; and Johnson, D. R. The P53 Antioncogene Induces Guanine Nucleotide-Dependent Stem Cell Division Kinetics. *Proc. Natl. Acad. Sci. USA* **1995**, *92*, 136-140.
30. Liu, Y.; Riley, L. B.; Bohn, S. A.; Boice, J. A.; Stadler, P. B.; Sherley, J. L. Comparison Of Bax, Waf1, And IMP Dehydrogenase Regulation In Response To Wild-Type P53 Expression Under Normal Growth Conditions. *J. Cell. Physiol.* **1998**, *177*, 364-376.
31. Lane, D. P. P53, Guardian of the Genome. *Nature* **1992**, *358*, 15-16.
32. Ko, L. J.; Prives, C. P53: Puzzle and Paradigm. *Genes & Devel.* **1996**, *10*, 1054-1072.
33. Levine, A. J. P53, The Cellular Gatekeeper for Growth and Division. *Cell* **1997**, *88*, 323-331.
34. Vogelstein, B.; Kinzler, K. W. P53 Function and Dysfunction. *Cell* **1992**, *70*, 523-526.
35. Hainaut, P. The Tumor Suppressor Protein p53: A Receptor to Genotoxic Stress That Controls Cell Growth and Survival. *Curr. Opin. Oncol.* **1995**, *7*, 76-82.

36. Giaccia, A. J.; Kastan, M. B. The Complexity of p53 Modulation: Emerging Patterns From Divergent Signals. *Genes & Devel.* **1998**, *12*, 2973-2983.
37. Finlay, C. A.; Hinds, P. W.; Levine, A. J. The P53 Proto-Oncogene Can Act As A Suppressor Of Transformation. *Cell* **1989**, *57*, 1083-1093.
38. Lane, D. P.; Benchimol, S. P53: Oncogene or Anti-oncogene. *Genes & Devel.* **1990**, *4*, 1-8.
39. Levine, A. J.; Momand, J.; Finlay, C. A. The p53 Tumour Suppressor Gene. *Nature* **1991**, *351*, 453-456.
40. Rambhatla, L.; Bohn, S. A.; Stadler, P. B.; Boyd, J. T.; Coss, R. A.; Sherley, J. L. Cellular Senescence: *Ex vivo* P53-Dependent Asymmetric Cell Kinetics. *J. Biomed. Biotech.* **2001**, in press.
41. Sherley, J. L. The P53 Tumor Suppressor Gene As Regulator Of Somatic Stem Cell Renewal Division. *Cope* **1996**, *12*, 9-10.
42. Potten, C. S.; Morris, R. J. Epithelial Stem Cells *In Vivo*. *J. Cell Sci. Suppl.* **1988**, *10*, 45-62.
43. Cairns, J. Mutation Selection And The Natural History Of Cancer. *Nature* **1975**, *255*, 197-200.
44. Rambhatla, L; Sherley, J. L., unpublished results.
45. Cohen, M. B.; Maybaum, J.; Sadée. W. Guanine Nucleotide Depletion And Toxicity In Mouse T Lymphoma (S-49) Cells. *J. Biol. Chem.* **1981**, *256*, 8713-8717.
46. Cohen, M. B.; Sadée, W. Contributions Of The Depletions Of Guanine And Adenine Nucleotides To The Toxicity Of Purine Starvation In The Mouse T Lymphoma Cell Line. *Cancer Res.* **1983**, *43*, 1587-1591.
47. Lui, M. S.; Faderan, M. A.; Liepnieks, J. J.; Natsumeda, Y.; Olah, E.; Jayaram, H. N.; Weber, G. Modulation of IMP Dehydrogenase Activity and Guanylate Metabolism by Tiazofurin (2-b-D-Ribofuranosylthiazole-4-carboxamide). *J. Biol. Chem.* **1984**, *259*, 5078-5082.
48. Jayaram, H. N.; Zhen, W.; Gharehbaghi, K. Biochemical Consequences of Resistance to Tiazofurin in Human Myelogenous Leukemic K562 Cells. *Cancer Res.* **1993**, *53*, 2344-2348.
49. Lin, D.; Fiscella, M; O'Connor, P. M.; Jackman, J.; Chen, M.; Luo, L. L.; Sala, A.; Travali, S.; Appella, E.; Mercer, W. E. (1994). Constitutive Expression Of B-Myb Can Bypass P53-Induced Wafl/Cip1-Mediated G1 Arrest. *Proc. Natl. Acad. Sci. USA* **1994**, *91*, 10079-10083.
50. Latham, K.M., Eastman, S. W.; Wong, A.; Hinds, P. W. Inhibition Of P53-Mediated Growth Arrest By Overexpression Of Cyclin-Dependent Kinases. *Mol. Cell. Biol.* **1996**, *16*, 4445-4455.
51. Sherr, C.J. Cancer Cell Cycles. *Science* **1996**, *274*, 1672-1677.
52. Liu, Y.; Sherley, J. L. unpublished data.

53. Milne, D. M.; Campbell, D. G.; Caudwell, F. B.; Meek, D. W. Phosphorylation of the Tumor Suppressor Protein p53 by Mitogen-Activated Protein Kinases. *J. Biol. Chem.* **1994**, *269*, 9253-9260.

54. Stephanopoulos, G. N.; Aristidou, A. A.; Nielsen, J. *Metabolic Engineering: Principles and Methodologies*; Academic Press: San Diego, 1998.

55. Hershfield, M. S; Seegmiller, J. E. Regulation of De Novo Purine Biosynthesis in Human Lymphoblasts: Coordinate Control of Proximal (Rate-Determining) Steps and the Inosinic Acid Branch Point. *J. Biol. Chem.* **1976**, *251*, 7348-7354.

56. Tunstead, J. R.; Merok, J. R.; Sherley, J. L. unpublished data.

57. Evans, D. R. CAD, A Chimeric Protein That Initiates De Novo Pyrimidine Biosynthesis In Higher Eukaryotes. In *Multidomain Proteins - Structure and Evolution*; Hardie, D. G., Coggins, J. R., Eds.; Elsevier: Amsterdam, New York, Oxford, 1986; pp 283-331.

58. Carrey, E. A. Key Enzymes In The Biosynthesis Of Purines And Pyrimidines: Their Regulation By Allosteric Effectors And By Phosphorylation. *Biochem. Soc. Trans.* **1995**, *23*, 899-902.

59. Fuhrmann, G.; Rosenberger, G.; Grusch, M.; Klein, N.; Hofmann, J.; Krupitza, G. The MYC Dualism In Growth And Death. *Mutat. Res.* **1999**, *437*, 205-217.

60. Land, H.; Parada, L. F.; Weinberg, R. A. Tumorigenic Conversion Of Primary Embryo Fibroblasts Requires At Least Two Cooperating Oncogenes. *Nature* **1983**, *304*, 596-602.

61. Obaya, A. J.; Mateyak, M. K.; Sedivy, J. M. Mysterious Liaisons: The Relationship Between C-Myc And The Cell Cycle. *Oncogene* **1999**, *18*, 2934-2941.

62. Persson, H.; Gray, H. E.; Godeau, F. Growth-Dependent Synthesis Of C-Myc-Encoded Proteins: Early Stimulation By Serum Factors In Synchronized Mouse 3T3 Cells. *Mol. Cell. Biol.* **1985**, *5*, 2903-2912.

63. Thompson, C. B.; Challoner, P. B.; Neiman, P. E.; Groudine, M. Levels Of C-Myc Oncogene mRNA Are Invariant Throughout The Cell Cycle. *Nature* **1985**, *314*, 363-366.

64. Hann, S. R.; Thompson, C. B.; Eisenman, R. N. C-Myc Oncogene Protein Synthesis Is Independent Of The Cell Cycle In Human And Avian Cells. *Nature* **1985**, *314*, 366-369.

65. Maruta, H.; Burgess, A. W. Regulation Of The Ras Signalling Network. *Bioessays* **1994**, *16*, 489-496.

66. Mateyak, M. K.; Obaya, A. J.; Adachi, S.; Sedivy, J. M. Phenotypes Of C-Myc-Deficient Rat Fibroblasts Isolated By Targeted Homologous Recombination. *Cell. Growth Differ.* **1997**, *8*, 1039-1048.

67. Prendergast, G. C. Mechanisms Of Apoptosis By C-Myc. *Oncogene* **1999**, *18*, 2967-2987.

68. Blackwood, E. M.; Eisenman, R. N. Max: A Helix-Loop-Helix Zipper Protein That Forms A Sequence-Specific DNA-Binding Complex With Myc. *Science* **1991**, *251*, 1211-1217.
69. Cole, M. D.; McMahon, S. B. The Myc Oncoprotein: A Critical Evaluation Of Transactivation And Target Gene Regulation. *Oncogene* **1999**, *18*, 2916-2924.
70. Schmidt, E. V. The Role Of C-Myc In Cellular Growth Control. *Oncogene* **1999**, *18*, 2988-2996.
71. Bush, A.; Mateyak, M.; Dugan, K.; Obaya, A.; Adachi, S.; Sedivy, J.; Cole, M. C-Myc Null Cells Misregulate Cad And Gadd45 But Not Other Proposed C- Myc Targets. *Genes Dev.* **1998**, *12*, 3797-3802.
72. Mac, S. M.; Farnham, P. J. Cad, A C-Myc Target Gene, Is Not Deregulated In Burkitt's Lymphoma Cell Lines. *Mol. Carcinog.* **2000**, *27*, 84-96.
73. Graves, L. M.; Guy, H. I.; Kozlowski, P.; Huang, M.; Lazarowski, E.; Pope, R. M.; Collins, M. A.; Dahlstrand, E. N.; Earp, H. S., 3rd; Evans, D. R. Regulation Of Carbamoyl Phosphate Synthetase By MAP Kinase. *Nature* **2000**, *403*, 328-332.
74. Carrey, E. A.; Campbell, D. G.; Hardie, D. G. Phosphorylation And Activation Of Hamster Carbamyl Phosphate Synthetase II By cAMP-Dependent Protein Kinase. A Novel Mechanism Of Regulation Of Pyrimidine Nucleotide Biosynthesis. *EMBO. J.* **1985**, *4*, 3735-3742.
75. Buchanan, J. M. The Amidotransferases. *Adv. Enzymol.* **1973**, *39*, 91-183.
76. Iwahana, H.; Oka, J.; Mizusawa, N.; Kudo, E.; Ii, S.; Yoshimoto, K.; Holmes, E. W.; Itakura, M. Molecular Cloning Of Human Amidophosphoribosyltransferase. *Biochem. Biophys. Res. Commun.* **1993**, *190*, 192-200.
77. Iwahana, H.; Yamaoka, T.; Mizutani, M.; Mizusawa, N.; Ii, S.; Yoshimoto, K.; Itakura, M. Molecular Cloning Of Rat Amidophosphoribosyltransferase. *J. Biol. Chem.* **1993**, *268*, 7225-7237.
78. Itakura, M.; Yamaoka, T.; Yoshikawa, H.; Yamashita, K. Paradoxical Effects Of Glucagon On De Novo Purine Synthesis In Rat Liver: In Vitro Phosphorylation Of Amidophosphoribosyltransferase. *Adv Exp Med Biol.* **1989**, *253B*, 29-35.
79. Ingley, E.; Hemmings, B. A. PKB/Akt Interacts With Inosine-5' Monophosphate Dehydrogenase Through Its Pleckstrin Homology Domain. *FEBS Lett.* **2000**, *478*, 253-259.
80. Parada, L. F.; Land, H.; Weinberg, R. A.; Wolf, D.; Rotter, V. Cooperation Between Gene Encoding p53 Tumour Antigen And *Ras* In Cellular Transformation. *Nature* **1984**, *312*, 649-651.
81. Aoki, T.; Weber, G. Carbamoyl Phosphate Synthetase (Glutamine-Hydrolyzing): Increased Activity In Cancer Cells. *Science* **1981**, *212*, 463-465.

82. Jackson, R. C.; Weber, G.; Morris, H. P. (1975). IMP Dehydrogenase, An Enzyme Linked With Proliferation And Malignancy. *Nature* **1975**, *256*, 331-333.

83. Collart, F. R.; Chubb, C. B.; Mirkin, B. L.; Huberman, E. Increased Inosine-5'-Phosphate Dehydrogenase Gene Expression in Solid Tumor Tissues and Tumor Cell Lines. *Cancer Res.* **1992**, *52*, 5826-5828.

84. Peters, G. J.; van der Wilt, C. L. Thymidylate Synthase As A Target In Cancer Chemotherapy. *Biochem. Soc. Trans.* **1995**, *23*, 884-888.

85. Chu, E.; Koeller, D. M.; Casey, J. L.; Drake, J. C.; Chabner, B. A.; Elwood, P. C.; Zinn, S.; Allegra, C. J. Autoregulation Of Human Thymidylate Synthase Messenger RNA Translation By Thymidylate Synthase. *Proc. Natl. Acad. Sci. U.S.A.* **1991**, *88*, 8977-8981.

86. Chu, E.; Voeller, D.; Koeller, D. M.; Drake, J. C.; Takimoto, C. H.; Maley, G. F.; Maley, F.; Allegra, C. J. Identification Of An RNA Binding Site For Human Thymidylate Synthase. *Proc. Natl. Acad. Sci. U.S.A.* **1993**, *90*, 517-521.

87. Chu, E.; Takechi, T.; Jones, K. L.; Voeller, D. M.; Copur, S. M.; Maley, G. F.; Maley, F.; Segal, S.; Allegra, C. J. Thymidylate Synthase Binds To C-Myc Rna In Human Colon Cancer Cells And In Vitro. *Mol. Cell. Biol.* **1995**, *15*, 179-185.

88. Chu, E.; Cogliati, T.; Copur, S. M.; Borre, A.; Voeller, D. M.; Allegra, C. J.; Segal, S. Identification Of In Vivo Target RNA Sequences Bound By Thymidylate Synthase. *Nucleic Acids. Res.* **1996**, *24*, 3222-3228.

89. Chu, E.; Copur, S. M.; Ju, J.; Chen, T. M.; Khleif, S.; Voeller, D. M.; Mizunuma, N.; Patel, M.; Maley, G. F.; Maley, F.; Allegra, C. J. Thymidylate Synthase Protein And P53 mRNA Form An In Vivo Ribonucleoprotein Complex. *Mol. Cell Biol.* **1999**, *19*, 1582-1594.

90. Ju, J.; Pedersen-Lane, J.; Maley, F.; Chu, E. Regulation Of P53 Expression By Thymidylate Synthase. *Proc. Natl. Acad. Sci. U.S.A.* **1999**, *96*, 3769-3774.

91. Keyomarsi, K.; Samet, J.; Molnar, G.; Pardee, A. B. The Thymidylate Synthase Inhibitor, ICI D1694, Overcomes Translational Detainment of the Enzyme. *J. Biol. Chem.* **1993**, *268*, 15142-15149.

92. Hauschka, P. V. Analysis of Nucleotide Pools in Animal Cells. In *Methods in Cell Biology*, Vol. VII; Prescott, D. M. Ed.; Academic Press: New York, 1973; pp 361-462.

93. Szekeres, T.; Fritzer, M.; Pillwein, K.; Felzmann, T.; Chiba, P. Cell Cycle Dependent Regulation of IIMP Dehydrogenase Activity and Effect of Tiazofurin. *Life Sciences* **1992**, *51*, 1309-1315.

94. Sweet, S.; Singh, G. Changes in Mitochondrial Mass, Membrane Potential, and Cellular Adenosine Triphosphate Content During the Cell Cycle of Human Leukemic (HL-60) Cells. *J. Cell. Physiol.* **1999**, *180*, 91-96.

95. Bianchi, V.; Borella, S.; Rampazzo, C.; Ferraro, P.; Calderazzo, F.; Bianchi, L. C.; Skog, S.; Reichard, P. Cell Cycle-dependent Metabolism of Pyrimidine Deoxynucleoside Triphosphates in CEM Cells. *J. Biol. Chem.* **1997**, *272*, 16118-16124.
96. Bischoff, J.R.; Friedman, P. N.; Marshak, D. R.; Prives, C.; Beach, D. Human P53 Is Phosphorylated By P60-Cdc2 And Cyclin B-Cdc2. *Proc. Natl. Acad. Sci. U S A* **1990**, *87*, 4766-4770.
97. Fan, H.; Villegas, C.; Wright, J. A. Ribonucleotide Reductase R2 Component Is A Novel Malignancy Determinant That Cooperates With Activated Oncogenes To Determine Transformation And Malignant Potential. *Proc. Natl. Acad. Sci. U.S.A.* **1996**, *93*, 14036-14040.
98. Fan, H.; Huang, A.; Villegas, C.; Wright, J. A. The R1 Component Of Mammalian Ribonucleotide Reductase Has Malignancy- Suppressing Activity As Demonstrated By Gene Transfer Experiments. *Proc. Natl. Acad. Sci. U.S.A.* **1997**, *94*, 13181-13186.
99. Filatov, D.; Ingemarson, R.; Johansson, E.; Rova, U.; Thelander, L. Mouse Ribonucleotide Reductase: From Genes To Proteins. *Biochem. Soc. Trans.* **1995**, *23*, 903-905.

Chapter 5

Targeted Disruption of the IMPDH: Type II Gene: Effect on Mouse Development and Lymphocyte Activation

Jing Jin Gu[1], Sander Stegmann[1], and Beverly S. Mitchell[1,2]

[1]Lineberger Comprehensive Cancer Center, and [2]Departments of Medicine and Pharmacology, University of North Carolina, Chapel Hill, NC 27599

We have examined the consequences of knocking out the IMPDH type II enzyme by gene targeting in a mouse model. Loss of both alleles of the gene encoding this enzyme results in very early embryonic lethality despite the presence of IMPDH type I and HPRT activities. Mice heterozygous for IMPDH type II expression have normal lymphocyte subset distributions and responses to a variety of mitogenic stimuli with marginal decrease in proliferation. However, mice deficient in both IMPDH type II and HPRT (IMPDH II +/-; HPRT -/o) demonstrated significantly decreased lymphocyte responsiveness to stimulation with anti-CD3 plus anti-CD28 antibodies and have a 30% mean reduction in GTP levels in lymphocytes activated by these antibodies. Furthermore, the cytolytic activity of T cells against allogeneic target cells is significantly impaired. These results demonstrate that a moderate decrease in the ability of murine lymphocytes to synthesize guanine nucleotides during stimulation results in a significant impairment in T cell activation and function.

Inosine 5'-monophosphate dehydrogenase (IMPDH; EC1.1.1.205) catalyzes the NAD-dependent conversion of IMP to XMP, a rate-limiting step in the *de novo* synthesis of guanine nucleotides which is critical for many essential cellular functions, such as DNA and RNA biosynthesis and signal transduction leading to cell proliferation and differentiation. The only alternative pathway for guanine nucleotide biosynthesis is a salvage pathway catalyzing the conversion of guanine to GMP by hypoxanthine-guanine phosphoribosyltransferase (HPRT), an enzyme encoded on the X-chromosome. The relative contributions of these two pathways in maintaining cellular guanine nucleotide pool have not been definitively elucidated, although deficiency in the salvage pathway has been linked to a human neurological disorder, Lesch-Nyhan Syndrome. No disease state has been directly linked to IMPDH deficiency.

IMPDH enzymatic activity results from two isoforms, termed type I and type II *[1]*, that are derived from two distinct genes under very different regulation *[2, 3]*. These two isoforms share 84% identity at the amino acid level and are indistinguishable in catalytic activity. Although increased expression of type II IMPDH is strongly correlated with cell proliferation and malignant transformation *[4-7]*, expression of both type I and type II are elevated in activated lymphocytes *[8]*. A study of the distribution of type I and type II IMPDH in various tissues has demonstrated that type II is expressed at higher levels in most tissues, whereas expression of type I, in general, is at a low level, but varies significantly in different tissues *[9]*.

Given the central role of IMPDH in guanine nucleotide synthesis, and the fact that high expression of type II is linked to neoplastic transformation, inhibitors of the enzyme have been sought as potential pharmacological agents for the treatment of malignancies and for immunosuppression. Inhibitors of IMPDH activity cause inhibition of cell proliferation that is reversible with the repletion of guanine nucleotides by exogenous guanine *[10, 11]*. Inhibition of IMPDH activity has been associated with the induction of cell differentiation in a variety of leukemic and tumor cell lines *[12-14]*, results in suppression of both T and B lymphocyte proliferation *[8, 10]*, and has been used, in conjunction with other immunosuppressive agents, to improve efficacy in tissue transplantation *[15, 16]*.

In order to acquire a better understanding of how guanine nucleotide biosynthesis is regulated by *de novo* and alternative pathways, and the relative biological importance of the two IMPDH isoforms in lymphocyte activation and function, we have used a genetic approach to inactivate the IMPDH type II gene using a specific gene targeting construct. The goals of our studies are to answer the following questions: (I) what is the role of IMPDH type II in mouse development; (II) will increased expression of IMPDH type I or HPRT compensate for the loss of expression of the type II gene; and (III) to what extent is lymphocyte activation affected by loss of IMPDH type II expression.

Methods

IMPDH type II gene targeting vector construction and embryonic stem cell selection. The pJNS2 vector (provided by Beverly Koller, Department of Medicine, University of North Carolina), containing *Neo* and *HSV-TK* genes, was used for making the knock-out construct. A 2.4 kb fragment containing IMPDH type II gene exons 1 to 5 and a 4.6 kb fragment containing exons 10 through 14 were inserted into *NotI*, *XhoI* and *XbaI* sites 5' and 3' to the *Neo* gene, respectively (Figure 1). Mouse 129 strain embryonic stem (ES) cells were transfected using electroporation and selected in the presence of 200 µg/ml G418 and 0.5 µg /ml ganciclovir. Genomic DNA from selected ES clones was digested with *Bgl II* and separated on an agarose gel. Southern hybridization was performed using a 0.6-kb DNA probe which binds to the sequences 5' to the ATG site. Two ES clones heterozygous for the presence of the targeting construct were microinjected into C57BL/6 blastocysts to generate chimeric mice.

Mouse breeding and offspring genotyping. Chimeric male mice with 65-100% chimerism were bred with C57BL/6 females to generate IMPDH type II heterozygous F1 mice. Genotyping was performed on tail DNA from 3-4 weeks old offspring. Heterozygous male and female mice were interbred to obtain homozygous mice. To obtain the IMPDH heterozygous plus HPRT deficient mice, IMPDH type II heterozygous mice were bred with HPRT deficient mice (The Jackson Laboratory, Bar Harbor, Maine, USA).

Western analysis. To determine expression of IMPDH protein, various mouse tissues or cells were homogenized in extraction buffer (200 mM Tris-HCl, pH7.5; 200 mM NaCl; 2 mM DTT; 5 mM MgCl$_2$; 1 mM EGTA; 2 µg/ml Leupeptins; 0.1 mM PMSF; and 2 µg/ml of Pepstatin A), freeze-thawed for 3 cycles followed by centrifugation to remove the cell debris. Ten µg of total cell lysate was separated on 15% SDS polyacrylamide gel, and IMPDH proteins were detected using a polyclonal antibody specific for both type I and type II proteins.

Enzyme Assays. IMPDH enzyme activity was measured using 10 µg of total splenocyte lysate in a 50 µl of reaction containing 80 µM [^{14}C]-8-inosine monophosphate; 100 mM Tris-HCl (pH8.0); 100mM KCl; 0.25 mM NAD; and 3 mM EDTA. HPRT enzyme activity was measured using 10 µg of protein extracted from tail in a 50 µl of reaction containing 100 µM of ^{14}C-8-hypoxanthine (48 mci/mmole); 50 mM Tris-HCl (pH 7.4); 5 mM MgCl$_2$ and 5 mM PRPP. The reactions were carried out at 37° C for 30 min, and 20 µl of reaction product were separated by thin-layer chromatography.

94

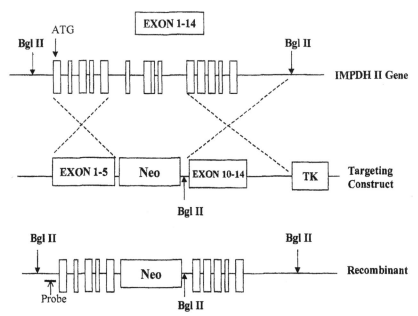

Figure 1: Targeted disruption of the IMPDH type II gene. Schematic diagram of IMPDH type II gene with 14 exons (open boxes); a knock-out construct with exons 1 to 5 and exons 10 to 14 cloned adjacent to the Neo gene; and the expected recombinant allele. The Bgl II sites are indicated, yielding an expected 11 kb fragment from the wild type allele and an expected 6 kb fragment from the recombinant allele. The probe used for Southern hybridization is indicated.

Splenocyte activation and proliferation. Mouse splenocytes were stimulated with Con A (5 μg/ml); LPS (20 μg/ml); PMA (10ng/ml) plus ionomycin (250 ng/ml); or added to the plate pre-coated with anti-CD3 (1 μg/ml) plus anti-CD28 (2.5 μg/ml) antibodies. Cells were incubated at 37° C for 48 hours, and 1 μCi of [methyl-^{3}H]-thymidine (80 Ci/mmol) was added at final 6 hours of culture. Cells were harvested onto glass fiber filters and measured in a scintillation counter. For testing drug sensitivity, cells were stimulated with either Con A or antibodies in the absence or presence of 10 nM mycophenolic acid for 72 hours.

^{51}Cr-release assay. To generate allogeneic-specific cytotoxic T lymphocytes, wild type or mutant splenocytes (C57BL/6; H-2^b) were mixed with irradiated (30 Gy) splenocytes from DBA/2 (H-2^d) mouse for 5 days (mixed lymphocyte reaction, MLR). Target cells (P815, H-2^d) were labeled with 150 μCi [^{51}Cr]-sodium chromate (250-500 mCi/mg Cr). The function of cytotoxic T lymphocytes generated from MLR was measured by mixing with labeled target cells at indicated effector to target cell ratios, and incubated at 37°C for 4 hours. The wells containing target cells with medium alone were used as spontaneous release controls, and wells containing target cells with 5% Triton X-100 were used as maximum release controls. Results are expressed as a percentage of specific lysis as calculated by the following formula:(release with effector cells-spontaneous release)/(maximum release- spontaneous release).

Interleukin-2 assay. Mouse splenocytes were stimulated with either Con A or antibodies, and culture supernatants were collected at time 0, 4, 8, 12, 24, 30, 36, 48, 56, 72 hr, filtered and stored at –80° C. To test relative amount of IL-2 in the culture supernatant, an IL-2 sensitive cell line, CTLL-2, was used. The proliferation of CTLL-2 cells in the presence of culture supernatant was measured by ^{3}H-thymidine incorporation.

Immunization of mice and serum immunoglobulin measurement. Wild type and mutant mice were injected peritoneally with 100 μg of NP32-Chicken γ globulin, and boosted with same amount of antigen after 14 days. Tail blood was collected at day 0, 7 and 19. Levels of IgG and IgM were measured by ELISA assays using NP5-BSA as antigen and alkaline phosphatase coupled secondary antibodies specific for mouse IgG and IgM.

Results

Targeting the IMPDH type II gene and generation of Type II deficient mice. To inactive the IMPDH type II gene in mice, we constructed a knockout vertor by cloning a 2.4 kb fragment containing exons 1 to 5 of the IMPDH type II gene and a 4.6 kb fragment containing exons 10 to 14 into the pJNS2 vector, 5' and 3' adjacent to the *Neo* gene, respectively (Figure 1). Homologous recombination will result in a new *Bgl II* site 3' to the *Neo* gene, generating a 6 kb *Bgl II* fragment from the recombinant allele as opposed to the 11 kb *Bgl II* fragment from the wild type allele. Mouse embryonic stem cells were transfected with this vector by electroporation and selected in the presence of both G418 and Ganciclovir. Cells with correctly targeted alleles were identified by Southern analysis of genomic DNA (data not shown) and constituted 10-15% of the clones analyzed.

Two IMPDH type II heterozygous ES clones were microinjected into C57BL/6 (B6) blastocysts to generate chimeric mice. Mice with high chimerism determined by coat color were further bred with B6 females with successful germ line transmission. IMPDH type II heterozygous mice were generated with no gross abnormal phenotype and were fertile. Therefore, they were further interbred to generate mice with homozygous deficiency of the IMPDH type II gene. Of 159 offspring analyzed, 57 were wild type (36%), and 102 were heterozygous (64%). No homozygous deficient mice was found, indicating that loss of both IMPDH type II alleles is embryonic lethal (Figure 2). Similarly, analysis of 122 blastocysts obtained from superovulated females revealed only eight that were homozygous deficient (6.5%). These data indicate that one allele of the type II gene is absolutely required for embryonic development.

Generation of IMPDH type II and HPRT mutant mice. In order to determine if loss of HPRT activity in addition to heterozygous deficiency of IMPDH type II will result in any phenotypic alterations, we crossed the IMPDH type II heterozygous mice with HPRT deficient mice and selected male offspring with four different genotypes (Figure 3): (1) wild type for both IMPDH type II and HPRT (IMPDH II +/+, HPRT +/o); (2) IMPDH II wild type, HPRT deficient (IMPDH II +/+, HPRT -/o); (3) IMPDH +/-, HPRT +/o; (4) IMPDH +/-, HPRT -/o. Mice containing each of these genotypes were born in an expected frequency and had no obvious developmental and behavioral abnormalities.

Regulation of expression of IMPDH type I in type II deficient mice. In order to determine whether expression of IMPDH type I, type II and HPRT, three major enzymes involved in guanine nucleotide synthesis, are coordinately regulated, we measured the IMPDH type I and type II protein levels in tissues from wild type and type II +/-, HPRT -/o mice (Figure 4). Expression of IMPDH type I protein is relatively higher in thymus, lung and brain. Similar levels were detected in the knock-out mouse, with a slightly increase in type I in the brain. Expression of the type II protein is moderately reduced in all tissues. Expression of both type I and type II proteins in splenocytes was greatly increased upon Con A stimulation, although to a lesser extent in the knock-out cells. An additional, higher molecular weight protein is consistently present in thymus and occasionally in lung and is not changed by loss of one IMPDH type

	# Analyzed	Wild Type	Heterozygous	Homozygous
Mice Born	159	57 (36%)	102 (64%)	0
Blastocysts	122	36 (30%)	78 (64%)	8 (6.5%)

Figure 2: Genotyping results from IMPDH type II heterozygous breeding.

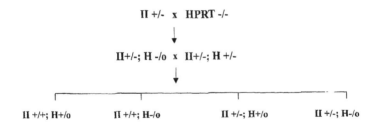

Figure 3: A schematic diagram of the breeding strategy for generating IMPDH type II and HPRT mutant mice.

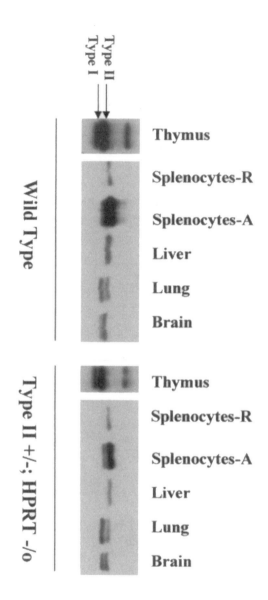

Figure 4: Expression of IMPDH proteins in wild type and knock-out mice tissues. Splenocytes were either resting (R) or Con A activated (A) for 48 hr.

II allele. IMPDH enzyme activity is significantly increased upon Con A stimulation (Figure 5A), and this increment is reduced by approximately 40% in IMPDH type II +/- mice regardless of HPRT status. Measurement of HPRT enzyme activity (Figure 5B) confirms the HPRT -/o genotype. We conclude from these findings that there are no compensatory increases in the expression of either IMPDH isoenzymes or HPRT as a result of the IMPDH II +/- or HPRT -/o genotype.

Effect of IMPDH and HPRT mutations on lymphocyte activation and drug sensitivity. To determine if alterations in the guanine nucleotide synthetic pathway affect lymphocyte activation, splenocytes from mice with the four different genotypes were cultured in the absence or presence of either B or T cell mitogens for 48 hours. Cell proliferation was measured by ^{3}H-thymidine incorporation (Figure 6). Lymphocytes from IMPDH II heterozygous mice or HPRT deficient mice showed a significant decrease in proliferative response to anti-CD3 plus anti-CD28 antibodies, while proliferation in response to other T and B cell activators was similar to that of wild type mice. Lymphocytes from mice with both mutations (IMPDH II +/-; HPRT -/o) demonstrated a more profound decrease in T cell proliferation, especially as a function of antibody stimulation, suggesting that receptor-activated signaling pathway might be more sensitive to guanine nucleotide level depletion in the cell.

To test whether lymphocytes from mice with both mutations would be more sensitive to the specific IMPDH inhibitor, mycophenolic acid (MPA), splenocytes were stimulated with either Con A or anti-CD3 plus anti-CD28 antibodies in the absence or presence of MPA (Figure 7). Proliferation of lymphocytes from wild type mice was slightly decreased in the presence of a very low dose of MPA (10 nM), while the proliferation of lymphocytes from mice containing both mutations was far more significantly decreased.

Effect of mutations on lymphocyte function. To determine whether the function of cytotoxic T lymphocytes (CTLs) is affected by guanine nucleotide depletion, cytolytic activity against allogeneic target cells was assayed after 5 days in a MLR, as described in the methods, by a standard ^{51}Cr-release assay (Figure 8). There was a small (20-30%), but consistent decrease in killing activity by T cells from mutant mice compared with cells from wild type mice. These differences are statistically significant at all the effector and target ratios.

To test if these mutations would also affect lymphokine secretion by CD4 positive T cells, splenocytes from wild type or mutant mice were cultured in the presence of Con A or antibodies and the culture supernatant was collected and assayed for presence of interleukin-2 (IL-2) by using an IL-2 sensitive cell line, CTLL-2. As indicated in Figure 9, proliferation of CTLL-2 cells cultured in supernatant from either wild type or mutant mice was very similar at all the time points, indicating no significant difference in IL-2 secretion.

B cell function was determined by their ability to secrete immunoglobulin (Ig) upon antigen stimulation. Wild type and mutant mice were injected with

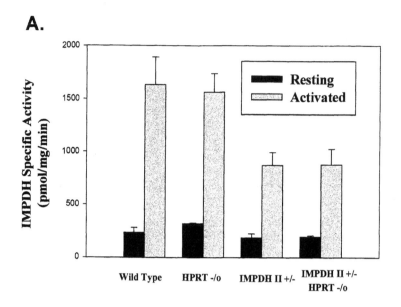

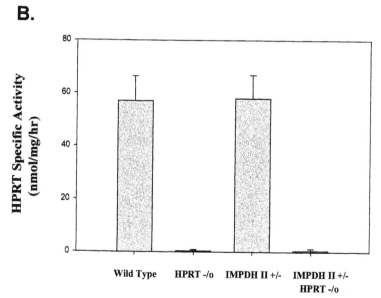

Figure 5: IMPDH and HPRT enzyme activities in wild type and mutant mice. (A) IMPDH enzyme activity of resting (black bar) and Con A activated (gray bar) splenocytes from mice with four different genotypes. (B) HPRT enzyme activity from mice with four different genotypes.

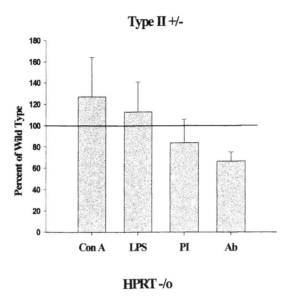

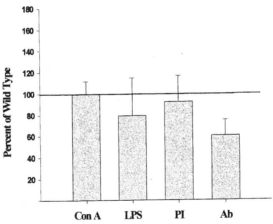

Figure 6: Effect of loss of IMPDH II and/or HPRT activities on mouse splenocyte proliferation. Splenocytes were stimulated with Con A or LPS or PMA plus Ionomycin (PI) or anti-CD3 plus anti-CD28 (Ab) for 48 hr. The proliferation was measured by ^{3}H-thymidine incorporation. Values are mean of three independent experiments performed in triplicate, and plotted as percentage of wild type splenocytes.

Continued on next page.

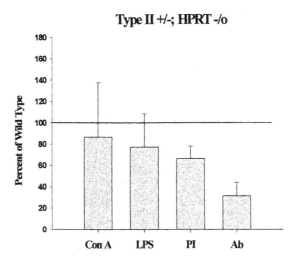

Figure 6. *Continued.*

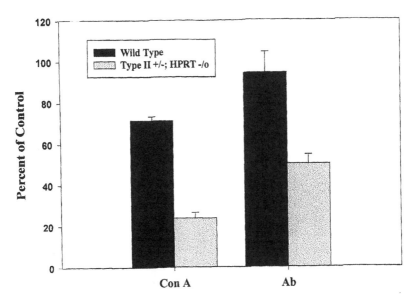

Figure 7: Effect of loss of IMPDH II and HPRT activity on drug sensitivity. Wild type (black bar) or mutant (gray bar) splenocytes were stimulated with either Con A or anti-CD3 plus anti-CD28 antibodies in the absence or presence of 10 nM of MPA for 72 hr. Values are the mean of three independent experiments performed in triplicate, and plotted as the percentage of proliferation in the absence of drug.

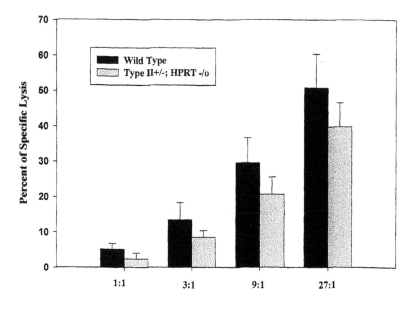

Effector : Target Ratio

*Figure 8: Effect of mutations on the cytolytic activity of cytotoxic T
lymphocytes. The function of cytotoxic T lymphocytes from wild type (black
bar) and mutant (gray bar) mice was analyzed in a ^{51}Cr-release assay. Values
are mean of 4 independent experiments performed in triplicate.* P<0.05 at
E:T=1:1; P<0.015 at E:T=3:1; P<0.012 at E:T=9:1 and 27:1.

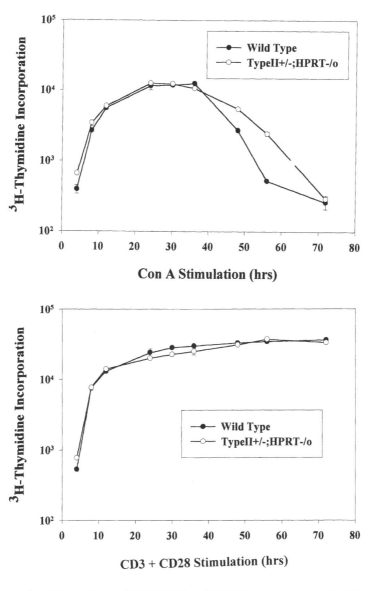

Figure 9: Effect of loss of IMPDH II and HPRT activity on interleukin 2 secretion. Splenocytes were stimulated with either Con A or anti-CD3 plus anti-CD28 antibodies and the amount of IL-2 in the supernatant was measured by CTLL-2 proliferation, as outlined in Methods. Wild type mice (closed circle) and mutant mice (opened circle).

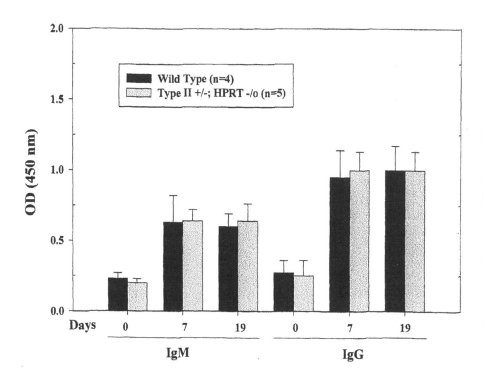

Figure 10: Effect of loss of IMPDH II and HPRT activity on serum immunoglobulin levels. Wild type (black bar) or mutant (gray bar) mice were immunized with NP32-Chicken γ globulin at day 0, and boosted at day 14. Serum collected at day 0, 7 and 19 was diluted 1350 fold for ELISA assay.

NP20-chicken γ globulin to elicit a T cell-dependent B cell response. Blood was collected before and after antigen stimulations and serum IgM and IgG levels were measured using specific antigen (NP5-BSA) by ELISA assays. No significant differences were found between wild type and mutant mice (Figure 10).

Discussion

The relative roles of the two IMPDH isoenzymes have not, to date, been elucidated. Although the marked up-regulation of the type II enzyme in proliferating and neoplastic tissues has argued for a major role for this isoenzyme in providing guanine nucleotides for cell proliferation, the gene encoding the type I enzyme has been shown to have a complex promoter region yielding three alternative transcripts [3] and also to be up-regulated during lymphocyte activation [8]. Hence, the type I enzyme has also been postulated, in the absence of experimental data, to have a potentially significant role in proliferation and development. We have now shown that homozygous deficiency of type II IMPDH results in embryonic lethality to mice at a very early stage of development. This result clearly demonstrates that the presence of IMPDH type I and HPRT activity cannot substitute for IMPDH type II in early development. Data from Downs have demonstrated a requirement for purines in oocyte meiotic arrest [17] and shown that mycophenolic acid (MPA), an inhibitor of both IMPDH enzymes, resulted in loss of implantation capacity of developing oocytes [18]. These results, in conjunction with our own, would indicate that IMPDH type II activity is required to generate guanine nucleotides during early maturation of the fertilized oocyte and would argue against a role for IMPDH type II protein that is independent of its enzymatic activity.

Loss of one allele of IMPDH type II resulted in a reduction in total IMPDH activity of roughly 50% in activated lymphocytes, indicating that this enzyme is responsible for the majority of the IMPDH activity in these cells. The lack of up-regulation of type I IMPDH, as judged by Western blot, or of HPRT activity argues against a compensatory effect of a reduction of IMPDH type II on the expression or activity of either of these proteins, again supporting a major role for the type II enzyme in intracellular guanine nucleotide production. Previous work has demonstrated that inhibition of IMPDH activity with MPA induces in increase in IMPDH type II mRNA production [19], but neither IMPDH type I or HPRT appears to be similarly regulated. Of interest is the observation that loss of one IMPDH type II allele impairs lymphocyte activation mediated by anti-CD3 and CD28 stimulation (figure 6). This result was also found in HPRT-deficient lymphocytes and was significantly enhanced in splenocytes from mice with both genetic defects. In contrast, proliferative responses to activation by Concanavalin A or LPS were not significantly reduced by the loss of these activities. These results would argue that signaling through the T cell receptor and/or CD28 receptor is reduced by the availability of guanine nucleotides. It does not support the argument that guanine nucleotides are required for cellular proliferation *per se*, since the extent of thymidine incorporation into DNA was similar in both mitogen and antibody-stimulated splenocytes.

We further investigated the biological response of cytotoxic T cells (CTL) and of activated B lymphocytes in response to loss of a single allele of IMPDH and of HPRT. CTL activity was reduced by approximately 20% at increasing effector:target ratios and these results were highly statistically significant. In contrast, the ability or T lymphocytes to secrete IL-2 in response to either ConA or CD3 plus CD28 stimulation was unimpaired while the ability of B lymphocytes to secrete either IgG or IgM in response to antigen stimulation *in vivo* was also not reduced. These data indicate that a moderate reduction in the ability to synthesize guanine nucleotides resulting from a 50% reduction in IMPDH activity and a complete loss of the salvage enzyme activity of HPRT have a very selective effect on T cell activation in the murine model, resulting in an impaired proliferative response to activation through the antigen receptors. CTL activity also appears to be highly dependent on guanine nucleotide production. However, other lymphocyte functions such as IL-2 production and T cell-dependent B cell activation appear less reliant on a moderate (approximately 30% in activated lymphocytes as compared to wild type control cells [20]) reduction in GTP levels. These observations provide a basis for the future delineation of GTP-dependent intracellular targets that are important in human lymphocyte activation and function. In turn, such studies may improve our ability to modulate lymphocyte responses *in vivo* in a more selective fashion than is currently feasible with standard IMPDH inhibitors such as Mycophenolate.

References

1. Natsumeda, Y., et al., *Two distinct cDNAs for human IMP dehydrogenase.* J. Biol. Chem., 1990. **265**: p. 5292-5295.
2. Zimmermann, A.G., et al., *Regulation of inosine-5'-monophosphate dehydrogenase type II gene expression in human T cells.* J. Biol. Chem., 1997. **272**: p. 22913-22923.
3. Gu, J.J., J. Spychala, and B.S. Mitchell, *Regulation of the human inosine monophosphate dehydrogenase type I gene.* J. Biol. Chem., 1997. **272**: p. 4458-4466.
4. Jackson, R.C. and G. Weber, *IMP dehydrogenase, an enzyme linked with proliferation and malignancy.* Nature, 1975. **256**: p. 331-333.
5. Collart, F.R., et al., *Increased inosine-5'-phosphate dehydrogenase gene expression in solid tumor tissues and tumor cell lines.* Cancer Research, 1992. **52**: p. 5826-5828.
6. Nagai, M., et al., *Selective up-regulation of type II inosine 5'-*

monophosphate dehydrogenase messenger RNA expression in human leukemias. Cancer Research, 1991. **51**: p. 3886-3890.

7. Nagai, M., Y. Natsumeda, and G. Weber, *Proliferation-linked regulation of type II IMP dehydrogenase gene in human normal lymphocytes and HL-60 leukemic cells.* Cancer Research, 1992. **52**: p. 258-261.

8. Dayton, J.S., et al., *Effects of human T lymphocyte activation on inosine monophosphate dehydrogenase expression.* Journal of Immunology, 1994. **152**: p. 984-991.

9. Senda, M. and Y. Natsumeda, *Tissue-differential expression of two distinct genes for human IMP dehydrogenase (E. C. 1. 1. 1. 205).* Life Sciences, 1994. **54**: p. 1917-1926.

10. Turka, L.A., et al., *Guanine ribonucleotide depletion inhibits T cell activation: mechanism of action of the immunosuppressive drug mizoribine.* Journal of Clinical Investigation, 1991. **87**: p. 940-948.

11. Dayton, J.S., et al., *Comparison of the effects of mizoribine with those of azathioprine, 6-mercaptopurine, and mycophenolic acid on T lymphocyte proliferation and purine ribonucleotide pools.* Molecular Pharmacology, 1992. **41**: p. 671-676.

12. Yu, J., et al., *Induction of erythroid differentiation in K562 cells by inhibitors of inosine monophosphate dehydrogenase.* Cancer Research, 1989. **49**: p. 5555-5560.

13. Sidi, Y., et al., *Growth inhibition and induction of phenotypic alterations in MCF-7 breast cancer cells by an IMP dehydrogenase inhibitor.* British Journal of Cancer, 1988. **58**: p. 61-63.

14. Kiguchi, K., et al., *Induction of cell differentiation in melanoma cells by inhibitors of IMP dehydrogenase: altered patterns of IMP dehydrogenase expression and activity.* Cell Growth Differ., 1990. **1**(6): p. 259-270.

15. Kokado, Y., et al., *A new triple-drug induction therapy with low dose cyclosporine, mizoribine and prednisolone in renal transplantation.* Transplantation Proceedings, 1989. **21**: p. 1575-1578.

16. Mita, K., et al., *Advantages of mizoribine over azathioprine in combination therapy with cyclosporine for renal transplantation.* Transplantation Proceedings, 1990. **22**: p. 1679-1681.

17. Downs, S.M., *Purine control of mouse oocyte maturation: Evidence that nonmetabolized hypoxanthine maintains meiotic arrest.* Molecular Reproduction and Development, 1993. **35**: p. 82-94.

18. Downs, S.M., *Induction of meiotic maturation in vivo in the mouse by IMP dehydrogenase inhibitors: Effects on the developmental capacity of ova.* Molecular Reproduction and Development, 1994. **38**: p. 293-302.

19. Glesne, D.A., F.R. Collart, and E. Huberman, *Regulation of IMP dehydrogenase gene expression by its end products, guanine nucleotides.* Mol. Cell. Biol., 1991. **11**(11): p. 5417-5425.

20. Gu, J.J., et al., *Inhibition of T lymphocyte activation in mice heterozygous for loss of the IMPDH II gene.* The Journal of Clinical Investigation, 2000. **106**: p. 599-606.

Chapter 6

Differential Splicing of *Pneumocystis carinii* Inosine 5′-Monophosphate Dehydrogenase mRNA: Implications for Producing Catalytically Active Protein

Dongjiu Ye, Jeniece Nott, and Sherry F. Queener[*]

Department of Pharmacology and Toxicology, Indiana University School of Medicine, Indianapolis, IN 46202

Pneumocystis carinii IMPDH was originally reported to be approximately 70 amino acids shorter at the amino terminus than IMPDH from other species[1], but recent work in our laboratory has demonstrated that four splicing variants of the protein are possible[2]. Three of these variants, including one identical to the originally described mRNA, contain a stop codon in the 5' region of the pre-mRNA that results in loss of regions of the amino terminus thought to be important for enzyme activity. The fourth variant codes for a form of IMPDH that contains 529 amino acids and retains key amino terminal sequences (GeneBank Accession No. 196975)[2]. In *P. carinii* isolated from infected rats, the splicing variant coding for the long form predominated and only the long form of the protein was detected; in organisms from culture, splicing variants coding for the short form of IMPDH predominated but only the long form of the protein was detected. The catalytic activity of this longer form of IMPDH from *P. carinii* was confirmed by a complementation study in bacteria lacking native IMPDH. *P. carinii* IMPDH long form was expressed with a his-tag on the amino terminal end and purified by metal-affinity chromatography. The protein retains catalytic activity and, like the native protein in *P. carinii*, exists as a tetramer in solution. The his-tagged protein showed a Km for NAD of 24 ± 1 micromolar and a Km for IMP of 51 ± 8 micromolar, similar to forms of the enzyme from other species. Substrate inhibition was observed at millimolar concentrations of both NAD and IMP. The enzyme activity was strongly inhibited by mycophenolic acid (IC50 0.066 micromolar) and showed a different pattern of inhibition with other inhibitors than the mammalian enzyme, suggesting selective drug design is possible.

Pneumocystis carinii

Taxonomy

Pneumocystis carinii is an pathogen whose taxonomic classification has been difficult since the organism was first described in 1909, when it was thought to be a life stage of trypanosomes[3]. The organism was quickly recognized to be distinct from trypanosomes[4,5], but continued to be considered a protozoan on the basis first of morphology and later on the basis of susceptibility to antiprotozoal agents[6-8]. *P. carinii* is now considered to be more related to fungi than to protozoans, primarily on the basis of homologies in nucleic acids and proteins[9,10].

Relation to human disease

P. carinii caused only sporadic outbreaks of human disease before 1980, with most cases occuring among protein-malnourished children living in crowded conditions[8,11]. The dramatic increase in *P. carinii* pneumonia that accompanied the onset of AIDS illustrated that *P. carinii* was primarily an opportunistic pathogen, proliferating and causing disease in those whose immune function was compromised. Other severely immunocompromised patients, such as those receiving immunosuppressive drugs for organ transplantation or other conditions are also susceptible to severe infections caused by this organism[12-14].

Before highly active anti-retroviral therapy (HAART) became widely used, *P. carinii* pneumonia was a leading cause of death in AIDS patients[15,16] but five years after HAART became available, the situation had changed. For many patients HAART allows the immune system to remain sufficiently functional to protect patients from many of the more common opportunistic infections, including *P. carinii* pneumonia. The use of prophylaxis for *P. carinii* pneumonia is also wide-spread and has lowered the incidence of this disease. In spite of these advances, significant numbers of AIDS patients and organ transplant patients continue to suffer from serious infections caused by this organism[12,17]. *P. carinii* infections are still a common finding in patients on first diagnosis with AIDS and the occurence of *P. carinii* pneumonia diminishes the prospects for long term survival.

Therapy of *P. carinii* infections

Therapy for *P. carinii* pneumonia has depended primarily upon older antimicrobial agents, such as pentamidine and the combination of trimethoprim with sulfamethoxazole. The use of pentamidine is often limited by toxicity unless it is administered in an aerosol form; this latter route is used more for prophylaxis than for therapy and is associated with more breakthrough infections than other routes[18,19]. Alternative therapies for *P. carinii* pneumonia exist and there are several regimens for prophylaxis, but many of the drugs used are associated with significant toxicity and failures of therapy or prophylaxis are not uncommon[17,20-22].

The first choice for both prophylaxis and therapy of *P. carinii* pneumonia is trimethoprim/sulfamethoxazole in the same fixed combination used for bacterial infections. Whereas this therapy is very effective and well tolerated in non-AIDS patients, side effects occur frequently in HIV-infected individuals[23]. Of more concern are recent reports that prophylaxis or therapy with trimethoprim/sulfamethoxazole may fail as drug-resistance mutations accumulate in dihydropteroate synthase, the target of sulfonamide drugs[24-26]. Questions have existed for some time about the role of trimethoprim in the combination when it is used against *P. carinii*, because the target enzyme dihydrofolate reductase in *P. carinii* is only weakly inhibited by trimethoprim[27]. Moreover, trimethoprim alone shows little or no activity against the organism in culture or animal models[28]. Thus, the extensive use of this combination for prophylaxis may have facilitated the development of drug resistance, since only one of the drugs in the combination seems to be active. If the drug-resistant forms of *P. carinii* spread, trimethoprim/ sulfamethoxazole, which is currently the primary clinical agent for this disease, will cease to be effective. Given the slow growth of the organism and the length of time it has taken for the first drug-resistant forms to be detected, the argument might be made that failure of the primary drug is a long way off, but no one can predict how rapidly drug-resistant forms may spread, given the new information on transmission of the organism. Although it was originally thought that humans carried cysts of *P. carinii* in their lungs and infections arose from organisms that might have been picked up long before the illness, current evidence suggests that *P. carinii* can be passed from human to human in certain settings, probably by the airborne route[29,30]. If this is the case, it might be expected that drug-resistance forms would spread more rapidly than if independent mutations were occurring in individual patients.

In order to maintain effective control of *P. carinii* infections in humans, new drugs must be developed to fill the void left by the failure of established agents. Several lines of study show promise, including the development of new generation inhibitors of dihydrofolate reductase that are selective for the *P. carinii* enzyme[31-35] and the development of chemical relatives of pentamidine with more selectivity and fewer side effects[36]. New drug targets also need to be identified in *P. carinii*. One exploitable target may be inosine 5'-monophosphate dehydrogenase (IMPDH, EC 1.1.1.205)[1].

Studies of purine metabolism in *P. carinii*

IMPDH catalyzes the NAD^+-dependent conversion of inosine 5'-monophosphate (IMP) to xanthosine 5'-monophosphate (XMP). IMPDH has been extensively studied as a drug target because the reaction catalyzed by this enzyme is a required and often rate-limiting step in *de novo* synthesis of guanine nucleotides (Figure 1). For example, the immunosuppressant mycophenolate mofetil, the anticancer drug tiazofurin, and the antiviral agent ribavirin are all inhibitors of IMPDH[37-40].

The potential of IMPDH to be a drug target in *P. carinii* was suggested by the fact that growth of the organism in short term culture was blocked by low concentrations of mycophenolic acid or tiazofurin[1] (Figure 2). The level of activity shown in the dose responses for these compounds places them among the most potent agents tested in the 24-well plate culture model, which has been widely used to demonstrate antipneumocystis activity[41-47]. By way of comparison, trimethoprim/sulfamethoxazole at 170/990 μM, respectively, produce inhibition similar to that shown by 1 μM mycophenolic acid.

One way some organisms offset inhibition of IMPDH is to salvage xanthine, guanine or guanosine. These metabolites enter the pathway after the IMPDH reaction and thus can drive synthesis of guanine nucleotides in the absence of significant IMPDH activity. Limited studies exist on metabolism in *P. carinii*

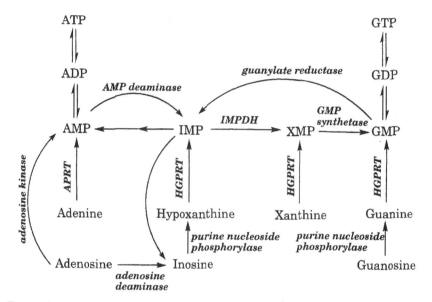

Figure 1. Purine metabolic pathways. A generalized sequence of metabolic reactions is shown, because the pathways and enzymes involved in P. carinii have not been elucidated. The process of de novo synthesis, not shown in the figure, would produce IMP. Salvage pathways are mediated by APRT, HGPRT, purine nucleoside phosphorylase and adenosine kinase. APRT, adenine phosphoribosyl transferase; HGPRT, hypoxanthine-guanosine phosphoribosyl transferase.

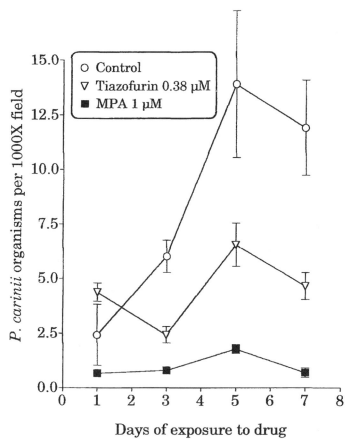

Figure 2. Inhibition of P. carinii *growth in short term cultures by IMPDH inhibitors. These studies were performed using the short term 24-well plate culture system designed for testing drug susceptibility of* P. carinii, *where the organism is grown on monolayers of human embryonic lung cells. Data shown are means ± SEM for quadruplicate samples. Error bars on the curve for 1 μM MPA (mycophenolic acid) are small and are within the range covered by the symbol.*

116

and no studies have focused on purine metabolism. Early reports suggested that PABA, glucose, pyruvate, amino acids and uridine were taken up by *P. carinii*, but that hypoxanthine and thymidine were not[48]. This result was interpreted as evidence that the organisms did not proliferate when they were isolated from rat lung and suspended in medium. An alternate explanation is that these organisms accurately reflected the naturally low state of purine and pyrimidine salvage in *P. carinii*. The critical enzyme for salvage of xanthine and guanine is HGPRT (hypoxanthine, guanosine phosphoribosyltransferase, EC 2.4.2.8), which still has not been identified in *P. carinii* .

The culture data (Figure 2) indicated that inhibition of IMPDH alone was sufficient to impair growth of *P. carinii*, consistent with the suggestion that salvage pathways in this organism may be of minor importance. In fact, adding excess guanine to the growth media of *P. carinii* in short term culture did not overcome the inhibition produced by mycophenolic acid (Figure 3); guanosine and xanthine also failed to reverse the growth inhibition. In separate studies, direct uptake of guanine, guanosine and xanthine could not be demonstrated with *P. carinii* isolated from infected rat lungs (data not shown). These results were consistent with the earlier studies by other laboratories that failed to show salvage of purines by *P. carinii*[48].

To date the evidence suggests that salvage of purines in *P. carinii* is weak or nonexistent. A low rate of purine salvage would tend to make the effects of inhibition of IMPDH more devastating to *P. carinii* than to other organisms, where salvage of significant amounts of purines could potentially bypass the blockade of IMPDH and allow synthesis of GMP from xanthine, guanine, or guanosine. That such a situation exists is suggested by the results of the culture studies which show profound growth inhibition at micromolar or submicromolar concentrations of IMPDH inhibitors. IMPDH thus seems to satisfy one criterion for being a potential drug target in *P. carinii*, namely that it seems essential to the organism.

Cloning IMPDH from *P. carinii*

cDNA library from cultured *P. carinii*

The difficulties inherent in acquiring sufficient supplies of *P. carinii* for biochemical studies drove us to explore recombinant methods to produce IMPDH from this organism. The original cloning was performed several years ago[1] using degenerate primers designed to highly conserved regions of known IMPDH proteins. The most highly purified organisms available at that time

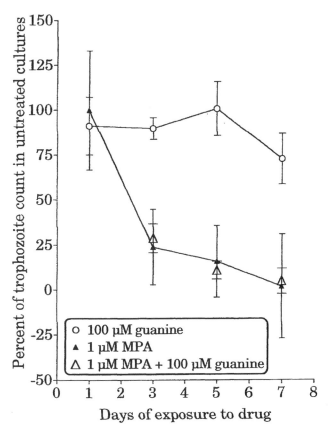

Figure 3. Effect of guanine on mycophenolic acid inhibition of P. carinii *growth in short term culture. The culture method is the same as for Figure 2, but the data for each experimental condition in this figure were expressed as percent of untreated control, means ± SEM. The results show that 100 µM guanine alone in the culture media had little effect on growth but, as expected, 1 µM mycophenolic acid produced a dramatic decline in the number of detected organisms; 100 µM guanine had no effect on this inhibition of growth.*

were from short term spinner flask culture and these were used for the DNA isolations[49]. This strategy yielded a 1.3 kb product that lacked both the 5' and the 3' end of the gene. This 1.3 kb product was used to screen a cDNA library also prepared from cultured *P. carinii*. No full length coding sequence was found in the library, but 3' and 5'-RACE were successfully employed to generate the cDNA for *P. carinii* IMPDH.

The protein sequence that was deduced from the cDNA was homologous to other forms of IMPDH, but was about 70 amino acids shorter at the amino terminus than IMPDH of other eukaryotic species[1]. Expression of this protein in *Escherichia coli* was difficult, but limited catalytic activity was observed in crude extracts of this recombinant bacteria and growth complementation studies suggested some IMPDH was present[1]. Redesign of the expression system failed to improve yields of this protein; when it was expressed in the IMPDH-deficient LH3 *E. coli*, soluble catalytic activity could not be detected and most of the recombinant protein was found in the particulate fractions, presumably in inclusion bodies.

Other laboratories have suggested that the amino terminus of IMPDH may be required for full catalytic activity, as that region is involved with forming the catalytic site and the potassium binding site[50]. The difficulties in expressing large amounts of catalytically active protein and the concerns about the effect of a 70 amino acid deletion at the N-terminus of the protein led us to reanalyze the 5' sequence of pre-mRNA transcripts of the *P. carinii* IMPDH gene.

The 5' sequence of the *P. carinii* IMPDH gene

Organisms were harvested from the lungs of *P. carinii*- infected immunosuppressed rats and purified from contaminating mammalian cells[51]. DNA purified from these organisms was used as template for PCR reactions; the forward primer was from nucleotide nine through twenty-eight of the previously reported sequence (GeneBank Accession number U42442), which was well upstream of the putative translation start site. The reverse primer was the reverse complement to nucleotide positions 246 through 265, a region about 60 nucleotides 3' of the putative start sequence. Thus, the design was intended to amplify only a small segment of the gene, starting within the coding sequence and covering the 5' region beyond the end of the coding sequence. The PCR reaction yielded a 420 bp product that contained four introns (Figure 4), which were identified by comparison to the previously reported sequence for the mRNA[1] and by the typical splicing recognition sites 5'gt-3'ag. The introns varied in length from 41 to 44 nucleotides, a size similar to those reported for other *P. carinii* genes[52]. An identical PCR product was generated when DNA isolated from cultured organisms was used as template.

Sequence determination of the 5' region of *P. carinii* IMPDH mRNA

In order to confirm the 5' nucleotide sequence of the mRNA for *P. carinii* IMPDH, we performed RT-PCR using freshly isolated total RNA preparations from organisms harvested from lungs of *P. carinii*- infected immunosuppressed rats and purified as noted above, as well as from organisms freshly harvested from short term spinner culture. The primers were the same as shown in Figure 4 for PCR. Based on the previously reported mRNA sequence the expected product size was 250 bp but in fact a range of sizes were seen, from 250 bp to about 350 bp.

Sequencing of the RT-PCR products showed four splicing variants, which lacked one, two, three or all four of the introns in this region of the gene[2]. The variant lacking all four introns was the same sequence as previously reported[1], which translates into an IMPDH lacking over 70 amino acids at the amino terminus. Two of the other splicing variants also gave rise to shortened deduced protein sequences, but one variant was different. The splicing variant that retained introns one and three gave rise to an open reading frame that led to a deduced protein sequence of 529 amino acids, 75 amino acids longer at the amino terminus than the previously reported form but continuous with it (Figure 5). This retention of putative introns in coding sequences is not unique. For example, in another fungus, *Cryptococcus neoformans*, an intron was identified as part of the coding sequence of the multidrug resistance gene 1[53]. When this intron was removed, 155 amino acids were lost from the amino terminus of that protein.

Implications of the splicing variants

Prevalence of the variants depends on the source of *P. carinii*

Because several independent RNA isolations were performed and the transformations were also repeated several times, it was possible to accumulate and sequence a total of seventeen clones representing a selection of the RT-PCR products. Nine of the clones arose from reactions that used RNA from organisms harvested from lungs of immunosuppressed rats. Among these nine clones, the splicing variant that gave rise to the 529 amino acid sequence was most prevalent, occuring in five of the nine; two clones lacked intron 2 and two clones retained only intron 1, but there were no products found that lacked all four introns. In contrast, among the eight clones derived from reactions that used RNA from cultured organisms, five of the clones lacked all four introns; one example of each of the other splicing variants was found. Thus, from

5' GTATCAAG CAGATACGGT GAAATACAGT ATGAAATTGG GTAATAAAAG ATATAAATAG 58

GAGGAATAGA Ggtataattt atgggggaag atagttattt atgtttatct gaagAAATTA 118
          Intron 1

AGAAGAAACT TGAAGAATAT TCGGAAAAGg tattaagatt ttttagatt taaattaaat 178
                              Intron 2

gattatatag GATGGATATG ATCTTGATAG TCTTATTTGT AGCCGACGAC ATGGGGGATT 238

AACgtataat gatattatta ttcttccggg atatattgat tttgaagTAA ATTCGGTTTG 298
Intron 3

TCTTGAAAGT CATATTACGA AAAAGATAGT ATTGAAGACG CCGTT**TATGA** Ggtatttta 358
                                                      Intron 4

gagagaataa gaaggaataa aggttaatt ttcagTTCAC CGATGGATAC GGTTACGGAA 418

TCGGATATG 3' 427

Figure 4. DNA sequence of the 5' region of the P. carinii IMPDH gene. Four
introns are designated with lower case letters. Primer sequences used for PCR
are underlined. The putative translation initiation codon identified in 1997
[1127] is in bold.

5′ CTGTATCAAGCAGATACGGTGAAATACAGTGATGAAATTGGGTAATAAAAGATATAAATAGGAGAGA

```
ATAGAGgtataatttatg ggg gaa gat agt tta tgt tta tct gaa gAA ATT    13
                    M   G   E   D   S   Y   L   C   L   S   E   E   I

AAG AAA CTT GAA GAA TAT TCG GAA AAG GAT GGA TAT GAT CTT GAT AGT       30
K   K   L   E   E   Y   S   E   K   D   G   Y   D   L   D   S

CTT ATT TGT AGC CGA CGA CAT GGG GGA TTA ACg tat aat gat att att att   47
L   I   C   S   R   R   H   G   G   L   T   Y   N   D   I   I   I

ctt ccg gga tat att gat ttt gaa gTA AAT TCG GTT TCT CTT GAA AGT CAT   64
L   P   G   Y   I   D   F   E   V   N   S   V   S   L   E   S   H

ATT ACG AAA AAG ATA GTA TTG AAG ACG CCG TTT ATG AGT TCA CCG ATG GAT   81
I   T   K   K   I   V   L   K   T   P   F   M   S   S   P   M   D

ACG GTT ACG GAA TCG GAT ATG GCG ATT AAT TTG                           92
T   V   T   E   S   D   M   A   I   N   L
```

Figure 5. cDNA sequence based on 5′ region of P. carinii IMPDH mRNA. The primers used for the RT-PCR reactions were the same as those used for PCR (Figure 4). Introns one and three, which are retained as coding sequence, are designated with lower case letters and underlined. Arrows designate the sites where introns two and four were spliced out. The methionine originally considered to be the translation start site is in bold face type.

cultured organisms seven of the eight clones gave rise to a shortened version of IMPDH and only one gave rise to the 529 amino acid deduced sequence. In the original studies reported in 1997, two independent clones gave rise to the 454 amino acid version of the protein[1].

Analysis of homology of short and long versions of *P. carinii* IMPDH

The sequence at the amino terminal end of the protein was compared amongst several eukaryotic species (Figure 6). The short version of *P. carinii* IMPDH lacks several regions that show high homology among the other eukaryotic proteins but the 529 amino acid version of the *P. carinii* IMPDH retains these regions of homology. Because these amino terminal sequences have been implicated in IMPDH function, we concluded that the long form of the protein is likely to be the active form in *P. carinii*. We therefore set out to express the longer version, assess its catalytic activity, and confirm its presence in intact *P. carinii*.

Expression of the long form of *P. carinii* IMPDH

Confirmation of catalytic activity

Catalytic activity of the long form of *P. carinii* IMPDH was confirmed by a complementation study in which the cDNA for the full coding sequence of the protein was cloned into a pTactac vector and the protein was expressed in a bacterial strain lacking IMPDH, LH3 *E. coli*. Because it lacks IMPDH activity, this strain of bacteria cannot grow without supplementation with guanine (Table 1). When the bacteria contains the plasmid expressing the long form of *P. carinii* IMPDH, it regains the ability to grow in the absence of guanine, implying that the expressed IMPDH is functional and supplies the guanine needs of the *E. coli*.

In order to confirm that the activity of IMPDH being expressed in LH3 *E. coli* was the *P. carinii* enzyme, we took advantage of the differing susceptibilities of the *E. coli* and the *P. carinii* proteins for mycophenolic acid. Mycophenolic acid has little effect on bacterial IMPDH but the culture susceptibilitiy data with *P. carinii* suggested that the IMPDH from *P. carinii*

```
        1                                                          50
PCL     MGEDSYLCLS  EEIKKKLEEY  SEKDGYDLDS  LICSRRHGGL  TYNDIIILPG
PCS     ..........  ..........  ..........  ..........  ..........
 CA     .....MVFET  SKATSYLKDY  PKKDGLSVKE  LIDSTNFGGL  TYNDFLILPG
 CH     ........MA  DYLISGGTSY  VPDDGLTAQQ  LFNCGD..GL  TYNDFLILPG
 HM     ........MA  DYLISGGTSY  VPDDGLTAQQ  LFNCGD..GL  TYNDFLILPG
 LD     ........M  ATNNANYRIK  TIKDGCTAEE  LFRGD...GL  TYNDFIILPG
                               **          *           **  ****    ****
        51                                                         100
PCL     YIDFEVNSVS  LESHITKKIV  LKTPFMSSPM  DTVTESDMAI  NLALLGGIGV
PCS     ..........  ..........  .....MSSPM  DTVTESDMAI  NLALLGGIGV
 CA     LINFPSSAVS  LETKLTKKIT  LKSPFVSSPM  DTVTEENMAI  HMALLGGIGI
 CH     YIDFTADQVD  LTSALTKKIT  LKTPLVSSPM  DTVTEAGMAI  AMALTGGIGF
 HM     YIDFTADQVD  LTSALTKKIT  LKTPLVSSPM  DTVTEAGMAI  AMALTGGIGF
 LD     FIDFGAADVN  ISGQFTKRIR  LHIPIVSSPM  DTITENEMAK  TMALMGGVGV
         *  *    *     **  *   *   *  ****  ** **  **   ** **  *
```

Figure 6. Homology of amino terminal region for IMPDH from several eukaryotic species. The first 100 amino acids at the amino terminus of the long form of P. carinii IMPDH showed strong homology to the amino terminal region of several other eukaryotic forms of IMPDH. Identical amino acids are indicated by an asterisk. PCL, 529 amino acid long form of P. carinii IMPDH; PCS, 454 amino acid short form of P. carinii IMPDH; CA, Candida albicans; CH, Chinese hamster; HM, human type I enzyme; LD, Leishmania donovani.

Table I Complementation of bacterial growth by recombinant *P. carinii* IMPDH

E. coli strain	Native IMPDH	IMPDH insert	OD600nm G+ media	OD600nm G- media
DH5α	intact	No	1.79	1.76
LH3	absent	No	1.11	0.02
LH3	absent	Long form	1.13	0.76

G+ media = media supplemented with guanine

G- media = media containing no guanine

might be quite sensitive to mycophenolic acid. We therefore conducted a simple growth experiment in which the recombinant LH3 bacteria expressed the *P. carinii* protein. We confirmed again that only the recombinant bacteria grew in the absence of guanine, but more importantly the growth in the absence of guanine was highly sensitive to the presence of mycophenolic acid (Figure 7). The IC50 for growth inhibition was 0.17 µM, a value far below that required for inhibition of bacterial IMPDH. These results suggested that the expressed IMPDH was the eukaryotic protein.

Confirmation of the sequence of the expressed protein

In order to produce adequate amounts of *P. carinii* IMPDH for purification the full length cDNA (GeneBank Accession Number 196975) was reengineered into a pET28a vector which added a 20 amino acid extension on the amino terminal end of the expressed protein. Included within this 20 amino acid region was a sequence of six histidines that allowed the protein to be purified by metal affinity chromatography. *P. carinii* IMPDH was thus expressed in BL21(DE3) *E. coli* and purified in a single step from soluble extracts of the recombinant bacteria. The protein was identified as a single band on polyacrylamide gel electrophoresis in the presence of SDS, isolated from the gel, digested with trypsin, and subjected to LC-ESI-MS. With this technology the peptides are separated by HPLC, each peptide is identified by its mass/charge ratio (m/z), and further identified by fragmentation. Thus, a peptide sequence is suggested by the m/z value and confirmed by the direct fragmentation that removes and identifies each amino acid in turn. Using this technique, 38% of the predicted sequence was directly confirmed (Figure 8). The conclusion is that the protein expressed in this system is the predicted *P. carinii* IMPDH.

What form of IMPDH exists in *P. carinii*?

IMPDH activity exists in *P. carinii*

The presence of IMPDH activity in *P. carinii* was implied by the susceptibility of the organism to mycophenolic acid, but had not been confirmed by direct assay of catalytic activity. We therefore screened four soluble protein extracts that had been prepared from organisms harvested from lungs of immunosuppressed rats. These extracts when prepared 7 to 9 years previously were demonstrated to contain *P. carinii* dihydrofolate reductase and

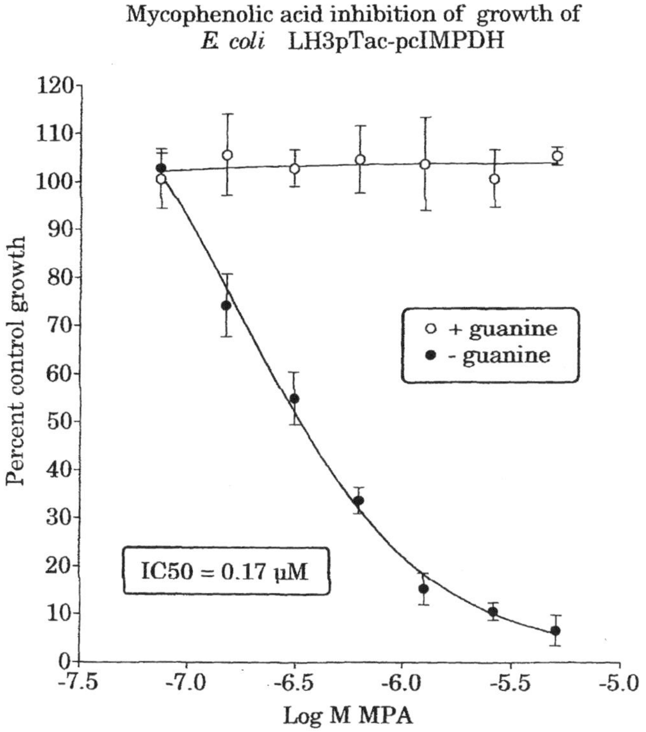

Figure 7. Mycophenolic acid inhibition of growth of E. coli LH3 pTac-pcIMPDH. IMPDH-deficient E. coli strain LH3 was transformed with a pTactac plasmid containing cDNA for the full coding sequence of the 529 amino acid form of P. carinii IMPDH. Growth in the presence of ampicillin confirmed the presence of the plasmid, which carries an ampicillin resistance marker. This recombinant bacteria also grew in the absence of guanine, suggesting functional IMPDH was being expressed. Data on the graph are expressed as percent of growth (OD600) observed in the absence of mycophenolic acid (MPA). Symbols denote mean ± SEM; open circles are for frowth in media with excess guanine, closed circles for media lacking guanine. The LH3 E. coli and the pTactac plasmid were both generous gifts from Dr. L. Hedstrom.

MGEDSYLCLSEEIKKKLEEYSEKDGYDLDSLICSRR
HGGLTYNDIIILPGYIDFEVNSVSLESHITKKIVLKTP
FMSSPMDTVTESDMAINLALLGGIGVIHHNCTIEEQTE
MVRKVK**KFENGFITSPIVLSLNHRVRDVRRIKEELGFS**
GIPITDTGQLNGKLLGIVTSRDIQFHNNDESFLSEVMT
KDLVTGSEGIR**LEEANEILR**SCKKGK**LPIVDKEGNLTA**
LLSRSDLMK**NLHFPLASKLPDSK**QLICAAAVGTRPDDR
IRLKHLVEAGLDIVVLDSSQGNSIYQINMIKWIKK**EFP**
NLEVIAGNVVTREQAANLISAGADALRVGMGSGSICIT
QEIMAVGRPQATAVYAVSEFASK**FGVPTIADGGIENIG**
HITKALALGASAVMMGNLLAGTTESPGQYYYRDGQRLK
SYRGMGSIDAMEHLSGKNKGDNAASSR**YFGEADTIRVA**
QGVSGSVIDKGSLHVYVPYLRTGLQHSLQDIGVQNLTE
LRKQVKEKNIR**FEFR**TVASQLEGNVHGLDSYQKKLWS

Figure 8. Amino acid sequence of P. carinii *IMPDH long form confirmed by
mass spectrometry. The amino acid sequence deduced from the cDNA sequence
is shown. The regions shown in bold face are those where the trypsin fragment
was identified by its mass/charge ratio and its sequence confirmed by
fragmentation analysis.*

dihydropteroate synthase activity, suggesting that other soluble enzymes from the organism might be measurable. Although all four extracts retained 71 to 108% of the original dihydrofolate reductase activity, only two showed measurable rates for IMPDH. To confirm that the rates were IMPDH, an attempt was made to purify the activity.

An aliquot of the soluble extract showing highest activity was subjected to gel filtration chromatography as a first purification step. Detectable activity was seen near the void volume for the G-100 column, which would be the expected elution pattern for a homotetrameric form of *P. carinii* IMPDH. The peak of activity near the void volume was further purified using an IMP affinity column. The activity was strongly retained by the affinity resin and eluted by 100 mM KCl and 5 mM IMP (Figure 9). An attempt was made to confirm the identity of the eluted protein by LS-ESI-MS, but the SDS-PAGE gels showed several proteins and only one, which was considerably smaller than the size expected for IMPDH, was present in amounts able to be identified by LS-ESI-MS. Surprisingly, this protein turned out to be rat mitochondrial malate dehydrogenase, which copurified on the IMP affinity column with IMPDH. Rat mitochondria are a known contaminant in the *P. carinii* preparations used to generate the extracts[27].

The IMPDH long form exists in *P. carinii*

Although there were weak protein bands on SDS-PAGE gels at the correct molecular weight for *P. carinii* IMPDH long form, the identity of the bands could not be confirmed by LS-ESI-MS. Therefore, we set out to confirm the presence of IMPDH by Western blotting. Polyclonal antibodies were generated against a his-tagged short form of *P. carinii* IMPDH for this purpose. This strategy was used because the short and long forms are identical except for the extra amino acids at the amino terminus. If the long version had been used to generate antibodies, it was possible that the antibodies might have recognized a region of protein missing in the short form. Using the short form as antigen assured that both short and long forms of the protein would be recognized, which was confirmed with the recombinant proteins. The polyclonal antibodies were purified by affinity methods to eliminate high background reactions. Rabbits are naturally infected with *P. carinii* at the time of weaning and therefore often show numerous reacting antibodies as background. The purified antibody reacted only to long and short form IMPDH.

P. carinii isolated from lungs of infected immunosuppressed rats or from spinner culture were lysed and the solubilized proteins were electrophoresed on SDS-PAGE gels under conditions to resolve proteins of sizes from 44 to 126 kD. After transfer to PVDF membranes, the proteins were reacted with the purified

128

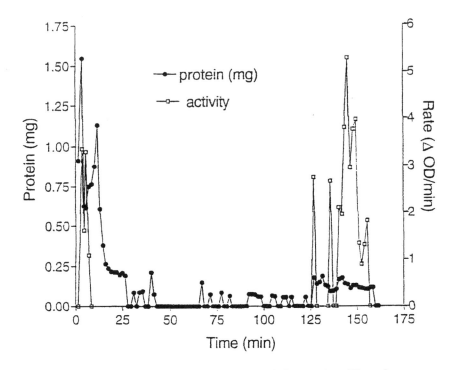

Figure 9. IMP-affinity chromatography of P. carinii *proteins. The column was prepared by published methods {151} and loaded with the large molecular weight proteins from a Sephadex G-100 column. The starting material for the G-100 column was a 100,000 x g supernate prepared from organisms harvested from lungs of immunosuppressed rats. The affinity column was washed for 130 minutes with 10 mM sodium phosphate buffer pH 6.5 containing 50 µM PMSF, 50 µg/ml trypsin inhibitor, and 20 µM 2-mercaptoethanol. Elution was with the same buffer adjusted to pH 7.5 and to which was added 100 mM KCl and 5 mM IMP. Protein was measured by the Coomassie blue method; each point on the protein curve represents the total protein per fraction. The total enzyme content for each fraction was determined by spectrophotometric assay performed at 37°C with 350 µM IMP and 1150 µM NAD+ in a Thermomax microplate reader.*

antibodies. The results showed that strong bands were visible in the samples from organisms harvested from either infected rat lungs or from culture (Figure 10). There was no reaction with control samples from uninfected rat lungs or from uninfected culture host cells. The size of the bands observed is consistent with the 529 amino acid form of the protein but not with the 454 amino acid form. We thus concluded that the long form of the protein is the functional form of IMPDH in *P. carinii*.

Catalytic properties of the expressed long form of *P. carinii* IMPDH

Kinetics of the expressed protein

The purified his-tagged protein was available in milligram quantities for biochemical studies. The kcat for the enzyme was 0.56 ± 0.032 1/sec, when assayed at 25 °C. This value is similar to that seen for other eukaryotic forms of IMPDH; for example, human type II enzyme is reported to have a value of 0.41 1/sec[54].

The kinetics for substrate and cofactor were also determined at 25°C. The Km for IMP was 51 ± 8 μM, which is somewhat higher than the value of 4 μM reported for human type II enzyme[55]. The Km for NAD was 24 ± 1 μM, which is similar to other eukaryotic forms but much lower than the Km shown by bacterial forms[56]. There was evidence of inhibition at higher levels of both IMP and NAD for the *P. carinii* IMPDH. The Ki for NAD was 2.6 ± 0.1 mM and for IMP was 7.9 ± 2 mM. This pattern is different from that seen with human IMPDH, where IMP seems not to cause substrate inhibition. Like *P. carinii* IMPDH, the human form is inhibited by NAD, with a Ki of 0.59 mM[54].

Susceptibility to inhibitors

The purified his-tagged *P. carinii* IMPDH was confirmed to be sensitive to mycophenolic acid (Table 2). The IC50 value was 0.066 μM when the assay was run with 250 μM NAD and 250 μM IMP, concentrations of cofactor and substrate well above the Km values but below concentrations that showed substrate inhibition. The pattern of inhibition for mycophenolic acid with this form of IMPDH appeared uncompetitive, which is the same pattern observed with other forms of IMPDH[55]. A Ki value was calculated at 0.06 μM, which is similar to the values calculated for human type II IMPDH[57].

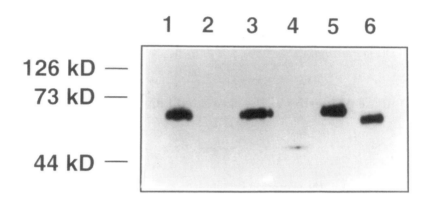

Figure 10. Identification of IMPDH in P. carinii *by Western blotting.* P. carinii *isolated from rat lung or culture were lysed with buffer containing 50 mM Tris pH 7, 2% SDS, 10% glycerol, and 5% 2-mercaptoethanol. Undissolved material was removed by centrifugation at 16,000 x g and the soluble proteins were resolved by SDS-polyacrylamide gel electrophoresis. The proteins on the gels were transferred to a PVDF membrane and reacted with purified antibody generated by immunization of rabbits with the pure his-tagged 454 amino acid version of IMPDH from* P. carinii. *The bands were visualized using anti-rabbit IgG conjugated with horse radish peroxidase. The proteins in each lane came from the following sources: Lane 1,* P. carinii *isolated from lungs of infected, immunosuppressed rats; Lane 2, lungs of uninfected rats (negative control); Lane 3,* P. carinii *isolated from short term spinner culture; Lane 4, human embryonic lung cells, the normal host cells for* P. carinii *in culture (negative control); Lane 5, four nanograms of the purified 529-amino acid, recombinant IMPDH from* P. carinii; *Lane 6, two nanograms of the purified 454-amino acid, recombinant IMPDH from* P. carinii.

Table II. Potential for selectivity against *Pneumocystis carinii* IMPDH

Compound name[3]	*P. carinii* IMPDH[1] IC50, μM	Human IMPDH[2] IC50, μM	IC50 Ratio, human/*P. carinii*
Mycophenolic acid	0.066 ± 0.003	(Ki = 0.01 - 0.037)	
C2-MAD 0.063 ± 0.005	(Ki = 0.33)		
7-O-C2-MAD	6.01 ± 1.55	inactive	>29
C6-MAD 79.7 ± 22.3	(Ki = 0.3)		
3'F-TAD	0.86 ± 0.09	0.7	1
F-ara-TAD 0.89 ± 0.76	2.6	3	
F-ara-CH2-TAD	0.95 ± 0.16	6	6
F-ara-CF2-TAD	0.33 ± 0.025	0.8	2
BAD	0.81 ± 0.25	0.8	1
F-ara-CH2-BAD	2.67 ± 0.66	175	66

[1]recombinant his-tagged long form assayed under standard conditions, 25°C. IC50 values are means ± SEM for 3 or 4 independent determinations with separate enzyme preparations.
[2]Type II enzyme; data supplied by K. Pankiewicz, except for mycophenolic acid (Carr et al., 1993)
[3]all compounds except mycophenolic acid were supplied by K. Pankiewicz

132

A preliminary series of inhibitors of IMPDH was tested with the purified *P. carinii* IMPDH to ascertain if any differences might be suggested between the mammalian enzyme and the enzyme from this eukaryotic pathogen. The most dramatic difference seems to be with 7-O-C2-MAD, a derivative of mycophenolic acid adenine dinucleotide. This compound is inactive against the type II human enzyme but has significant activity against *P. carinii* IMPDH in four independent trials, suggesting considerable selectivity might be developed toward the pathogen. Selectivity is also observed with F-ara-CH2-BAD. The *P. carinii* enzyme shows an IC50 of 2.67 μM with this compound, compared to a value of 175 μM for the human type II enzyme.

Conclusions and future directions

The success of the cloning and purification efforts have assured a ready supply of recombinant protein for continuing studies. Although work on the biology of *P. carinii* is very difficult, sufficient information has been gained to establish that IMPDH long form is present and that it is catalytically active. Moreover, the analysis of the purified protein suggests that the susceptibility of the cultured organism to mycophenolic acid is determined by the sensitivity of the IMPDH to that inhibitor; little salvage seems to exist. Thus, use of IMPDH inhibitors alone may be sufficient to control proliferation of the organism.

The suggestion for selectivity toward the *P. carinii* IMPDH revealed in the short series of inhibitors tested is an important justification for continuing work on this enzyme. A more complete series of compounds exploring some of the modifications that seem linked to selectivity will be examined, in collaboration with Dr. K. Pankiewicz (Pharmassett). In order to support design and synthesis of drugs, we are also exploring modified plasmid design to allow expression of IMPDH without the his-tag with the ultimate goal of cystallizing *P. carinii* IMPDH. X-ray crystallographic studies will offer strong support for selecting existing compounds for testing and for designing new, highly selective agents.

Acknowledgements

Special thanks to Dr. L. Hedstrom (Brandeis University) for supplying the pTactac plasmid and the LH3 *E. coli*. Thanks also to Dr. K. Pankiewicz (Pharmassett) for sending the array of inhibitors that were used to characterize the recombinant enzyme from *P. carinii*. The original culture work with *P. carinii* was done in the laboratory of Dr. J.W. Smith and Professor M. Bartlett (Department of Pathology and Laboratory Medicine, Indiana University School of Medicine, Indianapolis).

References

1. O'Gara, M. J., C. H. Lee, G. A. Weinberg, J. M. Nott, and S. F. Queener. IMP dehydrogenase from *Pneumocystis carinii* as a potential drug target. *Antimicrob. Agents Chemother.*1997,41, 40-48.
2. Ye, D., C. Lee, and S. F. Queener. Differential splicing of *Pneumocystis carinii* inosine 5'-monophosphate dehydrogenase pre-mRNA. *Gene* 2000, in review,
3. Chagas, C. Nova tripanozomiaza humana. *Mem. Inst. Oswaldo Cruz*1909,1,159-218.
4. Delanoe, P. and Mme. Delanoe. Sur les rapports des kystes des Carinii du poumon des rats avec le Trypanosoma lewisi. *C. R. Acad. Sci. (Paris)*1912,155,658-660.
5. Delanoe, P. and M. Delanoe. De la rarete de *Pneumocystis carinii* chez cobayes de la region de Paris; absense de cysts chez d'autres animauis lapin, grenouille, zanguilles. *Bull. Soc. Pathol. Expt.*1914,7,271-274.
6. Frenkel, J. K., M. S. Bartlett, and J. W. Smith. RNA homology and the reclassification of Pneumocystis. *Diagn. Microbiol. Infect. Dis.*1990,13,1-2.
7. Ivady, V. G., L. Paldy, and G. Unger. Weitere Derfahrungen bei der Behandlung der interstitiellen plasmacellularen Pneumonie mit Pentamidin. *Mschr. Kinderheilk.*1963,111,297-299.
8. Hughes, W. T., P. C. McNabb, T. D. Makres, and S. Feldman. Efficacy of trimethoprim and sulfamethoxazole in the prevention and treatment of *Pneumocystis carinii* pneumonitis. *Antimicrob. Agents Chemother.*1974,5,289-293.
9. Miller, R. F. and D. M. Mitchell. AIDS and the lung: Update 1995: 1-- *Pneumocystis carinii* pneumonia. *Thorax*1995,50,191-200.
10. Pixley, F. J., A. E. Wakefield, S. Banerji, and J. M. Hopkin. Mitochondrial gene sequences show fungal homology for *Pneumocystis carinii*. *Mol. Microbiol.*1991,5,1347-1351.
11. Burke, B. A. and R. A. Good. *Pneumocystis carinii* infection. *Medicine*1973,52,23-51.
12. Halme, M., I. Lautenschlager, S. Mattila, and P. Tukiainen. Breakthrough *Pneumocystis carinii* infections in lung and heart-lung transplant patients with chemoprophylaxis. *Transplant. Proc.*1999,31,197.
13. Raychaudhuri, S. P. and S. Siu. *Pneumocystis carinii* pneumonia in patients receiving immunosuppressive drugs for dermatological diseases. *Br. J. Dermatol.*1999,141,528-530.

14. Touzet, S., C. Pariset, M. Rabodonirina, and C. Pouteil-Noble. Nosocomial transmission of *Pneumocystis carinii* in renal transplantation. *Transplant. Proc.*2000,32,445.

15. Nüesch, R., C. Bellini, and W. Zimmerli. *Pneumocystis carinii* pneumonia in human immunodeficiency virus (HIV)-positive and HIV-negative immunocompromised patients. *Clin. Infect. Dis.*1999,29,1519-1523.

16. Boiselle, P. M., C. A. Crans,Jr., and M. A. Kaplan. The changing face of *Pneumocystis carinii* pneumonia in AIDS patients. *Am. J. Roentgenol.*1999,172,1301-1309.

17. Schliep, T. C. and R. L. Yarrish. *Pneumocystis carinii* pneumonia. *Seminars in Respiratory Infections*1999,14,333-343.

18. Schneider, M. M. E., A. I. M. Hoepelman, J. K. M. E. Schattenkerk, T. L. Nielsen, Y. Van Der Graaf, J. P. H. J. Frissen, I. M. E. Van Der Ende, A. F. P. Kolsters, J. C. C. Borleffs, and and the Dutch AIDS Treatment Group. A controlled trial of aerosolized pentamidine or trimethoprim-sulfamethoxazole as primary prophylaxis against *Pneumocystis carinii* pneumonia in patients with human immunodeficiency virus infection. *N. Eng. J. Med.*1992,327,1836-1841.

19. Hardy, W. D., J. Feinberg, D. M. Finkelstein, M. E. Power, W. He, C. Kaczka, P. T. Frame, M. Holmes, H. Waskin, R. J. Fass, W. G. Powderly, R. T. Steigbigel, A. Zuger, R. S. Holzman, and For the AIDS Clinical Trials Group. A controlled trial of trimethoprim-sulfamethoxazole or aerosolized pentamidine for secondary prophylaxis of *Pneumocystis carinii* pneumonia in patients with the acquired immunodeficiency syndrome. *N. Eng. J. Med.*1992,327,1842-1848.

20. Kovacs, J. A. and H. Masur. Drug therapy: Prophylaxis against opportunistic infections in patients with human immunodeficiency virus infection. *N. Engl. J. Med.*2000,342,1416-1429.

21. Masur, H., J. E. Kaplan, K. K. Holmes, B. L. Alston, N. Ampel, J. R. Anderson, A. C. Baker, D. Barr, J. G. Bartlett, J. E. Bennett, C. A. Benson, S. A. Bozzette, R. E. Chaisson, C. S. Crumpacker, J. S. Currier, L. Deyton, W. L. Drew, W. R. Duncan, R. W. Eisinger, W. El-Sadr, J. Feinberg, K. A. Freedberg, H. Furrer, and J. W. Gnann,Jr.. 1999 USPHS/IDSA guidelines for the prevention of opportunistic infections in persons infected with human immunodeficiency virus. *Clin. Infect. Dis.*2000,30 Suppl. 1,S29-S65.

22. Horowitz, H. W. and G. P. Wormser. Atovaquone compared with dapsone to prevent *Pneumocystis carinii* pneumonia. *N. Engl. J. Med.*1999,340,1512-1513.

23. Medina, I., J. Mills, G. Leoung, P. C. Hopewell, B. Lee, G. Modin, N. Benowitz, and C. B. Wofsy. Oral therapy for *Pneumocystis carinii*

pneumonia in the acquired immunodeficiency syndrome. *N. Eng. J. Med.*1990,323,776-782.

24. Meshnick, S. R. Drug-resistant *Pneumocystis carinii*. *Lancet*1999,354,1318-1319.

25. Bou, G., M. S. Figueroa, P. Martí-Belda, E. Navas, and N. Guerrero. Value of PCR for detection of *Toxoplasma gondii* in aqueous humor and blood samples from immunocompetent patients with ocular toxoplasmosis. *J. Clin. Microbiol.*1999,37,3465-3468.

26. Ma, L., L. Borio, H. Masur, and J. A. Kovacs. *Pneumocystis carinii* dihydropteroate synthase but not dihydrofolate reductase gene mutations correlate with prior trimethoprim sulfamethoxazole or dapsone use. *J. Infect. Dis.*1999,180,1969-1978.

27. Broughton, M. C. and S. F. Queener. *Pneumocystis carinii* dihydrofolate reductase used to screen potential antipneumocystis drugs. *Antimicrob. Agents Chemother.*1991,35,1348-1355.

28. Queener, S. F. New drug developments for opportunistic infections in immunosuppressed patients: *Pneumocystis carnii*. *J. Med. Chem.*1995,38,4739-4759.

29. Lundgren, B., K. Elvin, L. P. Rothman, I. Ljungström, C. Lidman, and J. D. Lundgren. Transmission of *Pneumocystis carinii* from patients to hospital staff. *Thorax*1997,52,422-424.

30. Vargas, S. L., W. T. Hughes, A. E. Wakefield, and H. S. Oz. Limited persistence in and subsequent elimination of *Pneumocystis carinii* from the lungs after *P. carinii* pneumonia. *J. Infect. Dis.*1995,172,506-510.

31. Robson, C., M. A. Meek, J. D. Grunwaldt, P. A. Lambert, S. F. Queener, D. Schmidt, and R. J. Griffin. Nonclassical 2,4-diamino-5-aryl-6-ethylpyrimidine antifolates: Activity as inhibitors of dihydrofolate reductase from *Pneumocystis carinii* and *Toxoplasma gondii* and as antitumor agents. *J. Med. Chem.*1997,40,3040-3048.

32. Gangjee, A., A. P. Vidwans, A. Vasudevan, S. F. Queener, R. L. Kisliuk, V. Cody, R. M. Li, N. Galitsky, J. R. Luft, and S. Pangborn. Structure-based design and synthesis of lipophilic 2,4-diamino-6-substituted quinazolines and their evaluation as inhibitors of dihydrofolate reductases and potential antitumor agents. *J. Med. Chem.*1998,41,3426-3434.

33. Rosowsky, A., V. Cody, N. Galitsky, H. N. Fu, A. T. Papoulis, and S. F. Queener. Structure-based design of selective inhibitors of dihydrofolate reductase: Synthesis and antiparasitic activity of 2,4-diaminopteridine analogues with a bridged diarylamine side chain. *J. Med. Chem.*1999,42,4853-4860.

34. Piper, J. R., C. A. Johnson, C. A. Krauth, R. L. Carter, C. A. Hosmer, S. F. Queener, S. E. Borotz, and E. R. Pfefferkorn. Lipophilic antifolates as agents against opportunistic infections .1. Agents superior to trimetrexate

and piritrexim against *Toxoplasma gondii* and *Pneumocystis carinii* in *in vitro* evaluations. *J. Med. Chem.*1996,39,1271-1280.

35. Cody, V., D. Chan, N. Galitsky, D. Rak, J. R. Luft, W. Pangborn, S. F. Queener, C. A. Laughton, and M. F. G. Stevens. Structural studies on bioactive compounds. 30. Crystal structure and molecular modeling studies on the *Pneumocystis carinii* dihydrofolate reductase cofactor complex with TAB, a highly selective antifolate. *Biochemistry*2000,39,3556-3564.

36. Boykin, D. W., A. Kumar, J. Spychala, M. Zhou, R. J. Lombardy, W. D. Wilson, C. C. Dykstra, S. K. Jones, J. E. Hall, R. R. Tidwell, C. Laughton, C. M. Nunn, and S. Neidle. Dicationic diarylfurans as anti-*Pneumocystis carinii* agents. *J. Med. Chem.*1995,38,912-916.

37. Warrens, A. N. The evolving role of mycophenolate mofetil in renal transplantation. *QJM*2000,93,15-20.

38. Vitale, M., L. Zamai, E. Falcieri, G. Zauli, P. Gobbi, S. Santi, C. Cinti, and G. Weber. IMP dehydrogenase inhibitor, tiazofurin, induces apoptosis in K562 human erythroleukemia cells. *Cytometry*1997,30,61-66.

39. Markland, W., T. J. McQuaid, J. Jain, and A. D. Kwong. Broad-spectrum antiviral activity of the IMP dehydrogenase inhibitor VX-497: a comparison with ribavirin and demonstration of antiviral additivity with alpha interferon. *Antimicrob. Agents Chemother.*2000,44,859-866.

40. Franchetti, P. and M. Grifantini. Nucleoside and non-nucleoside IMP dehydrogenase inhibitors as antitumor and antiviral agents. *Current Medicinal chemistry*1999,6,599-614.

41. Atzori, C., A. Bruno, G. Chichino, E. Bombardelli, M. Scaglia, and M. Ghione. Activity of bilobalide, a sesquiterpene from *Ginkgo biloba*, on *Pneumocystis carinii*. *Antimicrob. Agents Chemother.*1993,37,1492-1496.

42. Bartlett, M. S., T. D. Edlind, M. M. Durkin, M. M. Shaw, S. F. Queener, and J. W. Smith. Antimicrotubule benzimidazoles inhibit in vitro growth of *Pneumocystis carinii*. *Antimicrob. Agents Chemother.*1992,36,779-782.

43. Bartlett, M. S., M. M. Shaw, P. Navaran, J. W. Smith, and S. F. Queener. Evaluation of potent inhibitors of dihydrofolate reductase in a culture model for growth of *Pneumocystis carinii*. *Antimicrob. Agents Chemother.*1995,submitted,2436-2441.

44. Fishman, J. A., S. F. Queener, R. S. Roth, and M. S. Bartlett. Activity of topoisomerase inhibitors against *Pneumocystis carinii* in vitro and in an inoculated mouse model. *Antimicrob. Agents Chemother.*1993,37,1543-1546.

45. Merali, S. and S. R. Meshnick. Susceptibility of *Pneumocystis carinii* to artemisinin in vitro. *Antimicrob. Agents Chemother.*1991,35,1225-1227.

46. Queener, S. F., M. S. Bartlett, J. D. Richardson, M. M. Durkin, M. A. Jay, and J. W. Smith. Activity of clindamycin with primaquine against

Pneumocystis carinii in vitro and in vivo. *Antimicrob. Agents Chemother.*1988,32,807-813.

47. Rosowsky, A., J. H. Freisheim, J. B. Hynes, S. F. Queener, M. S. Bartlett, J. W. Smith, H. Lazarus, and E. J. Modest. Tricyclic 2,4-diaminopyrimidines with broad antifolate activity and the ability to inhibit *Pneumocystis carinii* growth in cultured human lung fibroblasts in the presence of leucovorin. *Biochem. Pharmacol.*1989,38,2677-2684.

48. Pesanti, E. L. and C. Cox. Metabolic and synthetic activities of *Pneumocystis carinii* in vitro. *Infection and Immunity*1981,34,908-914.

49. Lee, C. H., N. L. Bauer, M. M. Shaw, M. M. Durkin, M. S. Bartlett, S. F. Queener, and J. W. Smith. Proliferation of rat *Pneumocystis carinii* on cells sheeted on microcarrier beads in spinner flasks. *J. Clin. Microbiol.*1993,31,1659-1662.

50. Kerr, K. M. and L. Hedstrom. The roles of conserved carboxylate residues in IMP dehydrogenase and identification of a transition state analog. *Biochemistry*1997,36,13365-13373.

51. Merali, S. and A. B. Clarkson,Jr.. Polyamine content of *Pneumocystis carinii* and response to the ornithine decarboxylase inhibitor DL--difluoromethylornithine. *Antimicrob. Agents Chemother.*1996,40,973-978.

52. Fletcher, L. D., L. C. Berger, S. A. Peel, R. S. Baric, R. R. Tidwell, and C. C. Dykstra. Isolation and identification of six *Pneumocystis carinii* genes utilizing codon bias. *Gene*1993,129,167-174.

53. Thornewell, S. J., R. B. Peery, and P. L. Skatrud. Cloning and characterization of Cne MDR1: a *Cryptococcus neoformans* gene endocing a protein related to multidrug resistance proteins. *Gene*1997,201,21-29.

54. Wang, W. and L. Hedstrom. The kinetic mechanism of human inosine 5'-monophosphate type II: random addition of substrates, ordered release of products. *Biochemistry*1997,36,8479-8483.

55. Hedstrom, L. IMP dehydrogenase: Mechanism of action and inhibition. *Current Medicinal Chemistry*1999,6,545-560.

56. Hammerschmidt, D. E. and J. M. Beck. *Pneumocystis carinii* in the lung. *J. Lab. Clin. Med.*1996,128,524.

57. Carr, S. F., E. Papp, J. C. Wu, and Y. Natsumeda. Characterization of human type I and type II IMP dehydrogenases. *J. Biol. Chem.*1993,268,27286-27290.

Structure and Mechanism

Chapter 7

IMPDH Structure and Ligand Binding

Barry M. Goldstein*, Dipesh Risal, and Michael Strickler

Department of Biochemistry and Biophysics, University of Rochester
Medical Center, 601 Elmwood Avenue, Rochester, NY 14642
*Corresponding author: telephone: 716–275–5095; fax: 716–275–6007;
email: barry.goldstein@rochester.edu

Inosine monophosphate dehydrogenase (IMPDH, E.C.
1.1.1.205) is recognized as an important target for both
antileukemic and immunosuppressive therapy. IMPDH catalyzes the NAD-dependent oxidation of inosine 5' monophosphate (IMP) to xanthosine 5' monophosphate. The finding that
IMPDH exists as two isoforms, one of which (type II) is
induced in tumor cells, has led to the search for potentially
less toxic isoform-specific inhibitors. A number of crystal
structures of IMPDH have become available. These include
structures of the human type I and type II, hamster, *Tritrichomonas foetus*, *Streptococcus pyogenes* and *Borrelia burgdorferi* enzymes. Each structure crystallizes as a tetramer of α/β
barrels, the substrate and cofactor binding in a continuous cleft
on the C-terminal face of each barrel. The IMP base is well
positioned to stack against the NAD nicotinamide ring to
facilitate hydride transfer. The active site cleft is further
bounded by a highly flexible flap and loop. These structures
reveal enzyme-ligand interactions that suggest strategies for
the design of improved inhibitors. These include agents that
bind at either the substrate site or at the NAD site.

Abbreviations

BAD, benzamide adenine dinucleotide
CBS, cystathionine beta synthase
6-Cl-IMP, 6-chloro-9-β-D-ribofuranosylpurine 5'-monophosphate
EICAR, 5-ethylnyl-1-β-D-ribofuranosylimidazole-4-carboxamide
EICAR MP, EICAR 5'-monophosphate
FFAD, 5-β-D-ribofuranosylfuran-3-carboxamide adenine dinucleotide
IMP, inosine 5'-monophosphate
IMPDH, inosine 5'-monophosphate dehydrogenase
MAD, mycophenolic acid adenine dinucleotide
MPA, mycophenolic acid
NAD, nicotinamide adenine dinucleotide
PDB, protein data bank
SAD, 2-β-D-ribofuranosylselenazole-4-carboxamide adenine dinucleotide
β-SAD, β–methylene SAD
TAD, 2-β-D-ribofuranosylthiazole-4-carboxamide adenine dinucleotide
β-TAD, β–methylene TAD
XMP, xanthosine 5'-monophosphate.

I. Introduction

Inosine monophosphate dehydrogenase (IMPDH, E.C. 1.1.1.205) has been identified as a key enzyme in the regulation of cell proliferation and differentiation *(1, 2)*. In addition to its useful clinical effects, inhibition of IMPDH also provides a means of selectively probing G-protein-mediated mechanisms of myelopoesis. Over 400 literature citations now address the characterization, mechanism and biological functions of IMPDH, its role as a target for both antileukemic and immunosuppressive therapy, and its inhibition by chemotherapeutic agents. Despite more than 25 years of interest in IMPDH as a drug target, structural studies of the enzyme were, until relatively recently, lacking. Within the past five years, single crystal structures of both pro- and eukaryotic IMPDH enzymes have been obtained. These structures have provided insight into enzyme mechanism, ligand binding specificity, and drug design. Here we review the present state of published structural work on IMPDH, consider current IMPDH inhibitors in light of these findings, and examine potential strategies for the design of new agents.

The value of IMPDH as a chemotherapeutic target results from the enzyme's role in catalyzing the rate-limiting step in the de novo synthesis of the guanine nucleotides. The irreversible step catalyzed by IMPDH is the nicotinamide adenine dinucleotide (NAD) -dependent hydroxylation of inosine 5'-monophosphate (IMP, Figure 1a) to xanthosine 5' monophosphate (XMP) *(3, 4)*. The mammalian enzyme is a homotetramer consisting of four 56 kD mono-mers. Although the enzyme had been thought to follow an ordered bi-bi mechanism in which substrate IMP binds first *(5)*, more recent studies suggest a

142

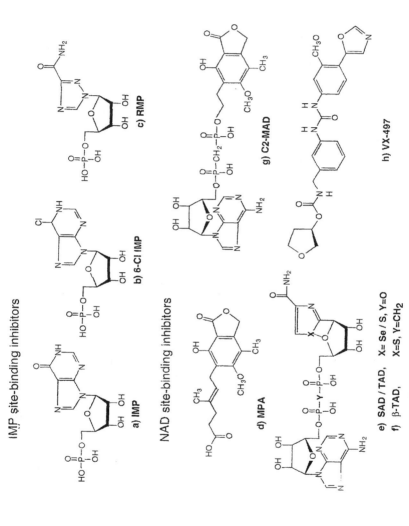

Figure 1. Selected IMPDH ligands and inhibitors.

random component to substrate addition *(6, 7)*. Nucleophilic attack of an active site cysteine (Cys 331) on the C2 position of the IMP base forms a covalent intermediate. Binding of cofactor NAD results in hydride transfer to the B-side of the nicotinamide ring to form product *(3, 5, 6, 8)*. Hydrolytic attack on the C2 tetrahedral carbon frees XMP, the precursor of GMP in the de novo pathway.

Inhibition of IMPDH produces a decline in guanine nucleotide pools, resulting in a variety of chemotherapeutically useful effects. Inhibition of IMPDH is associated with changes in nucleic acid synthesis, gene expression, signaling and, ultimately, cell proliferation and differentiation *(1, 2, 9, 10)*. IMPDH inhibitors have been exploited as antiviral, antileukemic and, more recently, immunosuppressive agents.

The inhibition of IMPDH and subsequent reduction in guanine nucleotides interrupts DNA and RNA synthesis in rapidly dividing tumor cells *(11)*. The inhibition of IMPDH also appears to compromise the ability of G proteins to function as transducers of intracellular signals *(10, 12-15)*. This is accompanied by down-regulation of c-myc and/or Ki-ras oncogenes in a variety of human tumor cell lines *(1, 16, 17)* and in blast cells of leukemic patients treated with the inhibitor *(1)*. IMPDH inhibitors have also been successfully employed in immunosuppressive therapy. The immune response represents normal lymphocyte proliferation and differentiation stimulated by antigen. Growth and differentiation of human lymphocytes are particularly dependent upon the IMPDH-catalyzed de novo pathway for purine nucleotide synthesis *(18)*. Inhibition of IMPDH leads to suppression of both T and B lymphocyte proliferation *(2)*.

Two distinct cDNA's encoding IMPDH have been isolated *(19)*. The transcripts, labeled type I and type II, are of identical size and share 84% sequence identity, but appear to serve different roles. The type I isoform is expressed at relatively constant levels in both normal and neoplastic cells, suggesting a constitutive "housekeeping" function *(20, 21)*. In contrast, expression of type II mRNA is preferentially up-regulated in human leukemic cell lines K562 and HL-60 *(22)*, as well as in human ovarian tumors *(19)* and leukemic cells from patients with chronic granulocytic, lymphocytic and acute myeloid leukemias *(20)*. This results in an 8-9-fold increase in activity of type II vs. type I IMPDH in in vitro leukemic lines *(22)* and a 1.5-5-fold increase in type II transcription in patient leukemic cells *(20)*. Conversely type II expression is down regulated over 90% in either HL-60 cells induced to differentiate *(21)*, or in patient leukemic cells treated in vitro *(20)*.

Thus, type II IMPDH is an inducible enzyme whose role is closely linked to cell differentiation and neoplastic transformation*(2)*. The disproportionate increases in IMPDH activity in malignant cells make this enzyme a key target for specific antileukemic chemotherapy *(1)*. The target for immune suppression is less clear. While the type II enzyme is the major isoform in normal human T lymphocytes, these cells appear to induce both type I and type II enzymes when stimulated by mitogen *(23, 24)*. Nevertheless, develop-ment of an isoform-specific immunosuppressant with an improved therapeutic ratio remains an actively sought goal (below).

As with all drugs, clinical use of IMPDH inhibitors is limited by metabolism, resistance and toxicity. Despite identification of a target and availability of a number of classes of inhibitors, these limitations still range from moderate to severe. The next step in the design of improved IMPDH inhibitors will be to improve resistance to metabolism, while simultaneously enhancing specificity for a particular isoform. An understanding of the detailed interactions between IMPDH and its inhibitors will be an invaluable step in the design of improved agents.

II. Crystal Structures of IMPDH Complexes

A. Overview of Available IMPDH Structures

Until a few years ago, no structure of any form of IMPDH was known. The situation has improved dramatically within the past six years, and is summarized in Table I.

A 2.6 Å structure of Chinese hamster IMPDH was published by Sintchak et al. of Vertex Pharmaceuticals in 1996 (25). The hamster structure differs from that of the human type II enzyme by only six amino acids, and was crystallized as a complex containing Cys 331-ligated IMP (XMP*) and inhibitor mycophenolic acid (MPA, Figure 1d). This structure defined both the overall topology of the enzyme and the location and topology of much of the active site.

A 2.9 Å structure of the human type II enzyme appeared in 1999 (26). The human enzyme was crystallized as a ternary complex containing the substrate analogue 6-Cl purine ("6-Cl IMP", Figure 1b) and the NAD analogue 2-β-D-ribofuranosylselenazole-4-carboxamide adenine dinucleotide (SAD, Figure 1e), the active anabolite of the antitumor drug selenazofurin (27, 28). The presence of this NAD analogue permitted definition of the remainder of the cofactor-binding site.

Additional mammalian IMPDH ternary complexes have been recently solved. A 2.8 Å his 93ala mutant hamster type II complex and a 2.8 Å wild type hamster type II complex have been described by the Vertex group (29). The cofactor sites in these complexes contain the novel inhibitors Vertex compound 1 and VX-497 (Figure 1h) respectively. Like the original MPA complex, these structures contain an XMP adduct with Cys-331.

Three additional human complexes have also recently been solved (30). These include a 2.5 Å structure of the human type I enzyme, and two additional human type II ternary complexes. One of the new human type II complexes contains ribavirin 5' monophosphate (RMP, 2Ic), the active anabolite of the antiviral agent and substrate analogue ribavirin and the novel inhibitor C2 mycophenolic acid adenine dinucleotide (C2-MAD, Figure 1g). The second human type II complex contains 6-Cl IMP and NAD, the first structure to include the normal cofactor. These manuscripts are in preparation, and should be available by the time this volume goes to press (30).

TABLE I. IMPDH Crystal Structures

Source	Ligands / Resolution	Homology‡ PDB entry	Resolution / Citation	PDB entry	Citation
Human type II	6-Cl IMP* / SAD 100 1B3O (26)		2.9		
Human type II	6-Cl IMP* / NAD RMP / MAD		2.8 † 2.5	in preparation †	in preparation in preparation
Human type I	6-Cl IMP*	83	2.5	1JCN	1JCN
Chinese hamster	XMP* / MPA 99		2.6	1JR1	(25)
Chinese hamster	XMP* / VX-497		2.8	(29)	(29)
Chinese hamster (H93A)	XMP* / Vertex Comp. 1		2.8	†	(29)
T. foetus (apo)	XMP	33	2.6 (2.3	1AK5	(33)
T. foetus	IMP / β-TAD		2.2	1LRT	(34)
T. foetus	Mizoribine		2.0	†	**
S. pyogenes	IMP	40	1.9	1ZFJ	(31)
B. burgdorferi	SO4*	38	1.9	1EEP	(32)

* ligand observed as covalent adduct with Cys 331.
† To be submitted
‡ Homology relative to the human sequence based on BLASTP search and alignment of sequences in SWISSPRO databank (95). Percentage reflects sequence identity over aligned regions.
These comprise 514 residues for the hamster sequence, 494 for T. foetus, 455 for S. pyogenes and 294 for
B. burgdorferi, which lacks a homologous flanking domain.

In addition to the mammalian enzymes, IMPDH from several microbial sources has also been studied. A 1.9 Å structure of *Streptococcus pyogenes* IMPDH complexed with IMP appeared in 1999 *(31)*. The higher resolution of this complex permitted definition of a flanking domain that had been disordered to varying degrees in the previous mammalian structures. A 2.4 Å structure from the spirochete *Borrelia burgdorferi*, the causative agent for Lyme disease, has also been solved, with a sulfate ion bound at the IMP phosphate *(32)*. This structure demonstrated additional flexibility in the active site loop (below).

Recent activity has also focused on IMPDH from the protozoan *Tritrichomonas foetus*. *T. foetus* shares 25-30% sequence identity with known mammalian forms of the enzyme. This system has proven a useful model in the identification of structural features unique to microbial systems that may be exploited in the development of species-specific antimicrobial agents (below). The first *T. foetus* IMPDH structures to appear were a 2.3Å apo structure and 2.6 Å binary complex *(33)*. The binary complex contained product XMP but no inhibitor or cofactor. An additional *T. foetus* structures is now in press *(34)*. This is a 2.2 Å structure of a ternary complex containing IMP and the NAD analogue β-methylene 2- β -D-thiazole-4-carboxamide adenine dinucleotide (β - TAD, 2If), a phosphodiesterase-resistant analogue of TAD, the active anabolite of the antitumor drug tiazofurin *(35, 36)*. This is one of the first published IMPDH complexes to contain both the native substrate IMP and an intact dinucleotide analogue. As this volume was in press, a manuscript describing two additional *T. foetus* complexes appeared; a 1.9Å binary complex containing RMP, and a 2.2 Å ternary complex containing RMP and a partial occupancy MPA ligand *(37)*. These structures identify the same potassium binding site at the monomer-monomer interface seen in the hamster complex. Coordinates are on hold until August 2003.

In addition to those mentioned above, a number of structures of human and hamster enzymes has been solved by both Vertex and Bristol-Myers Squibb in complex with a variety of proprietary inhibitors, but these coordinates have not been made available. Undoubtedly, further structures will appear while this volume is in press. Nevertheless, there is now a wealth of structural data available for IMPDH. Comparison of these structures yields a number of general features. These are discussed below.

B. IMPDH Crystal and Molecular Structure

All IMPDH structures show a tetrameric organization (Figure 2). With the exception of the *B. burgdorferi* enzyme, each monomer in the tetramer is composed of two domains, with a combined molecular weight of ~55kDa (Figure 2). A larger catalytic domain (394 residues in the human and hamster enzymes) forms an eight-stranded parallel α/β barrel of approximate dimensions 40x40x50 Å (Figure 2). The active site is located on the C-terminal face of the barrel (Figure 2). As will be described below, the cofactor-binding cleft of the active site in the human enzyme consists of residues from adjacent catalytic

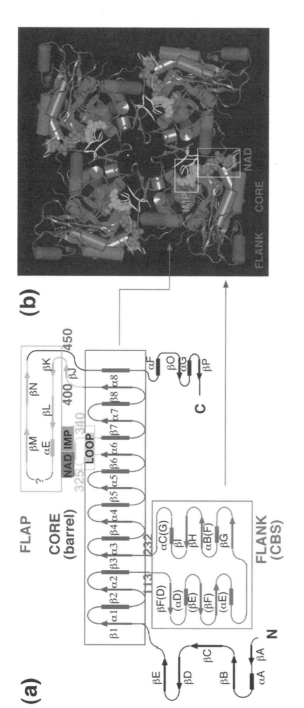

Figure 2. (a) Secondary structure of human and hamster IMPDH. The structure of the flanking CBS domain is partially disordered in these structures. The fold shown here is taken from the CBS domain defined in the S. Pyogenes structure (31). Nomenclature follows that used in the hamster structure (25). Labels in parentheses indicate S. pyogenes nomenclature. (b) IMPDH tetramer viewed down the crystallographic 4-fold. Elements of the structure are colored as follows: Core barrel domain (blue), flanking CBS domain (lavender), loop (yellow), flap (cyan), NAD (red), and IMP (green). Ligands IMP and NAD are circled. The cofactor site defined by NAD lies at the monomer–monomer interface.

(This figure is also in color insert.)

monomers (Figure 2), suggesting the necessity of at least a dimeric organization for activity. Potassium binding sites have also been identified at the domain interfaces in the hamster and *T. foetus* structures, *(25, 34)* suggesting an additional reason for a multimeric organization.

In each structure to date, the crystal lattice is composed of groups of octamers, each formed by the stacking of two tetramers along the 4-fold axis. Two different octamer structures are observed, depending upon which face of the tetramer forms the octamer interface. Octamers in the hamster and human structures are similar. These are formed by tetramers whose monomers are stacked against their N-terminal faces, with the active sites exposed to solvent. Octamers in the *T. foetus* and *B. burgdorferi* structures are formed by tetramers stacked against the opposite face. While there is some evidence that the *T. foetus* enzyme may form hexamers or octamers in solution *(33)*, the biological relevance of these higher order structures is unknown.

A smaller 120-residue flanking domain (residues 113-232 in the human and hamster enzymes) lies adjacent to the catalytic domain, inserted between the $\alpha2$ helix and $\beta3$ sheet of the barrel (Figure 2). The function of this domain is unknown. A deletion mutant of the human type II enzyme that substitutes a short linker for the flanking domain is apparently fully active *(25)*. A search of the flanking domain sequence in the ProDom domain database *(38)* identified two copies of a cystathionine beta synthase (CBS) -like domain *(39)* as the closest homologue. Both the human and *S. pyogenes* structures show approximate 2-fold symmetry in the β-sheet region of the flanking domain. This domain displays varying degrees of disorder, ranging from completely disordered in the *T. foetus* structure to fully ordered in the *S. pyogenes* complex *(31, 33)*. The latter provides the first complete structure of this domain. A 120° difference in the relative orientations of the flanking and catalytic domains observed between the hamster and human type II structures demonstrates that the linkages between these two domains are highly flexible *(26)*. In the human structure, the CBS domain does not participate in either the monomer–monomer or tetramer–tetramer interfaces.

C. Mobile Elements of the IMPDH Active Site

The IMPDH active site lies on the C-terminal face of the catalytic α/β barrel (Figure 2). This location is conserved in other barrel structures. The active site is composed of a long cleft containing a nucleotide binding pocket contiguous with an NAD binding groove (Figure 3). Substrate binds in the nucleotide pocket, forming a number of polar interactions that appear to be conserved across species. These involve both stationary residues forming one wall of the pocket, as well as residues in two mobile elements of the structure. The latter consist of a ~50-residue active site flap, and a 15-residue active site loop (Figure 3).

1. Active Site Loop

The active site loop (residues 325-340 in the human and hamster enzymes) contains a catalytically critical cysteine (Cys 331 in the hamster and human enzymes, Figure 3). Nucleophilic attack by this cysteine thiol on the C2 position of the inosine ring results in formation of the tetrahedral intermediate. Formation of this intermediate facilitates subsequent hydride transfer from the C2 position to the B-side of the NAD nicotinamide ring. Hydrolytic attack on the resulting thioimidate releases XMP product. In the hamster structure, the covalently bound thioimidate is observed directly *(25)*. This complex is trapped by the presence of MPA. It has been proposed that the MPA phenol hydroxyl oxygen displaces the catalytic water *(40)*. The covalent linkage between the IMP purine ring and the catalytic cysteine is not observed in the S. *pyogenes* binary complex, which lacks MPA, nor in the *T. foetus* IMP-β-TAD ternary complex, which lacks a hydride acceptor. In the S. *pyogenes* binary complex, two solvent molecules have been identified as candidates for the catalytic water *(31)*.

In addition to contributing the catalytic cysteine, the loop forms several polar interactions that close off one side of the IMP pocket and the nicotinamide end of the cofactor pocket (Figure 3). In the hamster, S. *pyogenes* and *T. foetus* structures, a conserved serine forms a hydrogen bond to the IMP phosphate group (Ser 329 in the hamster and human enzymes, Figure 3) *(25, 31, 34)*. In both the hamster and human MAD complexes, the main chain of conserved Gly 326 and the side chain of Thr 333 form H-bonds with the exocyclic carbonyl and hydroxyl oxygens of the MPA ring system *(25)* (Figure 3). Similar interactions are also seen in the *T. foetus* -β-TAD complex. Using the mammalian numbering scheme, the thiazole carboxamide group forms hydrogen bonds with the main chain nitrogen of Gly 326 and the main chain carbonyl oxygen of Gly 312. In *T. foetus*, Thr 333 is replaced structurally by an Arg, which also interacts with the thiazole carboxamide group. In the human SAD and NAD complexes, the position of the distal loop is distorted by the presence of 6-Cl IMP (see below). However, given that MPA and the β-TAD thiazole moieties bind in the nicotinamide end of the active site *(5, 25, 34)*, it is likely that at least some of these loop residue interactions are also formed with NAD.

The active site loop is highly mobile, even in the presence of bound ligand. This is demonstrated in the structure of the human type II complex *(26)* (Figure 4). This complex was crystallized with the IMP analogue 6-Cl IMP (Figure 1b) and the dinucleotide inhibitor SAD (Figure 1e). 6-Cl IMP undergoes IMPDH-catalyzed dehalogenation in the absence of NAD *(8)*. In the crystal structure, a covalent linkage is observed between the 6-position on the purine ring and Cys 331 (Figure 4), confirming that the dehalogenation reaction also proceeds via nucleophilic attack by the cysteine thiol *(8, 26)*. The nucleophilic attack by Cys 331 occurs on the opposite side of the purine ring from the C2 position; the site of nucleophilic attack on the normal IMP base (Figure 4). This is accomplished by reorientation of the loop, rather than by reorientation of the ligand (Figure 4). The position of the catalytic Cys on the loop in the *B. burgdorferi* apo structure is further shifted ~ 8 Å from the position observed in the human structure (Figure 4).

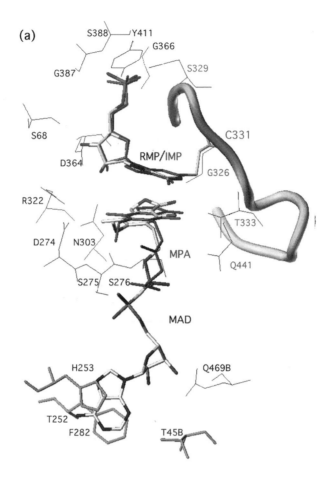

Figure 3. (a) Ligand environment in human type II- RMP- C2-MAD complex. IMP and MPA ligands from the hamster complex are shown in red for comparison. C2-MAD binds in the NAD site. Atoms in the RMP and C2-MAD ligands are color coded as follows: carbon, white; oxygen, red; nitrogen, blue; phosphorus, orange. Residues interacting with each ligand are labeled (residues 414 and 415 are omitted for clarity). The active site loop and the covalent linkage between substrate and catalytic residue Cys 331 are shown in yellow. Loop residues are labeled in yellow. Residues interacting with the adenosine end of NAD cofactor that are not conserved between the type I and II isoforms are highlighted in cyan. (b) Active site cleft of IMPDH showing the IMP and NAD binding sites. Residues of the barrel domains are illustrated as a solvent accessible surface. The active site loop (yellow) and flap model (white and cyan) are from the human type II RMP- C2-MAD complex. The terminal (cyan) end of the active site flap contains four residues not conserved between the types I and II isoforms. The NAD ligand is from the human type II – 6-Cl IMP–NAD complex. IMP is from the hamster complex (25). Ligand atoms are color coded as in (a). The sub-components of the NAD site, and selected residues are labeled for reference.

(This figure is also in color insert.)

(b)

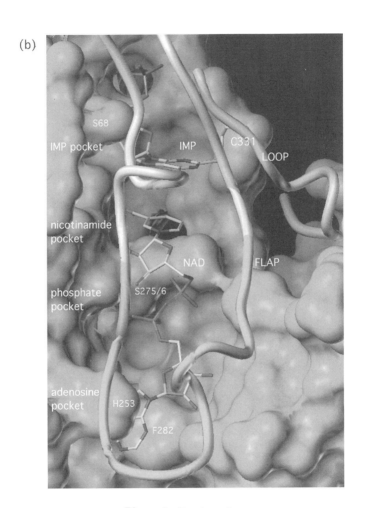

Figure 3. *Continued.*

(This figure is also in color insert.)

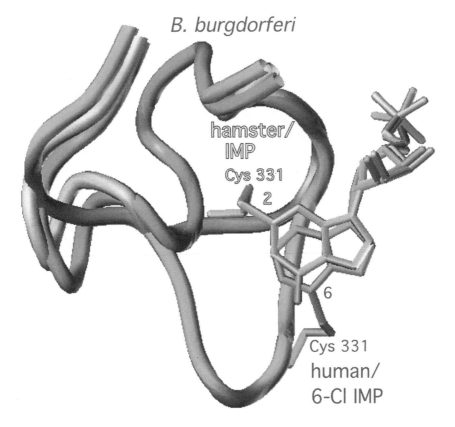

Figure 4. Comparison of loop positions in human (cyan) (26), hamster (yellow) (25) and
B. burgdorferi *(green) (32) complexes. The positions of the IMP and the dehalogenated
6-Cl IMP ligands are illustrated for the hamster-IMP and human 6-Cl IMP structures
respectively. Purine rings form covalent ligands with Cys 331 at the 2 and 6 positions for
IMP and 6-Cl IMP respectively.*
(This figure is also in color insert.)

2. Active Site Flap

The second mobile element of the enzyme that interacts with both ligands is the ~50-residue active site flap (residues 400-450 in the human and hamster enzymes). This element covers the active site cleft, apparently protecting binding of both substrate and cofactor (Figures 3 and 5).

In the hamster and *S. pyogenes* structures, polar interactions are observed between four flap residues and IMP (Figures 3 and 5) *(25, 31)*. A conserved Tyr (411 in the hamster enzyme) donates an H-bond to the phosphate group, and the main-chain atoms of three flap residues (414, 415, and 441) form H-bonds with the inosine base N1, N7 and carbonyl oxygen (Figure 3 and 5). In the hamster-MPA and human type II-RMP-MAD complexes, the side chain of flap residue Gln 441 also forms a direct hydrogen bond with the MPA hydroxyl group, and a water mediated H-bond with the end of the hexenoic acid tail (MPA) or a phosphate group (C2-MAD) (Figure 3 and 5). The distal portion of the flap, comprising residues 420-437, is disordered in both structures.

The series of interactions with IMP formed by the proximal part of the flap seen in the hamster and *S. pyogenes* complexes are extended in the human type II–ribavirin–MAD complex, which appears to be partially ordered by the presence of the dinucleotide ligand (Figure 3). Continuous electron density allows fitting of an α-carbon backbone model for the remainder residues 421-436. The terminal portion of the flap covers the phosphate and adenosine portions of the dinucleotide site, interacting with the dinucleotide both directly and via intervening residues. This segment of the flap is particularly rich in residues that are not conserved between the human type I and type II isoforms, and may be useful in the development of an isoform-specific inhibitor (see below).

Spectroscopic and denaturation studies suggest that initial binding of IMP induces a conformational change, most likely in the flap region, which facilitates cofactor binding *(41)*. Binding of IMP is the primary factor in stabilizing the flap against proteolysis *(41)*. Substrate-induced protection of the flap against proteolysis is further enhanced by cofactor binding *(41)*. These findings, combined with the crystallographic observations, suggest that the entire flap serves to stabilize both substrate and dinucleotide binding. Nevertheless, the flap remains a highly mobile element, showing significant disorder even in the presence of bound dinucleotide.

D. The Active Site Cleft

As noted, the IMP substrate and NAD cofactor bind in a continuous cleft on the C-terminal face of the α/β barrel (Figure 3). The active site cleft in the human enzyme also contains residues from adjacent catalytic monomers. This cleft can be further divided into IMP and NAD sites (Figure 3).

154

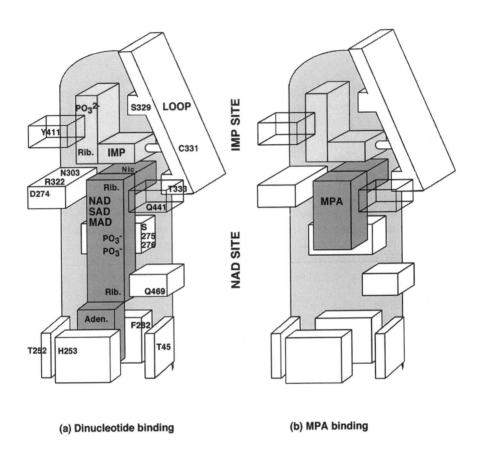

(a) Dinucleotide binding   **(b) MPA binding**

Figure 5. Schematic of the active site cleft showing binding of IMP (yellow) and (a) NAD and the dinucleotide inhibitors SAD and C2-MAD (red), (b) MPA (red), and (c) VX-497 (red). Flap residues interacting with ligands are drawn as transparent boxes. The flap itself is omitted for clarity in (a) and (b).

(This figure is also in color insert.)

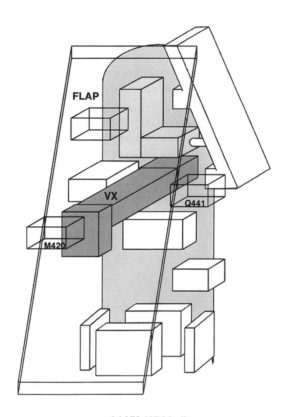

(c) VX-497 binding

Figure 3. *Continued.*

(This figure is also in color insert.)

1. The IMP site.

The IMP site has been well characterized in a number of structures, with complexes from hamster, human, *T. foetus and S. pyogenes* containing either IMP, the covalently modified thioimidate XMP*, or the IMP analogue RMP (Table I). As noted, upon binding of IMP, this end of the active site is covered by the active site loop (residues 325-340) (Figure 3). In the human enzyme, the loop contributes Ser 329, which hydrogen bonds to the IMP phosphate group, and the catalytic Cys 331, which attacks C2 on the inosine ring (Figure 3). Both residues are conserved across species.

The "back" wall of the IMP site contains a number of residues that stabilize binding of the IMP phosphate, ribose and inosine moieties (Figure 3 and 5). In the hamster and human enzymes, the phosphate group is locked into position by polar interactions with the main chain nitrogens of Gly 366, Gly 387, and Ser 388, and with the hydroxyl of Tyr 411. The ribose hydroxyls form H-bonds with the side chains of Ser 68 and Asp 364. The inosine base is stabilized by H-bonds with main chain atoms of Met 414, Gly 415, and Gln 441. Very similar interactions are observed in the *T. foetus* and *S. pyogenes* complexes.

2. The Cofactor Site

Characterization of the cofactor-binding groove has been aided by the hamster-MPA complex, as well as a number of recent structures containing NAD or cofactor analogues. These include human and *T. foetus* complexes containing NAD, SAD, β-TAD, and C2-MAD (Table 1).

The nicotinamide site lies in a deep pocket capped by the IMP hypoxanthine ring. The NAD nicotinamide ring stacks against the IMP base, favorably positioning this end of the cofactor for hydride transfer (Figure 3). The nicotinamide ring lies in the *anti* conformation relative to the ribose sugar, consistent with NMR studies suggesting that this conformation is adopted by IMPDH-bound NAD *(42)*. Further, in this position, hydride transfer occurs to the B-side of the nicotinamide ring, consistent with the fact that IMPDH is a B-side specific enzyme *(7)*.

The nicotinamide carboxamide group forms polar interactions with stationary residues forming the pocket walls (Figures 3 and 5). These include side chains of Asn 303, Arg 322, and Asp 274. Asn 303 and Asp 274 are conserved in 45 of 46 pro- and eukaryotic IMPDH sequences obtained to date. Arg 322 is conserved in all mouse, hamster and human sequences, but is substituted with Lys in a number of microbial enzymes, including *T. foetus*.

Like the IMP site, the nicotinamide end of the cofactor-binding pocket is covered by the active site loop, as well as the active site flap (Figure 3). The position of this segment of loop and flap in the human type II-NAD complex is distorted near the nicotinamide pocket by the presence of the 6-Cl IMP ligand (above, Figure 4). However, nicotinamide interactions with mobile elements of the site may be inferred from other structures.

The hamster-MPA complex contains an undistorted loop. In this structure, MPA binds at the nicotinamide end of the cofactor site, the substrate base stacking on the MPA bicyclic ring *(25)* (Figures 3 and 5). Here, the MPA bicyclic ring also forms H-bonds with stationary residues 303, 322 and 274, as observed in the human type II–NAD complex. In addition, MPA forms H-bonds with loop residues Gly 326 and Thr 333 on one side of the nicotinamide pocket and flap residue Gln 441, which covers the pocket. The same interactions are seen in the human type II-MAD complex (Figures 3 and 5). It is likely that at least some of these residues on the mobile elements of the active site interact with the NAD nicotinamide moiety.

The NAD ribose and phosphate groups continue at roughly a right angle to the nicotinamide pocket, binding in a long continuation of the active site cleft. The nicotinamide ribose hydrogen bonds with the same conserved Asp 274 that interacts with the nicotinamide carboxamide group (Figures 3 and 5). The phosphate groups form H-bonds with successive serines 275 and 276, which form the floor of the active site groove in this region (Figures 3 and 5). These phosphate–serine interactions are seen in the all IMPDH structures containing a dinucleotide to date, including human-NAD, SAD, and MAD complexes and the *T. foetus*–β-TAD complex. In the hamster-MPA complex, the MPA hexenoic "tail" binds in the phosphate groove, and also bonds with one of the two serines forming the floor (Ser 276) (Figures 3 and 5). These serines are conserved in all eukaryotic IMPDHs. As discussed above, the MPA hexenoic tail and one MAD phosphate also participate in a water-mediated H-bond with flap residue Gln 441.

In the human and hamster enzymes, the "right hand" wall of the NAD adenosine portion of the active site groove is formed by the neighboring symmetry-related catalytic monomer. Thus, the adenosine ribose hydroxyls form direct H-bonds with Gln 469 and a water-mediated bond with the main chain N of Ala 46, both residues being contributed by the neighboring monomer.

The solvent end of the cofactor groove is defined by residues that interact with the NAD adenine base. The adenosine end of the dinucleotide analogue is bound in a cleft between the α3 helix-β3 sheet junction of one monomer and the βC-βD sheet junction of the adjacent monomer.

In the human type II-NAD complex, the dinucleotide adenine ring is stacked between the side chains of Phe 282 and His 253 near the α3 helix. In this position, it makes two edge-on contacts between the adenine amino group and the side chain of residue Thr 252. On the opposite side of the ring, a contact is observed between adenine N3, and the side chain of Thr 45 on the adjacent monomer. Thus, the adenosine moiety mediates an unusual number of inter-monomer interactions *(26)*. The same adenosine-enzyme interactions are observed in the human type II-SAD and C2-MAD complexes (Figures 3 and 5). In the *T. foetus*-β-TAD complex, the adenosine moiety does not make intermonomer contacts, although its stacking interactions are conserved. In *T. foetus*, the adenine moiety stacks between Trp 269 and Arg 241, the residues that replace Phe 282 and His 253. Isoform and species differences at this end of the dinucleotide site may be exploited in the design of more selective chemotherapeutic agents (below).

III. Implications for Inhibitor Binding

Structural studies described above have defined an IMPDH active site cleft composed of contiguous IMP and NAD binding sites. The IMP site is covered by highly mobile flap and loop elements (Figure 3). The latter contains the catalytic residue Cys 331, which forms a covalent linkage with the substrate base in the catalytic intermediate. Additional residues donated by the flap, loop and stationery components of the IMP site participate in numerous polar interactions with the substrate phosphate, ribose and purine moieties. The NAD cleft may in turn be thought of as containing a nicotinamide pocket, a phosphate binding region, and an adenosine binding region. The nicotinamide pocket is capped by the substrate base, allowing stacking of the hydride donor and acceptor groups. The phosphate pocket is dominated by two conserved serines. The adenosine pocket is characterized by a tightly bound adenine ring sandwiched between two aromatic residues, and further constrained by two threonines contacting its edges. The cleft is covered by the highly mobile flap.

In principle, any ligand that binds to one of these locations so as to disrupt the catalytic mechanism is a potential chemotherapeutic agent. In practice, ligand design has focused on two major classes of compounds: those that bind at the IMP site and those that bind at the NAD site.

A. IMP Site-Binding Inhibitors

Development of IMP site-binders has focused on nucleotide analogues. These include the antiviral agents ribavirin *(43, 44)* and EICAR *(45, 46)*, the immunosuppressant mizoribine (bredinin) *(47, 48)*, and the "fat base" nucleoside imidazole[4,5-e][1,4]diazapine nucleoside *(49)*.

Ribavirin has seen widespread clinical use as an antiviral agent *(50-52)*. Ribavirin is converted to the 5' monophosphate (RMP, Figure 1c), which acts as a competitive inhibitor of human IMPDH with respect to IMP with Ki's of 0.4 and 0.7 μM for the type II and type I isoforms respectively *(53)*. Ribavirin has also been shown to act synergistically with NAD site inhibitors (below) as an antitumor agent in vitro *(54-57)*. The crystal structure of the human type II-RMP-MAD complex demonstrates that, as expected, RMP closely mimics binding of the normal substrate IMP.

Mizoribine is entering clinical use in Japan as an immunosuppressant agent in the treatment of acute allograft rejection following human renal transplantation *(58, 59)*. The active metabolite, mizoribine 5' monophosphate, acts as a competitive transition state analogue, binding IMPDH with a Ki of 0.5 nM with respect to IMP *(60)* (see also the chapter by Ishikawa et al.). The agent EICAR has shown significant antiviral activity in several in vitro and in vivo systems *(46, 61-63)*. The 5' monophosphate anabolites of EICAR and the "fat base" imidazole[4,5-e][1,4]diazapine nucleoside are also potent inhibitors of IMPDH, with Kd's of ~ 1 nM in the human enzyme *(49, 64)*. Like 6-Cl IMP, these agents inactivate IMPDH by forming covalent adducts with Cys 331 *(8, 49, 64)*.

Clearly, any of the IMP site binders can mimic the ground state substrate-enzyme interactions described above, particularly those involving the phosphate and ribose moieties. However, the more potent IMP site binders are also able to act as transition state analogues. These contain bases that either form covalent adducts with the active site cysteine, or otherwise mimic interactions specific to the transition state *(8, 49, 60, 64)*. Design of these agents may take advantage of the unusual flexibility of the active site loop containing Cys 331. Recall that formation of the covalent adduct with the inactivating agent 6-Cl IMP occurs via a shift in position of the active site loop in order to accommodate the 6-Cl purine moiety (Figure 4). This suggests that new IMP site binders may take advantage of the flexibility in the loop, and need not be limited by the stereochemical constraints of the native, ground state structure.

B. NAD Site-Binding Inhibitors

The second class of inhibitors comprises those that bind in the NAD site. These represent a structurally more diverse group, ranging from close isosteric analogues of NAD to distant analogues of mycophenolic acid.

Mycophenolate mofetil, a mycophenolic acid precursor, is approved in the United States for use as an immunosuppressant in the treatment of acute rejection in renal transplants *(65, 66)*. In vivo, mycophenolate mofetil is metabolized to mycophenolic acid (MPA, Figure 1d). The immunosuppressant activity of MPA is linked directly to its ability to reduce both T and B lymphocyte proliferation via inhibition of IMPDH *(2)*. The efficacy of MPA is limited by its rapid conversion to the glucuronide via uridine 5'-diphosphophoglucuronyl transferase*(2, 67, 68)*. Nevertheless, MPA remains the NAD site binding inhibitor with the highest affinity and specificity for type II IMPDH. MPA is an uncompetitive inhibitor with respect to NAD, binding with a Ki 's of 33 and 7 nM to the type I and II isoforms respectively *(2)*. MPA binds IMPDH with 10-100-fold greater affinity than the thiazole or selenazole dinucleotides (below), although some of this ~1-2 kcal/mol increase in binding may be due to the entropic advantage of a smaller ligand with fewer degrees of freedom.

Interestingly, MPA binds *T. foetus* IMPDH with an affinity over 1000-fold weaker than that which it binds the human enzyme. Hedstrom has suggested that this may be due to differences in the coupling between the nicotinamide and adenosine sites in the mammalian and *T. foetus* enzymes *(69)*. Binding of a ligand at the nicotinamide end of the cofactor site is thought to order the mobile components of the adenosine site to a greater degree in the *T. foetus* enzyme. Thus, while MPA binds only at the nicotinamide end of the cofactor site, some of this binding energy is required to order the adenosine site in the *T. foetus* enzyme *(69)*. This produces weaker binding in this system.

Although MPA shows some selectivity for the human type II IMPDH, its role as an antitumor agent has not been vigorously investigated. In vivo efficacy of MPA is limited by its rapid metabolism to the glucuronide *(67, 68)*. Active programs based upon the design of new MPA analogues to overcome these problems have been initiated by a number of pharmaceutical companies (below).

Benzamide riboside is a recently developed isosteric analogue of nicotinamide nucleoside *(70)*. In human myelogenous leukemia K562 cells, benzamide riboside is converted to benzamide adenine dinucleotide (BAD) via NAD pyrophosphorylase *(71, 72)*. BAD, a close isosteric analogue of NAD, is a competitive inhibitor of IMPDH from K562 cells, binding with a Ki of 0.118 μM with respect to NAD *(71)*. This agent is discussed in detail in the chapter by Jayaram et al.

One of the most widely studied IMPDH inhibitors is tiazofurin (2-β-D-ribofuranosylthiazole-4-carboxamide, NSC 286193), a C-glycosyl thiazole nucleoside originally synthesized as an analogue of ribavirin by Srivastava, Robins and co-workers more than 20 years ago *(73)*. Despite Initial Phase I clinical trials demonstrating a variety of dose-limiting toxic side effects, tiazofurin induced complete hematologic remissions in patients with end-stage acute nonlymphocytic leukemia or in myeloblastic crisis of chronic myeloid leukemia in phase II trials *(74, 75)*.

The efficacy of tiazofurin in the treatment of human leukemias appears to be related to two properties: the ability of tiazofurin to kill tumor cells and the ability to induce surviving cells to undergo maturation *(1, 74, 75)*. Both the antiproliferative and maturation-inducing effects of tiazofurin are attributed to the same biochemical effect: tiazofurin's ability to reduce intracellular guanine nucleotide pools via inhibition of IMPDH *(1, 75)*.

The major IMPDH inhibitor is a dinucleotide anabolite of tiazofurin. Tiazofurin is transported across the cell membrane and converted, via its 5'-phosphate, into an analogue of NAD. In this NAD analogue, called TAD (2-β-D-ribofuranosylthiazole-4-carboxamide adenine dinucleotide, Figure 1e), the nicotinamide ring is replaced by a thiazole-4-carboxamide moiety *(76)*. TAD is a noncompetitive inhibitor of IMPDH with respect to NAD. The Ki of TAD with respect NAD has been reported as ~0.5 μM in both isoforms. However, more recent analyses using tight-binding conditions suggest that these values may be 5-fold lower *(34)*.

Clinical use of tiazofurin is limited by toxicity and resistance. Resistance to tiazofurin is associated with a decline in the synthesis of TAD, accompanied by increased degradation of this active dinucleotide by a phosphodiesterase ("TADase") *(77, 78)*. Resistant cell lines, such as human colon carcinoma HT29, have lowered levels of anabolic enzymes as well as increased levels of TADase *(77, 78)*. Resistance has been a problem in the use of tiazofurin in the treatment of end stage myeloid leukemias, limiting the median response to 6 months *(74, 75)*.

One strategy to combat drug resistance has been the design of TADase-resistant analogs. The β-methylene analogue of TAD (β-TAD, Figure 1f) is resistant to this phosphodiesterase, due to replacement of the phosphodiester oxygen with a CH_2 bridge *(35)*. For this reason, β-TAD shows significant cytotoxicity against tiazofurin-resistant p388, L1210 and human colon carcinoma HT29 cells *(35, 77, 78)* . These modifications are discussed in detail in the chapter by Pankiewicz et al.

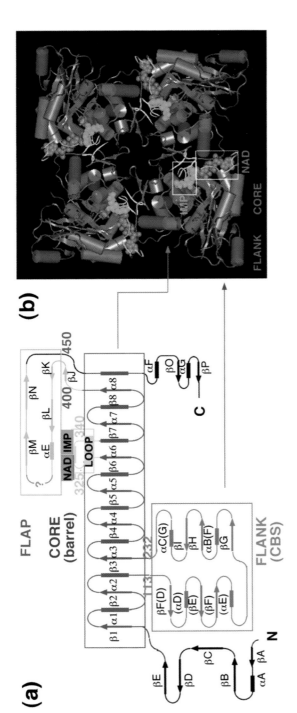

Figure 2. (a) Secondary structure of human and hamster IMPDH. The structure of the flanking CBS domain is partially disordered in these structures. The fold shown here is taken from the CBS domain defined in the S. Pyogenes structure (31). Nomenclature follows that used in the hamster structure (25). Labels in parentheses indicate S. pyogenes nomenclature. (b) IMPDH tetramer viewed down the crystallographic 4-fold. Elements of the structure are colored as follows: Core barrel domain (blue), flanking CBS domain (lavender), loop (yellow), flap (cyan), NAD (red), and IMP (green). Ligands IMP and NAD are circled. The cofactor site defined by NAD lies at the monomer-monomer interface.

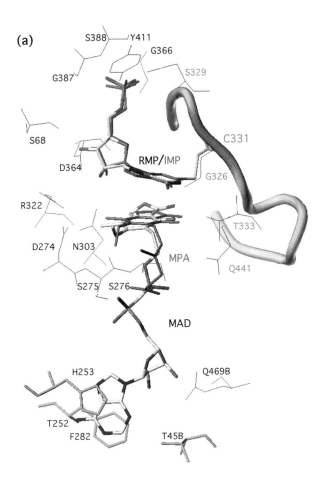

(a)

Figure 3. (a) Ligand environment in human type II- RMP- C2-MAD complex. IMP and MPA ligands from the hamster complex are shown in red for comparison. C2-MAD binds in the NAD site. Atoms in the RMP and C2-MAD ligands are color coded as follows: carbon, white; oxygen, red; nitrogen, blue; phosphorus, orange. Residues interacting with each ligand are labeled (residues 414 and 415 are omitted for clarity). The active site loop and the covalent linkage between substrate and catalytic residue Cys 331 are shown in yellow. Loop residues are labeled in yellow. Residues interacting with the adenosine end of NAD cofactor that are not conserved between the type I and II isoforms are highlighted in cyan. (b) Active site cleft of IMPDH showing the IMP and NAD binding sites. Residues of the barrel domains are illustrated as a solvent accessible surface. The active site loop (yellow) and flap model (white and cyan) are from the human type II RMP- C2-MAD complex. The terminal (cyan) end of the active site flap contains four residues not conserved between the types I and II isoforms. The NAD ligand is from the human type II – 6-Cl IMP–NAD complex. IMP is from the hamster complex (25). Ligand atoms are color coded as in (a). The sub-components of the NAD site, and selected residues are labeled for reference.

(b)

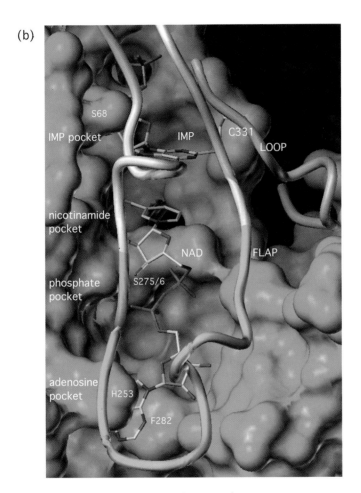

Figure 3. *Continued.*

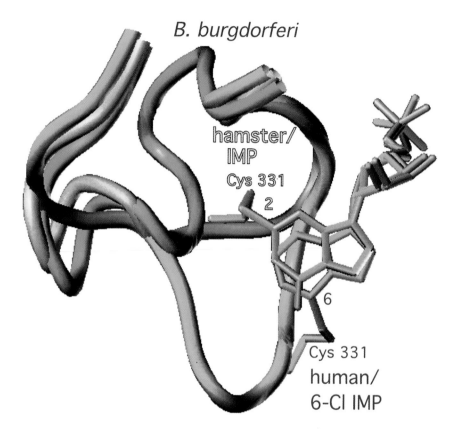

Figure 4. Comparison of loop positions in human (cyan) (26), hamster (yellow) (25) and B. burgdorferi *(green) (32) complexes. The positions of the IMP and the dehalogenated 6-Cl IMP ligands are illustrated for the hamster-IMP and human 6-Cl IMP structures respectively. Purine rings form covalent ligands with Cys 331 at the 2 and 6 positions for IMP and 6-Cl IMP respectively.*

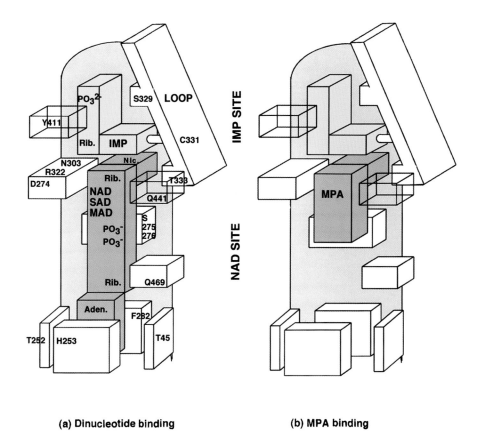

(a) Dinucleotide binding　　　　**(b) MPA binding**

Figure 5. Schematic of the active site cleft showing binding of IMP (yellow) and (a) NAD and the dinucleotide inhibitors SAD and C2-MAD (red), (b) MPA (red), and (c) VX-497 (red). Flap residues interacting with ligands are drawn as transparent boxes. The flap itself is omitted for clarity in (a) and (b).

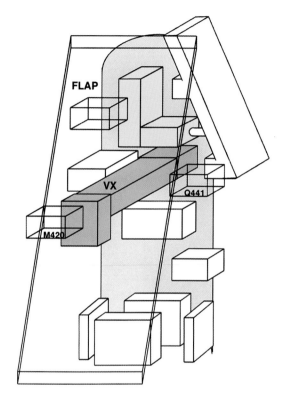

(c) VX-497 binding

Figure 5. *Continued.*

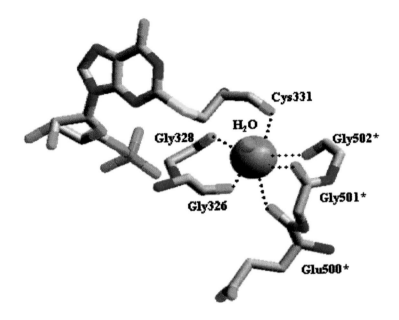

Plate 1. K⁺ and Environs in Hamster IMPDH (28). The large sphere in the center is K⁺ and the smaller foreground sphere is a water.

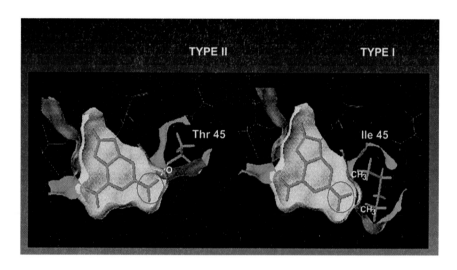

Plate 1. Model of IMPDH adenine-binding site occupied by 2,6,-diaminopurine containing dinucleotide. Van der Waals surfaces of adjacent residues are illustrated. At left: replacement of adenine with 2,6,-diaminopurine allows formation of hydrogen bonds with Thr-45 in IMPDH type II. At right: 2,6,-diamino purine exocylic nitrogen at the C2 forms unfavorable steric contacts with Ile-45 in IMPDH type I.

The crystal structure of human type II IMPDH with bound SAD, (the selenium analogue of TAD, Figure 1e) shows that this agent closely mimics NAD binding in the cofactor site. TAD and SAD show moderate specificity for IMPDH, binding to this target with 3-4 orders of magnitude greater affinity than to alcohol, glutamate or lactate dehydrogenases. The origins of this specificity have been attributed in part to entropic gains resulting from an intramolecular constraint. An intramolecular S/Se–O interaction constrains rotation about the C-glycosidic bond such that the S or Se heteroatom remains adjacent to the furanose oxygen, maintaining a *trans* conformation about the C-glycosyl bond *(79)*. In alcohol dehydrogenase, this distorts the conformation of the dinucleotide inhibitors relative to that required for NAD, and adversely influences binding *(26, 80)*. In IMPDH, the *trans* conformation mimics that required for NAD binding. Thus, the IMPDH-bound thiazole and selenazole dinucleotides appear to be constrained to a conformation favorable for target binding, gaining an entropic advantage over an unconstrained, more flexible ligand. Entropic contributions to ligand binding due to intramolecular constraints can be substantial *(81)*.

IV. New Strategies for Inhibitor Design

Despite the availability of IMPDH inhibitors, significant problems remain in their clinical application. Tiazofurin causes severe dose-limiting side effects requiring hospitalization and aggressive treatment *(74, 75)*. Some patients are resistant to tiazofurin. In sensitive patients the response is usually short lived (10-20 days), requiring multiple treatments leading to drug resistance *(74, 75)*. Thus, improvement in drug specificity and activity in resistant cells will be required to reduce toxicity and enhance efficacy *(74, 75)*. Gastrointestinal toxicity of MPA, although less severe than that of tiazofurin, is dose limiting *(65, 66)*. This is exacerbated by the rapid metabolism of MPA to its biologically inactive glucuronide *(65, 66)*. As noted above, MPA's role as an antitumor agent is also limited by its rapid metabolism to the glucuronide *(67, 68)*. Anemia and thyroid abnormalities can limit the use of ribavirin in the treatment of hepatitis *(82, 83)*. Ribavirin is also a known teratogen *(84)*, necessitating protection of health care workers administering the drug in its aerosolized form to treat respiratory syncytial virus infection *(85)*.

Improvement in drug specificity and activity will be required to reduce toxicity and enhance efficacy. High selectivity for type II IMPDH would be one of the more desirable attributes in an improved agent *(86)*. A type II-selective agent would target the dominant isoform in neoplastic cells while reducing interactions with the type I isoform in normal cells. Thus, a type II-specific agent would demonstrate both reduced toxicity and enhanced activity, providing a drug with an improved therapeutic ratio*(1, 2)*. Unfortunately, the 83% sequence identity between the constitutively expressed type I isoform and the inducible type II isoform complicates this task. Nevertheless, structural studies have identified residues at or near the ligand binding sites which may offer exploitable differences between isoforms.

Of the 514 residues in human IMPDH, 84 differ between the types I and II isoforms. Of these, ~25% are conservative substitutions. Thus, identification of any non-conserved residues that directly interact with either an IMP or NAD site-binding ligand is of value.

In the IMP site, the only residue that is not conserved between type I and II isoforms is Ser 327(II)->Cys(I). The closest distance between the side chain of this loop residue and the IMP phosphate group is 5.7Å. The distance between this ligand and the next closest non-conserved residue is 8.0 Å. Given the distance between IMP and the nearest non-conserved residue, as well as the conservative nature of the substitution, the IMP site appears less likely to provide opportunities for design of type II-specific agents. However, the ability of these agents to serve as transition state analogues insures their continued place among the more potent IMPDH inhibitors, with Kd's $\leq$ 1 nM (above). Given the continuity of the IMP and NAD sites (Figure 3), the linkage of a transition state analogue with an NAD site-binding moiety may offer one strategy for the design of potent, isoform-specific agents.

As noted, MPA show some selectivity for the type II enzyme (above) *(2)*. However, all residues identified as directly interacting with MPA in the hamster complex are conserved between the type I and type II isoforms. These observations suggest that subtle conformational changes in the MPA binding site and/or unobserved interactions with disordered residues contribute to differences in MPA binding between the two isoforms.

The NAD site does offer a number of potentially exploitable differences between the type I and II enzymes. Chief among these is a concentration of nonconserved residues at the terminal end of the flap region, comprising residues 421-436 . Four of the eight residues between 421 and 428 differ between the type I and II isoforms: Asp 421(II)->Glu(I), His 423(II)->Ser(I), Leu 424(II)->Ser(I) and Asn 428(II)->Lys(I). In the hamster complex, these residues are disordered. As described above, binding of a dinucleotide ligand in the NAD site appears to at least partially order this region of the flap through direct and indirect contacts with the ligand. In the human type II-RMP-MAD complex, the flap is sufficiently ordered to allow tracing of an α-carbon chain for terminal flap residues 420-437. This model indicates that, as expected, the terminal portion of the flap containing the non-conserved residues lies in proximity to the dinucleotide phosphate and adenosine binding regions (Figure 3b). Despite this, neither NAD nor the dinucleotide analogues show significant isoform specificity. One or more of the terminal flap residues may interact preferentially with MPA, particularly in the region of the hexenoic acid side chain, contributing to the modest isoform specificity of this agent.

In an effort to enhance the affinity and isoform selectivity of IMPDH inhibitors, several pharmaceutical companies have undertaken active programs in drug development, based on either MPA or de novo design. Most of this material remains proprietary, although a number of patents awarded to Syntex (now Roche) and Vertex Pharmaceuticals give some examples of the types of compounds under consideration *(87-92)*. These include 4-amino and 5- and 6-substituted derivatives of the MPA bicyclic ring system as well as derivatives of the hexenoic acid side-chain. A recent patent awarded to Vertex Pharmaceuticals also discloses a series of novel substituted ureas *(87)*. A manuscript describing complexes between the hamster enzyme and two of these de novo designed agents has also recently appeared *(29)*. The more potent of these agents, VX-497 inhibits IMPDH with a Ki of 7 nM (Figure 1h).

The VX-497 agent demonstrates a novel-binding mode (Figure 5c). The VX-497 phenyloxazole moiety binds in the nicotinamide pocket, stacking under the substrate hypoxanthine ring, and forming many of the same polar interactions as the NAD nicotinamide ring and the MPA bicyclic ring *(29)*. The VX-497 urea group hydrogen bonds with Asp 274, analogous to the nicotin-amide ribose hydroxyls. However, at this point, where a dinucleotide or MPA turns "down" into the groove, the remainder of the VX-497 molecule continues in roughly the same plane as the phenyl-oxazole moiety, coming "out" of the groove. In this orientation, VX-497 forms contacts with a number of flap residues, including Met 420 (Figure 5c). Thus, the furan end of VX-497 may be well positioned to exploit isoform differences in the flap. Unfortunately, the isoform specificity of these agents will take advantage of isoform differences in the flap region.

In addition to potential but as yet unobserved differences in the flap region, other interactions are directly observed between the dinucleotide ligand and non-conserved residues in the human complex *(26)*. Recall that the adenine base of SAD in this complex is stacked between aromatic residues His 253 and Phe 282, while forming edge-on interactions with Thr 252, as well as with Thr 45 from the adjacent monomer (Figure 6a). Three of these residues are not conserved between the type I and type II isoforms: His 253 (II) ->Arg(I), Phe 282(II)->Tyr(I) and Thr 45(II)->Ile(I). The crystal structure of the human type I-6-Cl IMP complex demonstrates that no significant structural differences in the adenosine end of the cofactor site are introduced by the type I substitutions. Thus, ligand modifications in the region of the dinucleotide pyrimidine base may well provide some degree of isoform specificity.

The adenosine region may also prove useful in the design of species-specific inhibitors. As noted above, His 253 and Phe 282 are replaced by Arg 241 and Trp 269 in the *T. foetus* enzyme. Despite this, stacking of the adenine moiety between the two residues is conserved (Figure 6b) *(34)*. As noted by Hedstrom, this opens the possibility for the design of a *T. foetus*-specific inhibitor *(34)*. More dramatic substitutions at the adenosine site in prokaryotic enzymes offer further opportunities for the design of specific antimicrobials.

New inhibitors have been developed which may exploit some of these findings *(93, 94)*. These compounds are "dinucleotide" NAD analogues in which the nicotinamide riboside is replaced by mycophenolic acid (Figure 1g). Variable components in compound design include the length of the linker and the presence of either a phosphodiester or methylene bridge. Three myco-phenolic-adenine "dinucleotide" hybrids have been constructed. These bind to type II IMPDH with Ki's = 0.2-0.6 μM and have IC_{50}'s comparable or superior to that of MPA in K562 cells in vitro (0.1-1.5 μM) *(93, 94)*. Further, these compounds are not metabolized to the glucuronide in K562 cells *(93, 94)* making them potentially useful chemotherapeutic leads.

The rationale behind the design of these agents was that the MPA end of the molecule would mimic MPA binding, with the adenine end mimicking

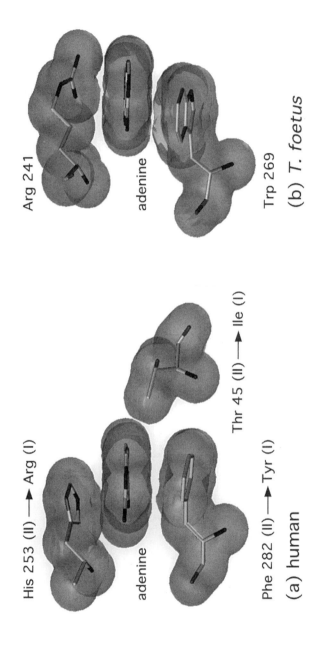

His 253 (II) ⟶ Arg (I)

adenine

Thr 45 (II) ⟶ Ile (I)

Phe 282 (II) ⟶ Tyr (I)

(a) human

Arg 241

adenine

Trp 269

(b) *T. foetus*

Figure 6. (a) Type I substitutions among residues interacting with the adenosine end of the NAD ligand in all human type II complexes. (b) Residues interacting with the adenosine end of the dinucleotide ligand in the T. foetus—IMP—β-TAD complex (34).

cofactor binding. This hypothesis has been confirmed by the structure of a complex between the human IMPDH type II enzyme, the substrate analogue RMP, and the C2-MAD hybrid (Figure 3a). Thus, in addition to these compounds' activity and resistance to metabolism, the interaction of the C2-MAD adenosine moiety with nonconserved residues may also be exploited in the design of new isoform-specific agents.

Perhaps the most intriguing development in the design of isoform specific agents has been the recent report of a novel series of 1,5-diaza-bicyclo[3.1.0]-hexane-2,4-diones *(95)*. To date, these agents show the clearest selectivity for the type II isoform. The most potent of these agents (Figure 7) is a competitive inhibitor of the type II isoform with respect to IMP, binding with a Ki of 5.1 µM. At this concentration, this agent shows no measurable inhibition of the type I isoform.

No crystal structure of an IMPDH complex containing these compounds has yet been reported. However, modeling studies suggest that these agents do not bind in either the IMP or the cofactor site. It has been suggested that the observed isoform selectivity results from interactions between the inhibitors and residues Ile 373 and Ala 374 in the α7 helix *(95)*. These non-conserved residues are replaced by Val's in the type I enzyme, and form part of a pocket adjacent to, but at some distance from, the IMP site (the α carbon of I 373 lies over 12 Å from the IMP phosphate). Subsequent biological studies have shown these compounds to be cytotoxic in a number of tumor cell lines, while showing no toxicity in the normal WI-38 cell line *(96)*. These novel compounds further demonstrate the wide variety of strategies now being pursued in the design of new IMPDH inhibitors.

V. Summary

IMPDH remains an attractive target for antitumor and immunosuppressive drug design. IMPDH inhibitors have been approved as immunosuppressants, and are under active investigation as antileukemic agents. However, clinical applications, particularly in antitumor therapy, remain limited by toxicity, metabolism, and resistance. Recent progress in crystal structure determinations of the enzyme, combined with active programs in drug design, has yielded a more thorough understanding of the determinants of ligand specificity in this system. Identification of residues involved in both substrate and NAD binding will aid the design of more potent inhibitors. Identification of potentially isoform-

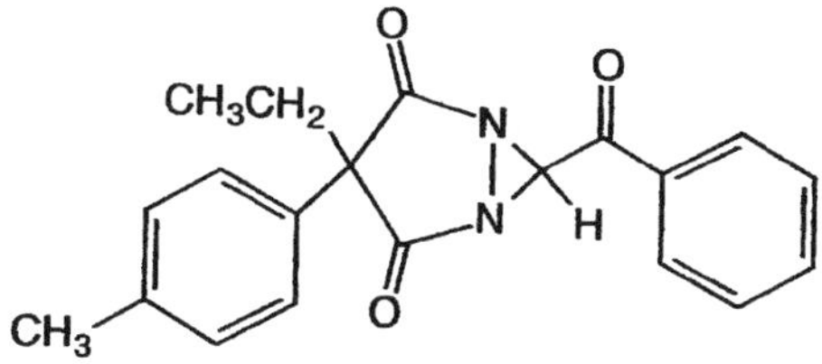

Figure 7. Novel type II-specific diazabicyclo[3.1.0]-hexane-2,4-dion inhibitor.

166

specific enzyme-ligand interactions holds the promise of more selective, less toxic agents.

References

1. Weber, G., Prajda, N., Abonyi, M., Look, K.Y.Tricot, G., *Anticancer Res.* **1996**, *16*, 3313-3322.
2. Wu, J.C., *Perspectives in Drug Discovery and Design* **1994**, *2*, 185-204.
3. Antonino, L.C.Wu, J.C., *Biochemistry* **1994**, *33*, 1753-1759.
4. Carr, S.F., Papp, E., Wu, J.C.Natsumeda, Y., *J. Biol. Chem.* **1993**, *268*, 27286-27290.
5. Hedstrom, L.Wang, C.C., *Biochemistry* **1990**, *29*, 849-854.
6. Wang, W.Hedstrom, L., *Biochemistry* **1997**, *36*, 8479-8483.
7. Xiang, B.Markham, G.D., *Arch. Biochem. Biophys.* **1997**, *348*, 378-382.
8. Antonino, L.C., Straub, K.Wu, J.C., *Biochemistry* **1994**, *33*, 1760-1765.
9. Pall, M.L., *GTP: A central regulator of Cellular Anabolism*, in *Current Topics in Cellular Regulation*, B.L. Horecker and E.R. Stadtman, Editors. 1985, Academic Press: New York. p. 1-20.
10. Mandanas, R.A., Leibowitz, D.S., Gharehbaghi, K., Tauchi, T., Burgess, G.S., Miyazawa, K., Jayaram, H.N.Boswell, H.S., *Blood* **1993**, *82*, 1838-1847.
11. Jayaram, H.N., Dion, R.L., Glazer, R.I., Johns, D.G., Robins, R.K., Srivastava, P.C.Cooney, D.A., *Biochem. Pharmacol.* **1982**, *31*, 2371-2380.
12. Manzoli, L., Billi, A.M., Gilmour, R.S., Martelli, A.M., Matteucci, A., Rubbini, S., Weber, G.Cocco, L., *Cancer Res.* **1995**, *55*, 2978-2980.
13. Kharbanda, S.M., Sherman, M.L.Kufe, D.W., *Blood* **1990**, *75*, 583-588.
14. Parandoosh, Z., Robins, R.K., Belei, M.Rubalcava, B., *Biochem. Biophys. Res. Commun.* **1989**, *164*, 869-874.
15. Parandoosh, Z., Rubalcava, B., Matsumoto, S.S., Jolley, W.B.Robins, R.K., *Life Sci.* **1990**, *46*, 315-320.
16. Vitale, M., Zamai, L., Falcieri, E., Zauli, G., Gobbi, P., Santi, S., Cinti, C.Weber, G., *Cytometry* **1997**, *30*, 61-66.
17. Olah, E., Csokay, B., Prajda, N., Kote-Jarai, Z., Yeh, Y.A.Weber, G., *Anticancer Res.* **1996**, *16*, 2469-2477.
18. Alison, A.C., Hovi, T., Watts, R.W.E.Webster, A.D.B., *CIBA Foundstion Symposium* **1977**, *48*, 207-224.
19. Natsumeda, Y., Ohno, S., Kawasaki, H., Konno, Y., Weber, G.Suzuki, K., *J. Biol. Chem.* **1990**, *265*, 5292-5295.
20. Nagai, M., Natsumeda, Y., Konno, Y., Hoffman, R., Irino, S.Weber, G., *Cancer Res.* **1991**, *51*, 3886-3890.
21. Nagai, M., Natsumeda, Y.Weber, G., *Cancer Res.* **1992**, *52*, 258-261.
22. Konno, Y., Natsumeda, Y., Nagai, M., Yamaji, Y., Ohno, S., Suzuki, K.Weber, G., *J. Biol. Chem.* **1991**, *266*, 506-509.
23. Dayton, J.S., Lindsten, T., Thompson, C.B.Mitchell, B.S., *J. Immunol.* **1994**, *152*, 984-991.
24. Gu, J.J., Spychala, J.Mitchell, B.S., *J. Biol. Chem.* **1997**, *272*, 4458-4466.
25. Sintchak, M.D., Fleming,' M.A., Futer, O., Raybuck, S.A., Chambers, S.P., Caron, P.R., Murcko, M.A.Wilson, K.P., *Cell* **1996**, *85*, 921-930.
26. Colby, T.D., Vanderveen, K., Strickler, M.D., Markham, G.D.Goldstein, B.M., *Proc Natl Acad Sci U S A* **1999**, *96*, 3531-3536.
27. Gebeyehu, G., Marquez, V.E., Van Cott, A., Cooney, D.A., Kelley, J.A., Jayaram, H.N., Ahluwalia, G.S., Dion, R.L., Wilson, Y.A.Johns, D.G., *J. Med. Chem.* **1985**, *28*, 99-105.
28. Boritzki, T.J., Berry, D.A., Besserer, J.A., Cook, P.D., Fry, D.W., Leopold, W.R.Jackson, R.C., *Biochem. Pharmacol.* **1985**, *34*, 1109-1114.
29. Sintchak, M.D.Nimmesgern, E., *Immunopharmacology.* **2000**, *47*, 163-184.
30. Risal, D.p., Strickler, M.D., Pankiewicz, K.W.Goldstein, B.M., *In preparation* **2003**.
31. Zhang, R., Evans, G., Rotella, F.J., Westbrook, E.M., Beno, D., Huberman, E., Joachimiak, A.Collart, F.R., *Biochemistry* **1999**, *38*, 4691-4700.

32. McMillan, F.M., Cahoon, M., White, A., Hedstrom, L., Petsko, G.A.Ringe, D., *Biochemistry* **2000**, *39*, 4533-4542.
33. Whitby, F.G., Luecke, H., Kuhn, P., Somoza, J.R., Huete-Perez, J.A., Phillips, J.D., Hill, C.P., Fletterick, R.J.Wang, C.C., *Biochemistry* **1997**, *36*, 10666-10674.
34. Gan, L., Petsko, G.A.Hedstrom, L., *Biochemistry* **2002**, in press.
35. Marquez, V.E., Tseng, C.K., Gebeyehu, G., Cooney, D.A., Ahluwalia, G.S., Kelley, J.A., Dalal, M., Fuller, R.W., Wilson, Y.A.Johns, D.G., *J. Med. Chem.* **1986**, *29*, 1726-1731.
36. Lesiak, K., Watanabe, K.A., Majumdar, A., Seidman, M., Vanderveen, K., Goldstein, B.M.Pankiewicz, K.W., *J. Med. Chem.* **1997**, *40*, 2533-2538.
37. Prosise, G.L., Wu, J.Z.Luecke, H., *J Biol Chem* **2002**, *13*, 13.
38. Corpet, F., Gouzy, J.Kahn, D., *Nucleic Acids Res.* **1998**, *26*, 323-326.
39. Kruger, W.D.Cox, D.R., *Proc. Natl. Acad. Sci. U.S.A.* **1994**, *91*, 6614-6618.
40. Link, J.O.Straub, K., *J. Am. Chem. Soc.* **1996**, *118*, 2091-2092.
41. Nimmesgern, E., Fox, T., Fleming, M.A.Thomson, J.A., *J. Biol. Chem.* **1996**, *271*, 19421-19427.
42. Schalk-Hihi, C.Markham, G.D., *Biochemistry* **1999**, *38*, 2542-2550.
43. Patterson, J.L.Fernandez-Larsson, R., *Reviews of Infectious Diseases* **1990**, *12*, 1139-1146.
44. Gilbert, B.E.Knight, V., *Antimicrobial Agents & Chemotherapy* **1986**, *30*, 201-205.
45. Minakawa, N., Takeda, T., Sasaki, T., Matsuda, A.Ueda, T., *J. Med. Chem.* **1991**, *34*, 778-786.
46. Balzarini, J., Karlsson, A., Wang, L., Bohman, C., Horska, K., Votruba, I., Fridland, A., Van Aerschot, A., Herdewijn, P.De Clercq, E., *J. Biol. Chem.* **1993**, *268*, 24591-24598.
47. Koyama, H.Tsuji, M., *Biochem. Pharmacol.* **1983**, *32*, 3547-3553.
48. Mizuno, K., Tsujino, M., Takada, M., Hayashi, M.Atsumi, K., *Journal of Antibiotics* **1974**, *27*, 775-782.
49. Wang, W.Hedstrom, L., *Biochemistry* **1998**, *37*, 11949-11952.
50. Wedemeyer, H., Jackel, E., Wedemeyer, J., Frank, H., Schuler, A., Trautwein, C.Manns, M.P., *Zeitschrift fur Gastroenterologie* **1998**, *36*, 819-827.
51. Reichard, O., Schvarcz, R.Weiland, O., *Hepatology* **1997**, *26*, 108S-111S.
52. Anonymous, *Pediatrics* **1996**, *97*, 137-140.
53. Hager, P.W., Collart, F.R., Huberman, E.Mitchell, B.S., *Biochem. Pharmacol.* **1995**, *49*, 1323-1329.
54. Yamada, Y., Goto, H., Yoshino, M.Ogasawara, N., *Biochimica et Biophysica Acta* **1990**, *1051*, 209-214.
55. Goldstein, B.M., Leary, J.F.Farley, B.A., *Blood* **1991**, *78*, 593-598.
56. Natsumeda, Y., Yamada, Y., Yamaji, Y.Weber, G., *Biochem. Biophys. Res. Commun.* **1988**, *153*, 321-327.
57. Prajda, N., Hata, Y., Abonyi, M., Singhal, R.L.Weber, G., *Cancer Res.* **1993**, *53*, 5982-5986.
58. Tanabe, K., Takahashi, K., Kawaguchi, H., Ito, K., Yamazaki, Y.Toma, H., *Journal of Urology* **1998**, *160*, 1212-1215.
59. Suthanthiran, M.Strom, T.B., *Surgical Clinics of North America* **1998**, *78*, 77-94.
60. Kerr, K.M.Hedstrom, L., *Biochemistry* **1997**, *36*, 13365-13373.
61. Neyts, J., Meerbach, A., McKenna, P.De Clercq, E., *Antiviral Research* **1996**, *30*, 125-132.
62. Jashes, M., Gonzalez, M., Lopez-Lastra, M., De Clercq, E.Sandino, A., *Antiviral Research* **1996**, *29*, 309-312.
63. Mori, S., Watanabe, W.Shigeta, S., *Tohoku Journal of Experimental Medicine* **1995**, *177*, 315-325.
64. Wang, W., Papov, V.V., Minakawa, N., Matsuda, A., Biemann, K.Hedstrom, L., *Biochemistry* **1996**, *35*, 95-101.
65. Behrend, M., *Clinical Nephrology* **1996**, *45*, 336-341.
66. Shaw, L.M., Sollinger, H.W., Halloran, P., Morris, R.E., Yatscoff, R.W., Ransom, J., Tsina, I., Keown, P., Holt, D.W., Lieberman, R., Jaklitsch, A.Potter, J., *Ther. Drug Monit.* **1995**, *17*, 690-699.
67. Franklin, T.J., Jacobs, V., Bruneau, P.Ple, P., *Adv. Enzyme Regul.* **1995**, *35*, 91-100.
68. Franklin, T.J., Jacobs, V., Jones, G., Ple, P.Bruneau, P., *Cancer Res.* **1996**, *56*, 984-987.
69. Digits, J.A.Hedstrom, L., *Biochemistry.* **2000**, *39*, 1771-1777.
70. Krohn, K., Heins, H.Wielckens, K., *J. Med. Chem.* **1992**, *35*, 511-517.
71. Gharehbaghi, K., Paull, K.D., Kelley, J.A., Barchi, J.J., Jr., Marquez, V.E., Cooney, D.A., Monks, A., Scudiero, D., Krohn, K.Jayaram, H.N., *International Journal of Cancer* **1994**, *56*, 892-899.

168

72. Jayaram, H.N., O'Connor, A., Grant, M.R., Yang, H., Grieco, P.A.Cooney, D.A., *J Exp Ther Oncol* **1996**, *1*, 278-285.
73. Srivastava, P.C., Pickering, M.V., Allen, L.B., Streeter, D.G., Campbell, M.T., Witkowski, J.T., Sidwell, R.W.Robins, R.K., *J. Med. Chem.* **1977**, *20*, 256-262.
74. Wright, D.G., Boosalis, M.S., Waraska, K., Oshry, L.J., Weintraub, L.R.Vosburgh, E., *Anticancer Res.* **1996**, *16*, 3349-3351.
75. Tricot, G.Weber, G., *Anticancer Res.* **1996**, *16*, 3341-3347.
76. Cooney, D.A., Jayaram, H.N., Glazer, R.I., Kelley, J.A., Marquez, V.E., Gebeyehu, G., Van Cott, A.C., Zwelling, L.A.Johns, D.G., *Adv. Enzyme Regul.* **1983**, *21*, 271-303.
77. Jayaram, H.N., Zhen, W.Gharehbaghi, K., *Cancer Res.* **1993**, *53*, 2344-2348.
78. Jayaram, H.N., Pillwein, K., Lui, M.S., Faderan, M.A.Weber, G., *Biochem. Pharmacol.* **1986**, *35*, 587-593.
79. Goldstein, B.M.Colby, T.D., *Advan. Enzym. Regul.* **2000**, *40*, 405-426.
80. Li, H., Hallows, W.H., Punzi, J.S., Marquez, V.E., Carrell, H.L., Pankiewicz, K.W., Watanabe, K.A.Goldstein, B.M., *Biochemistry* **1994**, *33*, 23-32.
81. Khan, A.R., Parrish, J.C., Fraser, M.E., Smith, W.W., Bartlett, P.A.James, M.N., *Biochemistry* **1998**, *37*, 16839-16845.
82. Thevenot, T., Mathurin, P., Moussalli, J., Perrin, M., Plassart, F., Blot, C., Opolon, P.Poynard, T., *Journal of Viral Hepatitis* **1997**, *4*, 243-253.
83. Sachithanandan, S., Clarke, G., Crowe, J.Fielding, J.F., *Journal of Interferon & Cytokine Research* **1997**, *17*, 409-411.
84. Kochhar, D.M., Penner, J.D.Knudsen, T.B., *Toxicology & Applied Pharmacology* **1980**, *52*, 99-112.
85. Shults, R.A., Baron, S., Decker, J., Deitchman, S.D.Connor, J.D., *Journal of Occupational & Environmental Medicine* **1996**, *38*, 257-263.
86. Natsumeda, Y.Carr, S.F., *Annals of the New York Academy of Sciences* **1993**, *696*, 88-93.
87. Armistead, D.M., et al., *Inhibitors of IMPDH Enzyme.* 1998, Vertex Pharmaceuticals Inc (Cambridge, Mass): United States Patent.
88. Morgans, J., et al., *5-substituted derivatives of mycophenolic acid.* 1997, Syntex (U.S.A.) Inc. (Palo Alto, CA): U.S.A.
89. Patterson, J.W., et al., *4-amino 6-substituted mycophenolic acid and derivatives*, in *Unites States Patent.* 1996, Syntex (U.S.A.) Inc. (Palo Alto, CA): Unites States.
90. Patterson, J.W., et al., *Method of using 4-amino 6-substituted mycophenolic acid and derivatives.* 1996, Syntex (U.S.A.) Inc. (Palo Alto, CA).
91. Patterson, J.W., et al., *6-substituted mycophenolic acid and derivatives*, in *United States Patent.* 1996, Syntex (U.S.A.) Inc. (Palo Alto, CA): Unites States.
92. Patterson, J.W., et al., *Method of using 4-amino mycophenolic acid and derivatives.* 1996, Syntex (U.S.A.) Inc. (Palo Alto, CA).
93. Lesiak, K., Watanabe, K.A., Majumdar, A., Powell, J., Seidman, M., Vanderveen, K., Goldstein, B.M.Pankiewicz, K.W., *J Med Chem* **1998**, *41*, 618-622.
94. Pankiewicz, K.W., Lesiak-Watanabe, K.B., Watanabe, K.A., Patterson, S.E., Jayaram, H.N., Yalowitz, J.A., Miller, M.D., Seidman, M., Majumdar, A., Prehna, G.Goldstein, B.M., *Journal of Medicinal Chemistry.* **2002**, *45*, 703-712.
95. Barnes, B.J., Eakin, A.E., Izydore, R.A.Hall, I.H., *Biochemistry.* **2000**, *39*, 13641-13650.
96. Barnes, B.J., Eakin, A.E., Izydore, R.A.Hall, I.H., *Biochemical Pharmacology.* **2001**, *62*, 91-100.
97. Altschul, S.F., Madden, T.L., Schaffer, A.A., Zhang, J., Zhang, Z., Miller, W.Lipman, D.J., *Nucleic Acids Res* **1997**, *25*, 3389-3402.

Acknowledgment: This work was supported in part by National Institutes of Health Grant GM62785.

Chapter 8

Monovalent Cation Activation of IMP Dehydrogenase

George D. Markham

Institute for Cancer Research, Fox Chase Cancer Society,
7701 Burholme Avenue, Philadelphia, PA 19111

IMP dehydrogenases are among the group of enzymes that require specific monovalent cations for maximal catalytic activity. These cations function in catalysis rather than in maintenance of global protein structure. In the IMP dehydrogenase catalyzed reaction, the ions specifically facilitate hydride transfer from the covalent enzyme-IMP intermediate to NAD, rather than either formation of enzyme-IMP or hydrolysis of the oxidized enzyme-XMP* intermediate. Possible mechanisms by which monovalent cations may fulfill this role are discussed.

The activation of IMP dehydrogenases by certain monovalent cations such as K^+ has been recognized since the pioneering work of Magasanik *(1)*. In concert with most monovalent cation (M^+) activated enzymes, elucidation of the specific roles of the cation in catalysis has been difficult. This article endeavors to summarize the state of knowledge of monovalent cation interactions with IMPDH from various organisms, and to place this information in the context of other well characterized monovalent cation activated enzymes.

The IMPDH-catalyzed reaction is shown in Figure 1. Also shown are the reactions in which a 2-halo-IMP is hydrolyzed to XMP and halide (Cl^- or F^-) in

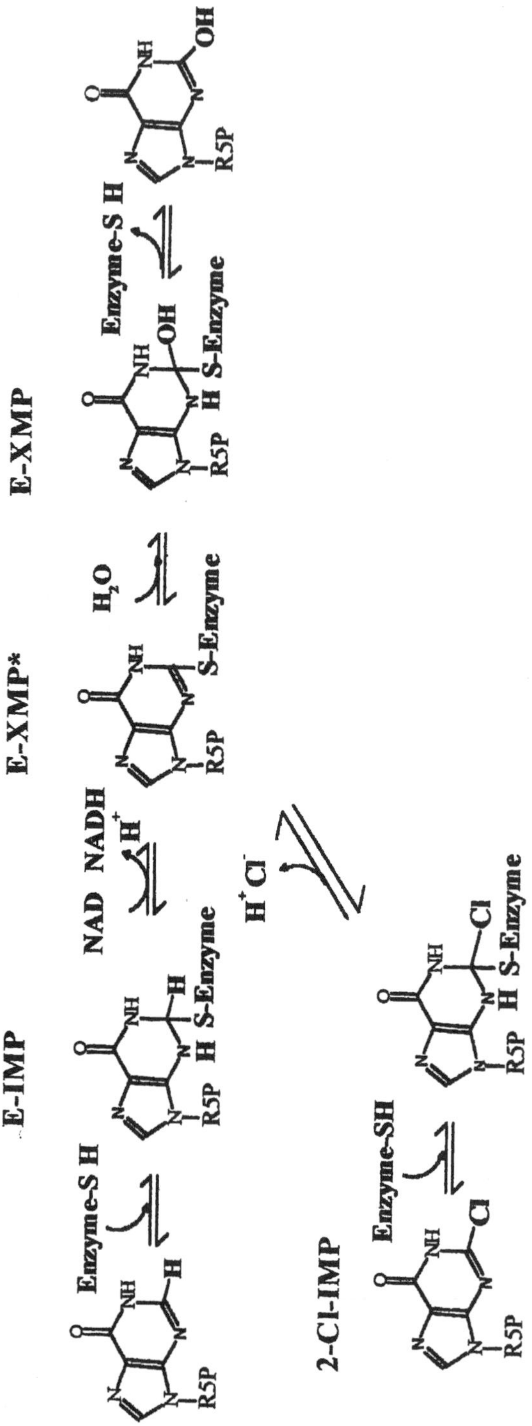
E-IMP
E-XMP*
E-XMP
2-Cl-IMP
Enzyme-S H
NAD NADH H+
H2O
H+Cl-
Enzyme-SH
S-Enzyme
R5P

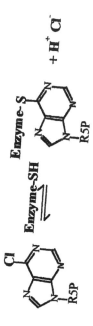

Figure 1. Reactions catalyzed by IMP dehydrogenase. Top: The IMPDH reaction. Middle: The 2-Cl-IMP hydrolytic reaction; the confluence of this reaction pathway and that for the IMPDH reaction is shown. Bottom: The reaction leading to inactivation by 6-Cl-IMP. The ribose 5'-monophosphate of the substrate is abbreviated R5P.

the absence of NAD *(2)*, and the inactivation of the enzyme by 6-Cl-IMP which irreversibly modifies the sulfhydryl group of the active site cysteine *(3)*. Other nucleotides that act as suicide inactivators by modifying the active site cysteine are known *(4)*.

Monovalent Cation Activation of Enzymes – General Considerations

Review articles by Suelter on the properties of enzymes activated by monovalent cations *(5,6)*, and more recently by Woehl and Dunn on monovalent cation participation in pyridoxal phosphate dependent enzymatic reactions *(7)* provide overviews of the topic. The general properties of metal ion coordination chemistry in the context of proteins have been discussed by Glusker *(8)*. A difficulty in assessing the literature on monovalent cation activation of any enzyme is that potentially interacting ions are commonly present as counter ions to the substrates and as buffer components. Thus, literature data can only be confidently amalgamated into an overview when the authors have noted that specific precautions were to control the ionic milieu.

The structural characterization of monovalent cation interactions with enzymes has been plagued by the relatively weak nature of the interactions and the uninformative spectroscopic properties of alkali metal ions. Furthermore, the variety of coordination numbers readily adopted by these ions, and their tendency toward binding to carbonyl groups of the polypeptide main chain, hinders deduction of the ligation scheme. Table I lists the coordination preferences for monovalent cations as established from high-resolution small molecule crystal structures, and the values reported for protein complexes. The deviation between the average coordination numbers and the number of ligands reported in protein structures with better than 2.5-Å resolution illustrates the limitations of protein crystallography in characterization of these cation-protein interactions.

The typically modest affinity of proteins for M^+ hinders determination the crucial parameter of binding stoichiometry. With a typical dissociation constant in the millimolar range, protein concentrations on the order of 1 mM binding sites would be required for accurate determination of the binding stoichiometry; unfortunately this solubility is not readily achieved with proteins such as IMPDH. The Tl^+ ion, which has a similar ionic radius to K^+, binds to some proteins with substantially higher affinity than other M^+ and has allowed determination of binding stoichiometry using the ^{204}Tl radioisotope *(cf. 9)*. Kinetic data for Tl^+ activation of IMPDH suggest that its affinity is not high enough for measurement of the binding stoichiometry.

Table I

Coordination Properties of Monovalent Cations

	Li^+	Na^+	K^+	Rb^+	Cs^+	Tl^+
Ionic Radius (Å)	0.60	0.95	1.33	1.48	1.69	1.49
Average Coordination Number[a]	4.9	6.4	7.9	8.0	8.8	6.9

Distribution of Protein Structures with Ions of Each Coordination Number

Coordination #			*Number of Structures*			
1	--	36	41	--	3	--
2	--	60	76	2	5	--
3	--	34	55	--	2	--
4	--	80	30	2	--	--
5	--	100	9	--	--	2
6	1	51	6	--	--	--
7	--	6	3	--	--	--
≥ 8	--	--	--	--	--	--

[a] In small molecule structures with oxygen or sulfur ligands; *(8)*.

[b] Structures at 2.5 Å or better resolution found using the Metalloprotein Database Web Site: http://www.scripps.edu/research/metallo/

When assessing methods for studying the structures of the cation binding sites in solution, the situation appears similarly grim. None of the monovalent cations that typically activate enzymes has distinctive optical spectroscopic properties, or is paramagnetic. For the alkali metal ions, which do generally have isotopes suitable for NMR studies, the NMR chemical shift ranges are too small to be of diagnostic value for determination of either coordination number or ligand type *(10)*. Furthermore, the quadrupole moments of most of these nuclei result in large line widths with attendant loss of both chemical shift discrimination and sensitivity. In a few cases NMR studies of such cations have provided information on the dynamics of interactions with proteins *(11)*. The spin $1/2$ $^{203/205}Tl$ nuclei do have chemical shift ranges large enough to provide a diagnostic parameter for both coordination number and ligand type *(12)*. However while the $^{203/205}Tl$ nuclei are potentially very sensitive NMR probes, their magnetogyric ratios (which are nearly equivalent) are the only ones between ^{31}P and ^{19}F; therefore their resonant frequencies are often inaccessible on commercial NMR probes. Nevertheless a few NMR studies of Tl^+ complexes with proteins have been reported *(13-15)*.

Thus, in most protein systems the deduction of M^+ binding sites and the structural basis for activation has relied on crystallographic data. The presence of large amounts of monovalent cations, particularly NH_4^+, in crystallization mother liquors requires that the functional significance of any deduced cation site must be taken with reservation until the site is tested by independent means. A mutagenesis approach is often not applicable since monovalent cation binding sites often contain immutable oxygen ligands the protein backbone. However in the case of pyruvate kinase where one K^+ ligand was identified as a glutamate side chain, the functional significance of site was verified by construction of a mutant that was active in the absence of a monovalent cation as a result of replacement of the glutamate by a lysine *(16,17)*. Spectroscopic data had demonstrated that the M^+ activator bound at the active site before the crystal structure was obtained, which provided additional confidence in the crystallographic assignment.

Thus, there are no generally applicable methods for characterization of interactions of monovalent cations with proteins in the absence of high-resolution crystallographic data. Thus far, IMPDH has been a difficult case.

IMP Dehydrogenase

Explicit tests of monovalent cation activation have been reported for IMPDH from the bacteria: *Aerobacter aerogenes, Bacillus subtillis, Borrelia burgdorferi, Escherichia coli, Streptococcus pyogenes*; the protozoa *Erwinia tenella,* and *Tritrichomonas foetus*; and the eukaryotic enzymes from mouse sarcoma 180 and the human type-2 isozyme *(18-26)*. The results indicate that different IMPDHs have distinctive tolerances for various cations, as listed in Table II.

The M^+ free activities of the enzymes vary considerably. The human type-2 enzyme has <1% of the maximal activity in the absence of added M^+ whereas the basal activity of the *E. coli* enzyme is ~5% of its maximal value. Although studies of IMPDH isolated from *T. foetus* did not note M^+ dependence, recent studies of the recombinant *T. foetus* IMPDH demonstrate substantial K^+ activation (L. Hedstrom, personal communication). The sequences and three dimensional structures are highly similar among the hamster, human, *S. pyogenes* and *B. burgdoferi* IMPDHs *(21,27-30)*. It is remarkable that the M^+ activation is observed across the majority of evolutionary forms of IMPDH. It would be instructive to determine whether monovalent cations activate the IMPDH from an archaeon such as the hyperthermophile *Pyrococcus furiosus*. The *P. furiosus* IMPDH is clearly related to IMPDH from other kingdoms, with 50% and 36% sequence identity to (72% and 61% similarity) to the *B. subtilis* and human enzymes despite evolutionary distance and dramatically different physiological constraints *(31)*. The phylogenetic relationships deduced by Collart et al. show the wide separation between the *T. foetus* IMPDH and that from other organisms *(31)*.

Table II.
Activation of IMP Dehydrogenases by Monovalent Cations

	Li$^+$	Na$^+$	K$^+$	NH$_4^+$	Rb$^+$	Cs$^+$	Tl$^+$	None
			*Relative Activity (% of maximal)*					
	Ionic Radius (Å)							
	0.60	0.95	1.33	1.48	1.48	1.69	1.40	
Organism								
A. aerogenes[a]	I	--	100	100	--	--	--	--
B. subtilis[b]	18	20	100	67	45	38	180	--
B. burgdorferi[c]	<0.5	<1	100	90	--	10	--	<0.5
E. coli[d]	I	I	100	61	82	13	--	4.5
S. pyogenes[e]	--	<5	100	98	94	--	--	<5
Human[g]	x	136	100	62	58	5	37	<1

-- Not reported
I = inhibitor
x = no effect at tested levels
[a](18); [b](20); [c](23); [d](22); [e](21); [f]Type-2 isozyme (26).

Table III[a]
Selectivity of IMPDH for Monovalent Cations

	Li$^+$	Na$^+$	K$^+$	NH$_4^+$	Rb$^+$	Cs$^+$	Tl$^+$
	Activator (Inhibition) Constant (mM)						
Critter							
A. aerogenes[a]	(33)	--	25	14	--	--	--
B. subtilis[b]	450	100	6.3	7.4	9.0	11	5.4
B. burgdorferi[c]	(51)	(61)	25	24	--	31	--
E. coli[d]	(55)	(76)	2.8	5.2	4.7	12	--
S. pyogenes[e]	--	--	--	--	--	--	--
Human[g]	x	42	3.3	9.7	11	--	0.8

[a] Designations and references as listed for Table II.
x = no effect at tested levels.

Selectivity for Cations

The kinetically determined activator constants (i.e., apparent dissociation constants) or inhibition constants for different M$^+$ are listed in Table III. Unfortunately the avidity of IMPDHs for M$^+$ is in the millimolar range and has not allowed direct determination of the number of cation binding sites. The significance of this limitation has become evident in light of the recent results with the *E. coli* IMPDH that support multiple kinetically significant interactions *(22)*.

In common with many other M^+ activated enzymes, the activation site(s) is selective for cationic radius, with both optimal activity and highest affinity being observed for K^+ and ions of similar size. The ability of NH_4^+ to replace K^+ despite their different binding modes, 4 hydrogen bonds from NH_4^+ rather than 6 to 8 typically oxygen ligands to K^+, suggests that the ion activates indirectly via a conformational modification rather than by direct involvement in the reaction. However a caveat is the plant (cowpea) IMPDH that was reported to be activated by K^+ but not NH_4^+ *(32)*. The ability of Mg^{2+} and Ca^{2+} to competitively inhibit K^+ activation of the *E. coli* enzyme is unusual for a monovalent cation binding site. Furthermore the discovery of mutants of the *E. coli* enzyme that are activated by divalent cations may allow use of a spectroscopically informative divalent cation to provide novel insight into the role of the metal ion *(22)*.

Structural Effects of M^+ Binding

Monovalent cation activation is not related to requirement for maintenance of the active tetrameric structure of human type-2 IMPDH, the *E. coli, T. foetus,* and *B. burgdoferi* enzymes *(22-24,26)*. For example, the human type-2 IMPDH remains tetrameric over a wide pH range in the presence of the non-interacting $(CH_3)_4N^+$ ion *(33)*.

The crystal structure of a hamster IMPDH in a covalent complex with oxidized IMP (XMP*) and the inhibitor mycophenolic acid (MPA) and K^+ (e.g., IMPDH-XMP*•MPA•K^+) was described at 2.6 Å resolution by Sintchak et al. *(28)*. This structure is thus far unique in reporting the position of an apparent K^+ ion (Plate 1).

The density ascribed to K^+ was described as surrounded at a distance of ~3.1 Å by five main chain carbonyl groups and a water molecule. Octahedral coordination is common for K^+ ions *(8)*. The only carbonyl group identified by the authors was that of the nucleophilic Cysteine-331 at a distance of 2.9 Å. The coordinates of this structure have not been deposited in a public data bank; coordinates for the protein were obtained from the authors and the positions of the putative K^+ ion and the waters were taken from a patent application *(34)*. Inspection of these coordinates indicates that other carbonyl ligands are donated by Glycine-328 (2.8 Å) or Glycine-326 (3.0 Å), Glutamate-500* (2.9 Å) Glycine-501* (3.1 Å), and Glycine-502* (2.7 Å) where residues with a * arise from a second subunit. However the nearest water appears at 3.5 Å from the K^+ and the carbonyls of six residues are closer, suggesting that the site might be considered 7 coordinate which is also common for K^+ *(8)*. The K^+ site is ~6 Å from the C2 of XMP* and $\geq$8 Å from MPA which binds at the nicotinamide region of the NAD site. Sinchak et al. propose that the hydroxyl

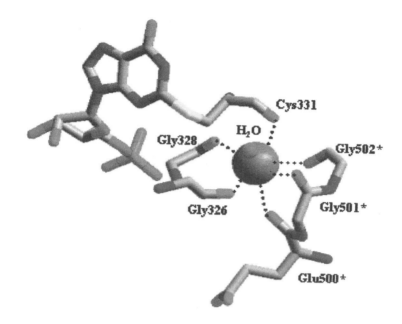

Plate 1. K⁺ and Environs in Hamster IMPDH (28). The large sphere in the center is K⁺ and the smaller foreground sphere is a water.

(This figure is also in color insert.)

of mycophenolic acid occupies the site of the nucleophilic water. Thus the structure implies that direct $M^+ \cdot OH_2$ coordination is not involved in the reaction, in accord with the M^+ independent rate of hydrolysis of 2-Cl-IMP (see below). Solution studies of conformational alterations on substrate binding to the hamster enzyme did not address the role of the M^+ *(35-37)*.

The 2.9Å resolution structure of the human type-2 IMPDH modified with the affinity label 6-Cl-IMP and complexed with the NAD analog inhibitor selenazofurin-adenine-dinucleotide (SAD) did not reveal the location of a M^+ *(29)*. Kerr et al. noted that conformational differences of the active site flap would not allow K^+ to bind in the human enzyme structure in the same place as observed in the hamster structure; they suggest that the NAD binding is not linked to the K^+ site seen in the hamster structure but rather to another M^+ site *(22)*.

A M^+ ion was not identified in the 2.4 Å crystal structure of the *B. burgdorferi* enzyme•sulfate complex *(30)*. The M^+ specificity of *B. burgdorferi* enzyme is similar to that of *E. coli* enzyme but different from the human enzyme *(23)*.

The crystal structure of the *Streptococcus pyogenes* enzyme was determined at 1.9 Å resolution in a non-covalent complex with IMP *(21)*. The enzyme was shown to have a requirement for monovalent cations, however even at this resolution a bound M^+ was not reported.

The structure of the *T. foetus* IMPDH was reported for the ligand-free enzyme at 2.3 Å resolution and the non-covalent E•XMP complex at 2.6 Å *(27)*. In both cases the structure was disordered near the probable NAD binding site. The active site arginine (Arg-382) uniquely present in the *T. foetus* enzyme interacts with the phosphate group of the substrate and appears unlikely to function as an aid to the M^+. The position corresponding to Arginine-382 has asparagine, glycine or serine in other isoforms.

In sum, the available crystallographic data can not be considered to conclusively reveal the M^+ activator binding site, much less how the cation functions.

Kinetic Studies

Bacterial IMPDH

The pioneering kinetic studies of Morrison and colleagues first demonstrated that K^+ activation of *Aerobacter aerogenes* (now *Klebsiella sp.*) was related to the NAD site *(18,19)*. The combination of steady state kinetic studies and isotope exchange at equilibrium results showed that IMP and K^+ bind randomly while NAD binds to the E•IMP or E•IMP•K^+ complex *(18,19)*. Binding of IMP to the enzyme was confirmed and K^+ did not significantly alter the affinity. The studies also showed that at a saturating K^+ concentration the mechanism has random addition of IMP and NAD to the E•K^+ complex. This arises because at the ~0.1 M concentration of K^+ need for saturation the (second order) binding rate is much faster than the binding rates for IMP or NAD at their typical millimolar concentrations. The random addition of IMP and NAD is consistent with that later deduced for the human and other IMPDH. Heyde et al. also concluded that there was a significant fraction of the enzyme in an E•K^+•XMP complex during steady state turnover, which was later observed for IMPDH from other organisms.

The kinetic data for the *A. aerogenes* enzyme were consistent with compulsory dissociation of M^+ during each turnover *(18,19)*. The observation that high concentrations of M^+ inhibit many IMPDHs, and at least for the human type-2 IMPDH ions that activate at lower concentrations also inhibit at lower concentrations, may also reflect that the M^+ must dissociate to regenerate the free enzyme in each catalytic cycle. This possibility has not been definitively resolved.

Recent studies with the *E. coli* enzyme demonstrated differences in cation selectivity from the human enzyme *(22)*. Inhibition by Li^+, Na^+, Mg^{2+} and Ca^{2+} was competitive with both K^+ and NAD, supporting interaction between the M^+

site and NAD site. However, the concentration dependence of the inhibition was non-linear leading to the proposal that multiple binding sites might be involved. Several mutants that affected M^+ activation were constructed in a systematic mutagenesis study of conserved acidic residues *(38)*. In one interesting mutant (Aspartate-50→Alanine), Mg^{2+} inhibition became non-completive with respect to K^+ and competitive with both IMP and NAD; this mutant was inactive in the absence of K^+. This change in inhibition pattern might be due to conformational alterations in the mutants or alternatively to the presence of a second cation binding site. A second site was proposed by Kerr et al. since the analogous hamster residue (Aspartate-71) lies > 15Å from the K^+ in the crystal structure *(22)*. The other mutants studied, Aspartate-13→Alanine and Glutamate-469→Alanine, were activated by all of the cations tested, while the K_m for K^+ is increased 38 and 15-fold, respectively. The K_m for Mg^{2+} of ~20 mM is comparable for the Aspartate-13→Alanine and Glutamate-469→Alanine mutants, and is 4-5 fold larger than the K_i for inhibition of the wild type enzyme. Interestingly M^+ activation of the Glutamate-469→Alanine mutant was cooperative which again suggests more than one interaction site.

Human IMPDH

Studies of human IMPDH have emphasized the type-2 isozyme that is a target for anticancer and immunosuppressive drugs *(39)*. Characterization of M^+ activation of the type-1 enzyme has not been reported. Equilibrium binding studies show that IMP and XMP bind to the human type-2 enzyme in the absence of an M^+ activator, and that the affinity for IMP is enhanced only ~2-fold by K^+ *(26)*. Kinetic isotope effect studies with 2-^{2}H-IMP showed that there was no ^{2}H effect on V_{max} in the IMPDH reaction with Na^+, K^+, NH_4^+ and Rb^+ as activators, indicating that hydride transfer is not rate limiting in turnover *(40,41)*. However, the presence of substantial ^{2}H effects on V_{max}/K_m for both IMP and NAD with each ion demonstrated that both substrates can dissociate from the E•IMP•NAD complex, i.e. binding is random. In the presence of each ion there is a smaller kinetic isotope effect on the V_{max}/K_m for IMP, indicating that IMP dissociates more slowly than NAD, although the magnitudes of the effects suggest that both substrates dissociate at rates comparable to the forward reaction rate. The comparable isotope effects with the various cations (1.9 to 2.4 fold on V_{max}/K_m of IMP, 2.7 to 3.5 fold for V_{max}/K_m of NAD) suggest that the cations do not directly participate in the hydride transfer step of the reaction. However until pre-steady state kinetics are compared with the various M^+ to ensure that the same steps in the mechanism are being altered in the reactions with the various ions, the isotope effects provide little insight into the mechanism of M^+ activation. Wang and Hedstrom demonstrated exchange reactions between NADH and thio-NAD in the presence of IMP, showing that hydride transfer was reversible *(40)*. Despite random substrate binding, product release is ordered due to the slow hydrolysis of E-XMP*, a presumably M^+ independent step *(40)*.

Important insights have been obtained from the reaction of 2-chloro-IMP which occurs in the absence of NAD *(2)*. This reaction provides a mechanism to study formation and hydrolysis of E-XMP* in the absence of hydride transfer reaction. The 2-Cl-IMP hydrolysis reaction catalyzed by human type-2 IMPDH was not activated by M^+ *(33)*. Inactivation of both the human and *E. coli* enzyme by 6-Cl-IMP was also independent of M^+ *(22,33)*. The cation-free reactions with 2-Cl-IMP and 6-Cl-IMP demonstrate formation of catalytically active M^+ independent complexes with the human type-2 and *E. coli* enzymes *(22,33)*.

Solvent ^2H_2O isotope effects were not observed on V_{max} for either the IMPDH reaction in the presence of K^+ or 2-Cl-IMP hydrolysis *(33,41)*. ^2H_2O did cause V_{max}/K_m reductions for both IMP (1.4-fold) and NAD (2.0-fold) indicating a kinetically significant proton transfer early in the IMPDH reaction. However, ^2H_2O did not alter the V_{max}/K_m for 2-Cl-IMP indicating that proton transfers are relatively fast in this reaction.

pH-rate profiles showed comparable pK values for the group that must be deprotonated for the IMPDH reaction and 6-Cl-IMP inactivation (8.1 in the free enzyme, 7.5 in the E•6-Cl-IMP complex) , suggesting that this pK reflects the active site cysteine-SH *(33)*. The pK of this group is in a typical range for cysteine thiols in peptides *(42)*. The K^+ independence of the pK indicates that the M^+ does not dramatically alter the acidity of the –SH which is surprising given the proximity of the K^+ in the hamster IMPDH structure. The comparable pK values for the group that must be protonated for both the IMPDH reaction and the 2-Cl-IMP hydrolysis (9.4 to 9.9) suggest that the M^+ does not determine the pK of this group, which is unidentified but may be the purine ring. There are therefore no indications that the M^+ has any role in the acid-base chemistry of IMPDH-catalyzed reactions.

M^+ and Mycophenolic Acid Inhibition

Mycophenolic acid was found to be an uncompetitive inhibitor with respect to K^+ as well as IMP and NAD with *Eimeria tenella* IMPDH *(25)*, indicating that MPA and K^+ bind to different enzyme forms. Since MPA binds with high affinity to E-XMP*, this result suggests that that K^+ binding to E-XMP* is not required for tight MPA binding. It is surprising that detailed studies of the effects of monovalent cations on the affinity of the potent inhibitor mycophenolic acid have not been reported.

The extensive kinetic studies of the *T. foetus* enzyme demonstrated random substrate addition, which was confirmed by equilibrium binding studies *(43)*. Mutagenesis studies showed than only half of the 400-fold lower affinity of MPA for the *T. foetus* enzyme that the humans enzyme was attributable to differences in amino acids located in the active site *(44)*. The residual difference was attributed to differences in interactions between the adenosine and nicotinamide portions of the NAD binding site *(45)*. The M^+ ion binding site in the hamster structure is ≥ 8 Å from MPA and may not make a substantial contribution to the difference in affinity among IMPDH isoforms.

Conclusions

In common with other M^+ activated enzymes, elucidation of the particular catalytic roles of the ions in IMPDH catalysis has been difficult. While the kinetic data clearly connect M^+ activation to the NAD site, structural data which explain this relationship are not yet available. The M^+ accelerates the hydride transfer step of the reaction with apparently minor, if any, roles in binding of the substrate, determining the nucleophilicity of the reactive cysteine sulfhydryl or in the hydrolysis of the covalent E-XMP* intermediate. It thus appears that a role of the M^+ in activating the nucleophilic water has been excluded. It is unclear whether the M^+ activated IMPDH isozymes use the M^+ to orient NAD in a particularly reactive conformation. Nevertheless, an elusive M^+ induced structural change involved in positioning NAD appears more probable than is direct interaction of the substrate with either the M^+ or one of its first coordination sphere ligands. Similar M^+ activation by adjustment of protein conformation has been described for pyruvate kinase, S-adenosylmethionine synthetase, dialkylglycine decarboxylase, tryptophan synthase and tyrosine phenol lyase *(16,46-48)*. The additional possibility of a through-space electrostatic enhancement of the electrophilicity of NAD by a properly positioned M^+ remains to be explored but M^+ binding at the Sintchak site is unlikely to provide this effect because it is located on the opposite side of the hypoxanthine ring.

Acknowledgements
This work was supported by National Institutes of Health Grants GM31186, CA06927, and also supported by an appropriation from the Commonwealth of Pennsylvania.

References

1. Magasanik, B.; Moyed, M. S.; Gehrig, L. B. *J. Biol. Chem.* **1957**, *226*, 229-350.

2. Antonino, L. C.; Wu, J. C. *Biochemistry* **1994**, *33*, 1753-1759.
3. Antonino, L. C.; Straub, K.; Wu, J. C. *Biochemistry* **1994**, *33*, 1760-1765.
4. Hedstrom, L. *Curr. Med. Chem.* **1999**, *6*, 545-560.
5. Suelter, C. H. *Science* **1970**, *168*, 789-795.
6. Suelter, C. H. In *Metal Ions in Biological Systems*; Segel, H., Ed.; Marcel Dekker, NY, 1974; Vol. 3, pp 201-251.
7. Woehl, E. U.; Dunn, M. F. *Coord. Chem. Rev.* **1995**, *144*, 147-197.
8. Glusker, J. P. *Adv. Protein Chem.* **1991**, *42*, 1-76.
9. Kayne, F. J. *Arch. Biochem. Biophys.* **1971**, *143*, 232-239.
10. Drakenberg, T., Johansson, C., Forsen, S. *Meth. Mol. Biol.* **1997**, *60*, 299-323.
11. Aramini, J. M.; Vogel, H. J. *Biochem. Cell Biol.* **1998**, *76*, 210-222.
12. Hinton, J. F., Metz, K.R., Briggs, R.W. *Progress Nucl. Magn. Res. Spectrosc.* **1988**, *20*, 423-513.
13. Reuben, J.; Kayne, F. J. *J. Biol. Chem.* **1971**, *246*, 6227-6234.
14. Markham, G. D. *J. Biol. Chem.* **1986**, *261*, 1507-1509.
15. Loria, J. P.; Nowak, T. *Biochemistry* **1998**, *37*, 6967-6974.
16. Larsen, T. M.; Laughlin, L. T.; Holden, H. M.; Rayment, I.; Reed, G. H. *Biochemistry* **1994**, *33*, 6301-6309.
17. Laughlin, L. T.; Reed, G. H. *Arch. Biochem. Biophys.* **1997**, *348*, 262-267.
18. Heyde, E.; Morrison, J. F. *Biochim. Biophys. Acta* **1976**, *429*, 661-671.
19. Heyde, E.; Nagabhushanam, A.; Vonarx, M.; Morrison, J. F. *Biochim. Biophys. Acta* **1976**, *429*, 645-660.
20. Wu, T. W.; Scrimgeour, K. G. *Can. J. Biochem.* **1973**, *51*, 1391-1398.
21. Zhang, R.; Evans, G.; Rotella, F. J.; Westbrook, E. M.; Beno, D.; Huberman, E.; Joachimiak, A.; Collart, F. R. *Biochemistry* **1999**, *38*, 4691-4700.
22. Kerr, K. M.; Cahoon, M.; Bosco, D. A.; Hedstrom, L. *Arch. Biochem. Biophys.* **2000**, *375*, 131-137.
23. Zhou, X.; Cahoon, M.; Rosa, P.; Hedstrom, L. *J. Biol. Chem.* **1997**, *272*, 21977-21981.
24. Verham, R.; Meek, T. D.; Hedstrom, L.; Wang, C. C. *Mol. Biochem. Parasitol.* **1987**, *24*, 1-12.
25. Hupe, D. J.; Azzolina, B. A.; Behrens, N. D. *J. Biol. Chem.* **1986**, *261*, 8363-8369.
26. Xiang, B.; Taylor, J. C.; Markham, G. D. *J. Biol. Chem.* **1996**, *271*, 1435-1440.
27. Whitby, F. G.; Luecke, H.; Kuhn, P.; Somoza, J. R.; Huete-Perez, J. A.; Phillips, J. D.; Hill, C. P.; Fletterick, R. J.; Wang, C. C. *Biochemistry* **1997**, *36*, 10666-10674.

28. Sintchak, M. D.; Fleming, M. A.; Futer, O.; Raybuck, S. A.; Chambers, S. P.; Caron, P. R.; Murcko, M. A.; Wilson, K. P. *Cell* **1996**, *85*, 921-930.
29. Colby, T. D.; Vanderveen, K.; Strickler, M. D.; Markham, G. D.; Goldstein, B. M. *Proc. Natl. Acad. Sci. U.S.A.* **1999**, *96*, 3531-3536.
30. McMillan, F. M.; Cahoon, M.; White, A.; Hedstrom, L.; Petsko, G. A.; Ringe, D. *Biochemistry* **2000**, *39*, 4533-4542.
31. Collart, F. R.; Osipiuk, J.; Trent, J.; Olsen, G. J.; Huberman, E. *Gene* **1996**, *174*, 209-216.
32. Atkins, C. A.; Shelp, B. J.; Storer, P. J. *Arch. Biochem. Biophys.* **1985**, *236*, 807-814.
33. Markham, G. D.; Bock, C. L.; Schalk-Hihi, C. *Biochemistry* **1999**, *38*, 4433-4440.
34. Vertex Pharmaceuticals Incorporated **1997**, World Patent Application WO97/41211.
35. Bruzzese, F. J.; Connelly, P. R. *Biochemistry* **1997**, *36*, 10428-10438.
36. Nimmesgern, E.; Fox, T.; Fleming, M. A.; Thomson, J. A. *J. Biol. Chem.* **1996**, *271*, 19421-19427.
37. Nimmesgern, E.; Black, J.; Futer, O.; Fulghum, J. R.; Chambers, S. P.; Brummel, C. L.; Raybuck, S. A.; Sintchak, M. D. *Protein Expr. Purif.* **1999**, *17*, 282-289.
38. Kerr, K. M.; Digits, J. A.; Kuperwasser, N.; Hedstrom, L. *Biochemistry* **2000**, *39*, 9804-9810.
39. Goldstein, B. M.; Colby, T. D. *Curr. Med. Chem.* **1999**, *6*, 519-536.
40. Wang, W.; Hedstrom, L. *Biochemistry* **1997**, *36*, 8479-8483.
41. Xiang, B.; Markham, G. D. *Arch. Biochem. Biophys.* **1997**, *348*, 378-382.
42. Peters, G. H.; Frimurer, T. M.; Olsen, O. H. *Biochemistry* **1998**, *37*, 5383-5393.
43. Digits, J. A.; Hedstrom, L. *Biochemistry* **1999**, *38*, 2295-2306.
44. Digits, J. A.; Hedstrom, L. *Biochemistry* **1999**, *38*, 15388-15397.
45. Digits, J. A.; Hedstrom, L. *Biochemistry* **2000**, *39*, 1771-1777.
46. Antson, A. A.; Demidkina, T. V.; Gollnick, P.; Dauter, Z.; von Tersch, R. L.; Long, J.; Berezhnoy, S. N.; Phillips, R. S.; Harutyunyan, E. H.; Wilson, K. S. *Biochemistry* **1993**, *32*, 4195-4206.
47. Toney, M. D.; Hohenester, E.; Keller, J. W.; Jansonius, J. N. *J. Mol. Biol.* **1995**, *245*, 151-179.
48. Rhee, S.; Parris, K. D.; Ahmed, S. A.; Miles, E. W.; Davies, D. R. *Biochemistry* **1996**, *35*, 4211-4221.

Chapter 9

IMP Dehydrogenase: Mechanism of Drug Selectivity

Lizbeth Hedstrom[1] and Jennifer A. Digits[1-3]

[1]Department of Biochemistry, MS 009, Brandeis University,
Waltham, MA 02454
[2]Current address: Department of Pharmacology and Cancer Biology, Duke
University Medical Center, Levine Science and Research Center, Box 3686,
Durham, NC 27710
[3]Supported by National Institutes of Health Molecular Structure and
Function Training Grant GM07956 (J.A.D.), NIH GM54403 (L.H.) and a
grant from the Markey Charitable Trust to Brandeis University

IMP dehydrogenase catalyzes the conversion of IMP to XMP
with the reduction of NAD to NADH. IMP dehydrogenase is
a potential antimicrobial target. Mycophenolic acid (MPA)
inhibits mammalian IMP dehydrogenases but not microbial
enzymes. This inhibitor specificity derives in part from the
residues that contact MPA. The remainder of drug selectivity
derives from coupling between the MPA site and the
adenosine subsite of the dinucleotide binding site. This
coupling appears to be mediated by a flexible flap that is
believed to cover the active site.

Introduction

Mycophenolic Acid

Isolated in the 1890's, mycophenolic acid (MPA) can be considered the first antibiotic (*1*). MPA's fortunes have waxed and waned over the last century, culminating recently in the use of the MPA derivative mycophenolate mofetil in immunosuppressive therapy. MPA inhibits guanine nucleotide metabolism, targeting IMP dehydrogenase. This enzyme catalyzes the conversion of IMP to XMP with the concomitant reduction of NAD. MPA is a rare uncompetitive inhibitor with respect to both substrates. It sequesters a covalent enzyme-IMP intermediate (termed E-XMP*, see below). MPA is a potent inhibitor of mammalian IMP dehydrogenases, but a poor inhibitor of the enzyme from microbial sources. Understanding the source of this selectivity is important for the design of IMP dehydrogenase inhibitors which could serve as antimicrobial therapy.

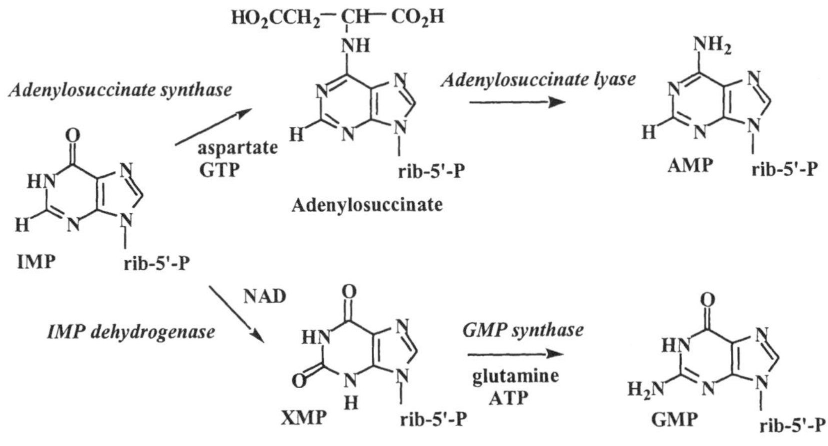

Figure 1. Purine Biosynthesis.

IMP dehydrogenase catalyzes a critical step in purine biosynthesis

IMP is located at the heart of purine metabolism (Figure 1). It can be converted into both adenine and guanine nucleotides. The IMP dehydrogenase reaction is the rate-limiting and first committed step of guanine nucleotide biosynthesis. IMP dehydrogenase levels are amplified in rapidly proliferating cells (2). IMP dehydrogenase inhibitors depress guanine nucleotide pools, disrupting many critical cellular functions and stopping cell growth. This antiproliferative activity is the rationale underlying IMP dehydrogenase-based chemotherapy.

The mechanism of the IMP dehydrogenase reaction

The IMP dehydrogenase reaction involves the attack of the active site Cys on the 2 position of IMP (Figure 2). This reaction is reminiscent of the aldehyde dehydrogenase reaction. A hydride is expelled from the resulting adduct, producing NADH and E-XMP*. Attack of the Cys and hydride transfer could be stepwise as shown in Figure 2, or concerted. The presence of E-XMP* is well established from the work of several laboratories (3-6). NADH is released from the E-XMP*•NADH complex, which allows water (or MPA) to access E-XMP*. Water is presumably activated by a basic residue on the enzyme, although this residue has not yet been identified. Hydrolysis of E-XMP* produces XMP. This hydrolysis reaction is at least partially rate limiting in both microbial and human IMP dehydrogenases.

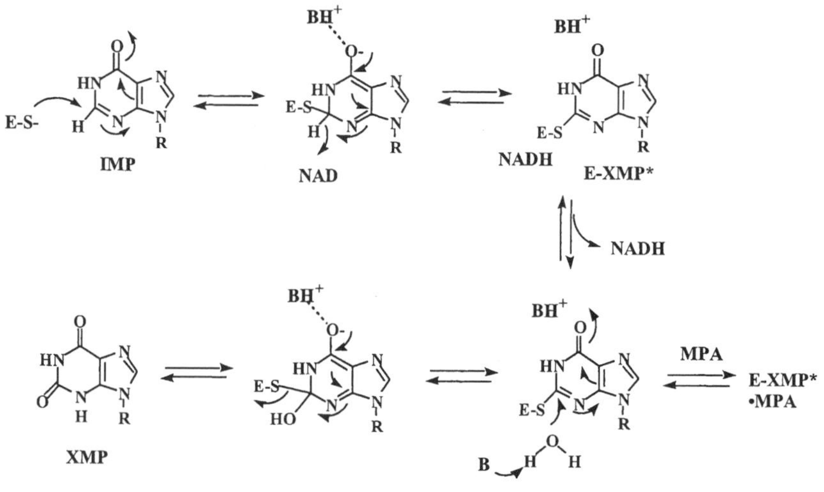

Figure 2. Chemical mechanism of the IMP dehydrogenase reaction.

The kinetic properties of IMP dehydrogenases

The kinetic mechanism of IMP dehydrogenase has been delineated for the *T. foetus* enzyme (*7*). The human type II and *E. coli* enzymes appear to follow the same mechanism (*8-10*), which suggests that all IMP dehydrogenase reactions may have the same kinetic features: substrates bind in a random fashion, hydride transfer is fast and reversible and NADH release precedes hydrolysis of E-XMP* (Figure 3). High concentrations of NAD inhibit IMP dehydrogenase by trapping E-XMP* (*6*). NAD inhibition may be a means of regulating IMP dehydrogenase activity *in vivo* (*11*). IMP binding is a two step process in the *T. foetus* enzyme. Similar two-step binding is expected to be observed in other enzymes.

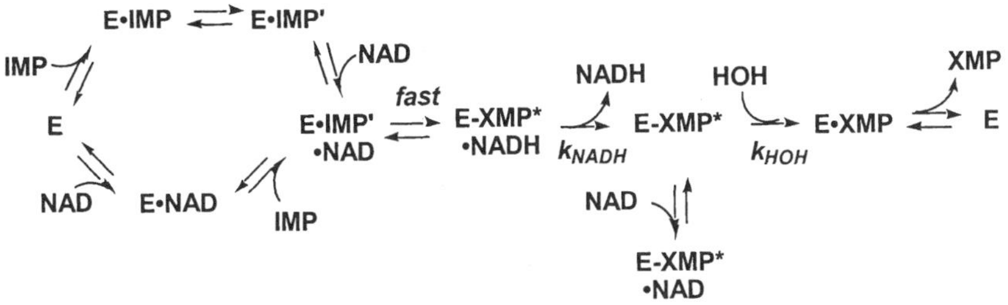

Figure 3. Kinetic mechanism of the IMP dehydrogenase reaction.

Table I. Kinetic parameters for IMP dehydrogenases

Enzyme source	MPA K_I (μM)	k_{cat} (s^{-1})	IMP K_m (μM)	NAD K_m (μM)
human type II[a]	0.02	0.4	13	6
Leishmania donovani[e]	0.2	n.d.	n.d.	700
Tritrichomonas foetus[b]	9	1.9	1.7	150
Borrelia burgdorferi[c]	8	2.6	29	1100
Escherichia coli	20[e]	13[d]	61[d]	2000[d]

NOTE: Enzymes were expressed in *E. coli*. a. Reference (*8, 12*) b. Reference (*7,22*) c. Reference (*20*) d. Reference (*13*) e. L.H., unpublished observations

The steady state kinetic properties of microbial and human type II IMP dehydrogenases differ significantly as shown in Table I. In general, microbial enzymes have higher values of k_{cat}, K_m for NAD and K_i for MPA. Intriguingly, these trends may arise from an underlying mechanistic link between catalysis and MPA sensitivity. The *E. coli* enzyme is the most striking example of this phenomenon: the values of the kinetic parameters of the *E. coli* enzyme are 30-1000-fold greater than the human type II isozyme. The higher value of k_{cat} of *E. coli* IMP dehydrogenase appears to derive from the acceleration of all of the unimolecular steps of the reaction: hydride transfer, NADH release and E-XMP* hydrolysis (*10*). These steps control the accumulation and availability of E-XMP*, and thus will also control MPA sensitivity.

The structural determinants of MPA sensitivity

Our goal is to identify the structural features which control the function of IMP dehydrogenase. We will focus our studies on the *T. foetus* and human type II enzymes. *T. foetus* IMP dehydrogenase, like most microbial enzymes, is resistant to MPA inhibition (Table I). It has a strong intrinsic protein fluorescence signal, relatively low values of K_m and comparatively little NAD substrate inhibition. These properties facilitate kinetic investigation. The human type II isozyme is amplified in tumor cells and appears to be the target for cancer and immunosuppressive therapy. Both enzymes are produced in high yields from an *E. coli* expression system (*12,13*). Our initial objective is to construct a variant of *T. foetus* IMP dehydrogenase which is sensitive to MPA in order to identify the structural features which control MPA sensitivity.

The mechanism of MPA inhibition

MPA has an unusual mode of inhibition. It binds to the E-XMP* intermediate and prevents its hydrolysis. Thus MPA is an uncompetitive inhibitor with respect to both substrates. MPA binds in the nicotinamide subsite of the dinucleotide site, stacking against the purine ring of E-XMP* (Figure 4) (*14,15*). Two limiting scenarios can account for differences in MPA sensitivity. The structure of the MPA binding site could be different in the *T. foetus* and human type II enzymes, i.e., the binding energy associated with the enzyme-MPA interactions could be different. Alternatively, the accessibility of the E-XMP* intermediate will also determine MPA sensitivity. Of course, these two mechanisms are not mutually exclusive: changes in the structure of the MPA binding site can also be expected to affect the accumulation of E-XMP*. Nevertheless, this dichotomy between the "structural" and "kinetic" origins of affinity provides a useful intellectual framework to approach the question of how MPA selectivity is derived.

The structure of IMP dehydrogenase

IMP dehydrogenase is an α/β barrel protein. Six structures have been reported to date, including complexes of Chinese hamster, human type II, *T. foetus*, *B. burgdorferi* and *Streptococcus pyogenes* enzymes (*14,16-19*). These structures have varying degrees of disorder depending on the bound ligand. Like all α/β barrel proteins, the active site is located in the loops C-terminal to the beta sheets. The active site Cys is found in the loop following $\beta 6$. This loop appears to be very mobile. It is disordered in the E•SO$_4^{-2}$ and E•XMP complexes of *T. foetus* IMPDH, and has different structures in the human type II, *S. pyogenes* and *B. burgdorferi* structures. The loop following $\beta 8$ forms a flap which covers the active site (Figure 4). The flap is at least partially disordered in all of the IMP dehydrogenase structures reported to date. IMP dehydrogenase contains an unusual subdomain protruding from the other end of the barrel. This subdomain is not required for IMP dehydrogenase activity (*20,21*). The function of the subdomain is unknown.

The MPA binding site is defined in the structure of the E-XMP*•MPA complex of Chinese hamster IMP dehydrogenase (*14*). Two residues in MPA site differ in human type II and *T. foetus* IMP dehydrogenases: Arg322 is Lys in the *T. foetus* enzyme (Chinese hamster and human type II enzyme numbering) and Gln441 is Glu (Figure 5). Moreover, all mammalian IMP dehydrogenases contain Arg322 and Gln441, while most microbial enzymes contain the Lys and Glu substitutions. However, the *L. donovani* IMP dehydrogenase is an exception to this rule: this enzyme contains Arg322 and Gln441 like mammalian IMP dehydrogenases, yet has a reduced sensitivity to MPA (200 nM versus 20 nM for human type II; Table I). Therefore residues 322 and 441 can not be the only structural determinants of MPA specificity.

Selectivity derives in part from the MPA binding site.

We constructed a double mutant of *T. foetus* IMP dehydrogenase (K310R/E431Q, *T. foetus* enzyme numbering) containing the mammalian

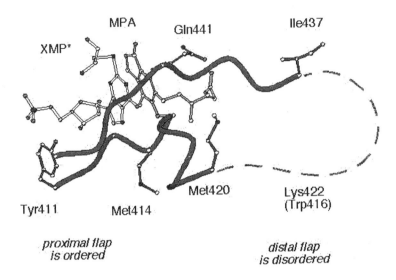

XMP* MPA Gln441 Ile437

Tyr411 Met414 Met420 Lys422 (Trp416)

proximal flap is ordered *distal flap is disordered*

Figure 4. Structure of the Chinese hamster E-XMP*•MPA complex.

Coordinates provided by M. Sintchak (14). E-XMP* is drawn in white bonds, MPA is drawn in light gray bonds and the flap is shown in dark gray. MPA stacks against the purine ring as would the nicotinamide ring of the substrate. Residues 421-436 of the flap are disordered. Chinese hamster numbering (equivalent to human type II numbering) is shown. The approximate position of Trp416 (*T. foetus* numbering) is shown in parentheses.

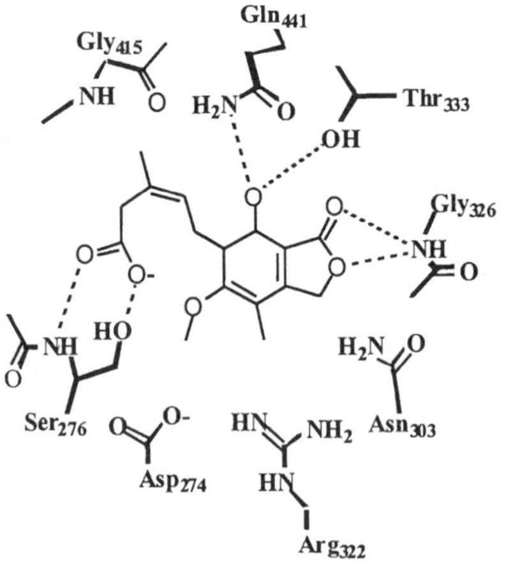

Figure 5. The MPA binding site.

residues at positions 322 and 441. These substitutions increase the MPA sensitivity of *T. foetus* IMP dehydrogenase to a level comparable to the *L donovani* enzyme (K_i decreases from 9 μM to 500 nM; Table II) (*22*). These results suggest that ~20-fold of the difference in MPA sensitivity can be attributed to residues 322 and 441. However, human type II enzyme is an additional 25-fold more sensitive than K310R/E431Q. These observations confirm that MPA sensitivity is not solely determined by the residues in the MPA binding site.

The differences in the kinetic properties between *T. foetus* and human type II IMP dehydrogenase also can not completely account for the difference in drug sensitivity. Both NADH release and hydride transfer are partially rate limiting in the *T. foetus* IMP dehydrogenase reaction (Figure 3 and Table II). In contrast, hydrolysis of E-XMP* is completely rate-limiting for the human type II isozyme. Thus E-XMP* is the major enzyme form in the human type II enzyme reaction (under conditions of saturating substrates) while E•IMP•NAD, E-XMP*•NADH, E-XMP* are all occupied in the *T. foetus* IMP dehydrogenase reaction. However, this difference in accumulation of E-XMP* can only account for a 3-fold decrease in MPA affinity.

The kinetic properties of K310R/E431Q resemble human IMP dehydrogenase type II. The value of k_{cat} is comparable to that of the human enzyme (0.6 s^{-1} versus 0.4 s^{-1}) and the hydrolysis of E-XMP* appears to be completely rate-limiting (Table II). This observation indicates that E-XMP* will be the major enzyme form in the K310R/E431Q reaction. Therefore, of the 20-fold increase in MPA sensitivity relative to wild-type *T. foetus* IMP dehydrogenase, 3-fold results from a change in accumulation of E-XMP* and 7-fold results directly from the interactions at the MPA binding site. These interactions presumably involve the newly introduced Arg and Gln residues. Interestingly, NADH release appears to be much faster in K310R/E431Q than in the human enzyme. This observation suggests that K310R/E431Q has a defect in dinucleotide binding.

Table II. Kinetic constants for mutant IMP dehydrogenases.

Enzyme	MPA K_i (μM)	k_{cat} (s^{-1})	k_{NADH} (s^{-1})	k_{HOH} (s^{-1})
T. foetus	9.0	1.9	6.5	3.8
K310R/E431Q	0.5	0.5	> 55	0.5
human type II	0.02	0.4	5.0	0.4

NOTE: Reference (*22*).

Selectivity is determined by residues outside the MPA site

The results cited above indicate that structural features outside the MPA binding site determine MPA sensitivity. Drug resistance often results from mutations outside the drug binding site. The effects of such mutations are puzzling, and the sites of such substitutions are difficult to identify a priori. In looking beyond the MPA binding site for potential residues which could influence MPA sensitivity, we considered two regions of the enzyme: the IMP binding site and the adenosine subsite. The IMP binding site is highly conserved in all IMP dehydrogenases. Therefore it seemed unlikely to be the source of MPA sensitivity. Unfortunately, the structure of the adenosine subsite is only partially characterized in the structure of a ternary complex of human type II IMP dehydrogenase (16). This complex contains the NAD analog SAD. In addition, the active site Cys is modified by 6-Cl-IMP, such that the Cys is attached to the 6 position of the purine ring. The IMP analog is found in the same position as E-XMP*, but the Cys loop has moved to the other side of the purine ring. The unusual position of the Cys loop displaces the proximal section of the flap, so that part of the IMP site is disordered. Density is observed in the distal section of the flap which interacts with NAD sites (16,23). Unfortunately, it has not been possible to fit this density to protein sequence. Nevertheless, it is apparent that the adenosine subsite varies widely among IMP dehydrogenases from different organisms. The residues which surround the adenine ring, His253, Phe282 and Thr45 in human type II IMP dehydrogenase, are not conserved. In addition, the flap residues are also not conserved. Therefore MPA selectivity might originate in the adenosine subsite although MPA does not bind in the adenosine subsite.

The coupling of the nicotinamide and adenosine subsites

The adenosine subsite must be coupled to the nicotinamide subsite if it is to influence MPA binding. We used a multiple inhibitor experiment to investigate this coupling (24). These experiments yield the "interaction constant" α which measures how the binding of one inhibitor influences the second. A value of $\alpha < 1$ indicates a synergistic interaction, $\alpha > 1$ indicates an antagonistic interaction while $\alpha = 1$ indicates that the two inhibitors are independent. If the two inhibitors are mutually exclusive, as would be the case if they bound to the same site, $\alpha = $. The NAD analog tiazofurin adenine dinucleotide (TAD) is a potent inhibitor of IMP dehydrogenases (25,26). We dissected TAD into two components: tiazofurin to probe the nicotinamide subsite and ADP to probe the adenosine subsite. Both tiazofurin and ADP are poor inhibitors of IMP dehydrogenase (Table III) (15,26). However, a mixture of tiazofurin and ADP strongly inhibits *T. foetus* IMP dehydrogenase. Thus tiazofurin and ADP interact synergistically ($\alpha = 0.007$, Table III), which demonstrates that the nicotinamide and adenosine sites are tightly coupled in the case of *T. foetus* IMP dehydrogenase. In contrast, no increase in inhibition is observed when a mixture of tiazofurin and ADP is used to inhibit human type II IMP dehydrogenase. The two inhibitors are independent (α = 1.0), which demonstrates that the nicotinamide and adenosine subsites are not coupled in the case of the human enzyme. The difference in coupling between the nicotinamide and adenosine subsites in the human type II and *T. foetus* enzymes is sufficient to account for the difference in MPA sensitivity.

We performed the same experiment with K310R/E431Q (Table III). Tiazofurin and ADP interact synergistically as in the *T. foetus* enzyme, although the value of α is greater (α = 0.02). Therefore the increased MPA sensitivity of this mutant enzyme can be attributed in part to a decrease in coupling between the nicotinamide and adenosine subsites. The remaining coupling energy is sufficient to account for the difference in MPA sensitivity of K310R/E431Q and human type II enzymes.

Table III. Multiple inhibitor experiments

Inhibitor	Constant	T. foetus	K310R/E431Q	Human
β-CH$_2$-TAD	K_{ii} (NC)	1.6 μM	4.3 μM	0.06 μM
	K_{is} (NC)	2.6 μM	14 μM	0.06 μM
Tiazofurin	K_{ii} (NC)	69 mM	75 mM	1.3 mM
	K_{is} (NC)	50 mM	75 mM	4.2 mM
ADP	K_{is} (C)	31 mM	1.4 mM	8.8 mM
Tiazofurin/ADP	α	0.007	0.02	1

NOTE: Reference (*27*)

The Distribution of Binding Energy

At first glance, tiazofurin inhibition appears to be incompatible with the more "human-like" properties of K310R/E431Q. Human type II IMP dehydrogenase is 12-fold more sensitive to tiazofurin than the *T. foetus* enzyme as measured by the value of K_i (Table III). The Lys210Arg and Glu431Gln mutations might naively be expected to decrease the value of K_i for tiazofurin inhibition of the *T. foetus* enzyme. This decrease in K_i is not observed (Table III). However, this view fails to account for the coupling energy associated with tiazofurin binding. The total, or intrinsic, binding energy associated with tiazofurin (ΔG_{ibe}) is the sum of the observed binding energy (ΔG_{obs}) and the coupling or interaction energy (ΔG_{int}) (28). For human type II IMP dehydrogenase, no coupling is observed between the nicotinamide and adenosine subsites. Therefore $\Delta G_{ibe} = \Delta G_{obs} = -4.0$ kcal/mol. In contrast, strong coupling is observed in the *T. foetus* enzyme, and $\Delta G_{ibe} = \Delta G_{obs} + \Delta G_{int} = -4.6$ kcal/mol. Therefore the intrinsic binding energy of tiazofurin is actually greater in the *T. foetus* IMP dehydrogenase than the human type II enzyme. The reduction in coupling observed in K310R/E431Q relative to the wild-type enzyme lowers the intrinsic binding energy of tiazofurin to the level of the human type II enzyme: ΔG_{ibe} -4.0 kcal/mol. This similarity in the values of ΔG_{ibe} is required if, as designed, the tiazofurin binding sites of the human type II and K310R/E431Q enzymes are identical.

The intrinsic binding energy associated with ADP is also greater for *T. foetus* IMP dehydrogenase than for the human type II enzyme. This result reflects the different structures of the adenosine subsites in the two enzymes. In addition, this result suggests that dinucleotide binding relies more on the adenosine subsite in the *T. foetus* enzyme. The total binding energy associated with ADP is greater in K310R/E431Q IMP dehydrogenase than the *T. foetus* enzyme. This result suggests that the mutations have also changed the ADP subsite. ADP extends into the MPA binding site, such that the β phosphate overlaps with the carboxyl group of MPA, so the effect of the mutations on ADP binding is not surprising.

The "Advantage of Connection"

The data of Table III can also be used to calculate the "advantage of connection", i.e., the gain in affinity for linking tiazofurin and ADP to form TAD. This "advantage" has the form of an effective concentration, derived from the product of the K_i values for tiazofurin and ADP divided by the K_i for TAD. The advantage of connectivity in our system is rather small: 24 M for K310R/E431Q, 190 M for human type II and 700 M for *T. foetus* IMP dehydrogenases. Advantages of connectivity of $>10^4$ M have been measured in other systems (29,30), and values as high as 10^8 M are theoretically possible (28). The comparatively small advantages that we observe may result from the apparent overlap of our component pieces: the 5' oxygen of tiazofurin will occupy the same position as one of the β-phosphate oxygens of ADP. The "double counting" of this oxygen would lower the apparent advantage of connectivity. In addition, ADP is more negatively charged than TAD, which could also lower the apparent advantage of connection.

The conformation of the flap

The results cited above demonstrate that coupling between the nicotinamide and adenosine subsites is an important determinant of MPA sensitivity. One model that can explain this observation is shown in Figure 6. The binding of tiazofurin to *T. foetus* IMP dehydrogenase (Figure 6, A) orders the adenosine binding site. This conformational change increases the affinity of ADP, but reduces the apparent affinity of tiazofurin. In contrast, no such conformational change is observed when tiazofurin binds to human type II IMP dehydrogenase, ADP affinity does not increase in the presence of tiazofurin, and all of the binding energy of tiazofurin is observed. Extending this model to MPA (Figure 6 B), the adenosine site is also becomes ordered when MPA binds to *T. foetus* IMP dehydrogenase; this conformational change decreases the apparent affinity of MPA. In contrast, no conformational change is induced when MPA binds to human type II IMP dehydrogenase, and all of the binding energy of MPA is observed.

The structure of the Chinese hamster E-XMP*•MPA complex is consistent with this model (Figure 4). The proximal section of the flap covers XMP* and MPA, but the distal section is disordered. The distal section is believed to be part of the adenosine subsite (*16*). Thus it appears that the nicotinamide and adenosine subsites are independent in mammalian IMP dehydrogenases. We believe that both the proximal and distal sections of the flap are ordered when MPA binds to *T. foetus* IMP dehydrogenase. We tested this hypothesis by constructing a version of the *T. foetus* enzyme containing a single Trp in the distal section of the flap. If MPA binding changes the conformation of the distal flap, then the fluorescence of this Trp residue should change. This mutant enzyme, "Trp416", has kinetic properties similar to the wild-type enzyme (*27*). MPA quenches the fluorescence of the E•IMP complex of "Trp416", which suggests that the distal flap undergoes a conformational change when MPA binds.

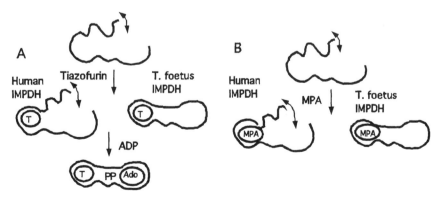

Figure 6. A model for coupling of the nicotinamide and adenosine subsites. Adapted from reference (27).

Summary

The origins of the 450-fold difference in MPA sensitivity between human type II and *T. foetus* IMP dehydrogenases can be apportioned as follows: 3-fold arises from the difference in accumulation of E-XMP*; 7-fold arises from the residues at the MPA binding site and 20-fold arises from coupling between the nicotinamide and adenosine subsites. We believe that the flap is the key structural element controlling this coupling.

References

1. Bentley, R. *Chemical Reviews* **2000** *100*, 3801.
2. Jackson, R.; Weber, G.; Morris, H.P. *Nature* **1975**. *256*, 331.
3. Huete-Perez, J.A.; Wu, J.C.; Witby, F.G.; Wang, C.C. *Biochemistry* **1995**. *34*, 13889.
4. Link, J.O.; Straub, K. *J. Am. Chem. Soc.* **1996**. *118*, 2091.
5. Fleming, M.A.; Chambers, S.P.; Connelly, P.R.; Nimmesgern, E.; Fox, T.; Bruzzese, F.J.; Hoe, S.T.; Fulghum, J.R.; Livingston, D.J.; Stuver, C.M.; Sintchak, M.D.; Wilson, K.P.; Thomson, J.A. *Biochemistry* **1996**. *35*, 6990.
6. Wu, J.C.; Carr, S.F.; Antonino, L.C.; Papp, E.; Pease, J.H. *FASEB J.* **1995**. *9*, A1337.
7. Digits, J.A.; Hedstrom, L. *Biochemistry* **1999**. *38*, 2295.
8. Wang, W.; Hedstrom, L. *Biochemistry* **1997**. *36*, 8479.
9. Xiang, B.; Markham, G.D. *Arch. Biochem. Biophys.* **1997**. *348*, 378.
10. Kerr, K.M.; Digits, J.A.; Kuperwasser, N.; Hedstrom, L. *Biochemistry* **2000** *32*, 9804.
11. Hupe, D.; Azzolina, B.; Behrens, N. *J. Biol. Chem.* **1986**. *261*, 8363-8369.
12. Farazi, T.; Leichman, J.; Harris, T.; Cahoon, M.; Hedstrom, L. *J. Biol. Chem.* **1997**. *272*, 961-965.
13. Kerr, K.M.; Hedstrom, L. *Biochemistry* **1997**. *36*, 13365-13373.
14. Sintchak, M.D.; Fleming, M.A.; Futer, O.; Raybuck, S.A.; Chambers, S.P.; Caron, P.R.; Murcko, M.; Wilson, K.P. *Cell* **1996**. *85*, 921-930.
15. Hedstrom, L.; Wang, C.C. *Biochemistry* **1990** 849-554.
16. Colby, T.D.; Vanderveen, K.; Strickler, M.D.; Markham, G.D.; Goldstein, B.M. *Proc. Natl. Acad. Sci., USA* **1999**. *96*, 3531-3536.

17. Whitby, F.G.; Luecke, H.; Kuhn, P.; Somoza, J.R.; Huete-Perez, J.A.; Philips, J.D.; Hill, C.P.; Fletterick, R.J.; Wang, C.C. *Biochemistry* **1997**. *36*, 10666-10674.
18. McMillan, F.M.; Cahoon, M.; White, A.; Hedstrom, L.; Petsko, G.A.; Ringe, D. *Biochemistry* **2000**. *39*, 4533-4542.
19. Zhang, R.-G.; Evans, G.; Rotella, F.J.; Westbrook, E.M.; Beno, D.; Huberman, E.; Joachimiak, A.; Collart, F.R. *Biochemistry* **1999**. *38*, 4691-4700.
20. Zhou, X.; Cahoon, M.; Rosa, P.; Hedstrom, L. *J. Biol. Chem.* **1997**. *272*, 21977-21981.
21. Nimmesgern, E.; Black, J.; Futer, O.; Fulghum, J.R.; Chambers, S.P.; Brummel, C.L.; Rayuck, S.A.; Sintchak, M.D. *Protein Expression and Purification* **1999**. *17*, 282-289.
22. Digits, J.A.; Hedstrom, L. *Biochemistry* **1999**. *38*, 15388-15397.
23. Goldstein, B.M.; Colby, T.D. *Current Med. Chem.* **1999**. *6*, 519-536.
24. Yonetani, T.; Theorell, H. *Arch. Biochem. Biophys.* **1964**. *106*, 2193-2196.
25. Gebeyehu, G.; Marquez, V.; Van Cott, A.; Cooney, D.; Kelley, J.; Jayaram, H.; Ahluwalia, G.; Dion, R.; Wilson, Y.; Johns, D. *J. Med. Chem.* **1985**. *28*, 99-105.
26. Marquez, V.; Tseng, C.K.H.; Gebeyehu, G.; Cooney, D.A.; Ahluwalia, G.S.; Kelley, J.A.; Dalal, M.; Fuller, R.W.; Wilson, Y.A.; Johns, D.G. *J. Med. Chem.* **1986**. *29*, 1726-1731.
27. Digits, J.A.; Hedstrom, L. *Biochemistry* **2000**. *39*, 1771-1777.
28. Jencks, W.P. *Proc. Natl. Acad. Sci. USA* **1981**. *78*, 4046-4050.
29. Nakamura, C.E.; Abeles, R.H. *Biochemistry* **1985**. *24*, 1364-1376.
30. Carlow, D.A.; Wolfenden, R. *Biochemistry* **1998**. *37*, 11873-11878.

Chapter 10

The Discovery of Thiazole-4-Carboxamide Adenine Dinucleotide (TAD) and a Recent Synthetic Approach for the Construction of a Hydrolytically Resistant Surrogate

Victor E. Marquez

National Cancer Institute, National Institutes of Health, Laboratory of Medicinal Chemistry, Center for Cancer Research, Building 376 Boyles Street, Room 104, NCI–Frederick, P.O. Box B, Frederick, MD 21702 (marquezv@dc37a.nci.nih.gov)

A short account of the mechanism of action of the oncolytic nucleoside, tiazofurin, and the discovery of its active metabolite, TAD, is described. The chapter also includes a brief history of the early chemical approaches used for the synthesis of TAD and other analogues, including the phosphodiesterase-resistant β-methylene TAD phosphonate. A new synthetic approach to the latter compound is highlighted as a viable way to obtain this class of compounds in larger quantities.

The C -nucleoside tiazofurin (2-β-D-ribofuranosylthiazole-4-carboxamide, **1a**, X = S) was first synthesized by ICN chemists as one in a series of potential antiviral compounds (*1,2*). Although the compound exhibited moderate antiviral activity, it was not sufficient to warrant further development as an antiviral agent (*3*). However, initial results from tests conducted at the National Cancer Institute revealed that tiazofurin was an effective antitumor agent as judged by the percent increase in life span (ILS) in murine models (Table 1). Furthermore, tiazofurin proved to be effective and curative against an extremely resistant tumor line such as Lewis Lung carcinoma (Table 2) (*4*).

1a, X = S (tiazofurin)
1b, X = Se (selenazofurin)

Table 1. Effect of tiazofurin (1a) on several mouse tumor models

Tumor System[a]	Treatment schedule and route[b]	%ILS (median)
IP L1210 leukemia	QD 1-9 (IP)	134
IP L1210 leukemia	QD 1-9 (PO)	127
IP L1210 leukemia	QD 1-5 (IV)	126
IP P388 leukemia	QD 1-9 (IP)	127
IV Lewis Lung	QD 1-9 (IP)	224
IC Lewis Lung	QD 1-9 (IP)	108

[a] tumor implated IP = intraperitoneal, IV =intravenous
[b] treatment IP = intraperitoneal, PO = oral, IV = intravenous

Table 2. Effect of tiazofurin on the life span of mice inoculated intravenously with Lewis Lung carcinoma

Drug	Dose (mg/Kg)	60-day survivors
Control		0/40
Tiazofurin (1a)	400	7/10
QD 1-9 (IP)	200	10/10
	100	9/10
	50	10/10
	25	10/10

The complex studies that ultimately deciphered the mechanism of action of tiazofurin have been summarized in two excellent reviews (5,6). Briefly, two laboratories almost simultaneously confirmed that the oncolytic effect of tiazofurin was dependent on its conversion to a phosphodiester analogue of NAD, the ordinary cofactor in the inosine monophosphate dehydrogenase (IMPD) reaction (7,8). Formation of this metabolite, thiazole-4-carboxamide adenine dinucleotide (2, TAD), was responsible for the profound dose-dependent fall in the intratumoral concentration of GTP resulting from the potent inhibition of IMPD (5,6). IMPD catalyzes a unique step in the *de novo* biosynthesis of guanine nucleotides, which is considered essential for supporting cell proliferation. This NAD-dependent reaction is responsible for the conversion of IMP to XMP, which is then aminated to GMP by GMP synthetase (Figure 1). Through the successive action of several enzymes, GMP is then converted to some of the building blocks of DNA (dGTP) and RNA (GTP) synthesis.

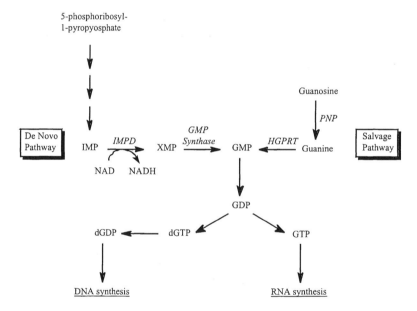

Figure 1. Schematic representation of the de novo and salvage pathways of purine nucleotide biosynthesis converging on GMP

The isolation and characterization of TAD (7,8) was soon followed by its first total synthesis, which firmly confirmed the identity of the metabolite (9).

With NAD as the variable substrate, TAD showed non-competitive inhibition of IMPD with a K_i in the submicromolar range ($K_i = 0.12$ μM). TAD did not have substrate activity as a coenzyme, i.e. the molecule did not undergo reduction, and its inhibitory activity was very specific as shown by its inability to inhibit other dehydrogenases at concentrations capable of inhibiting IMPD (5,6).

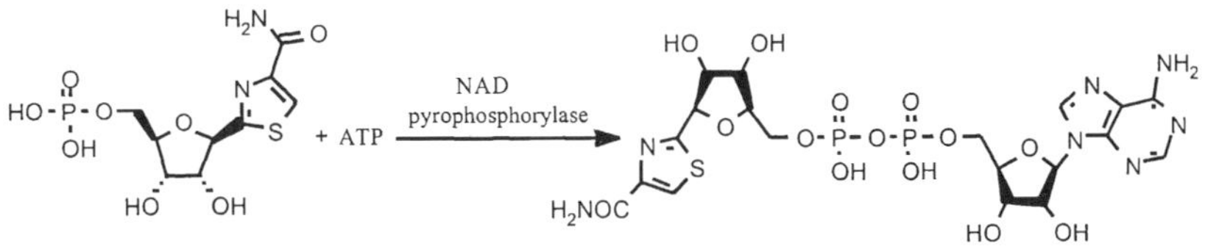

Figure 2. Biosynthesis of TAD

The biosynthesis of TAD starts with the generation of the 5'-monophosphate of tiazofurin (**1a**-MP) by adenosine kinase and/or by 5'-nucleotidase (*10*). The ensuing formation of TAD from **1a**-MP and ATP is mediated by NAD pyrophosphorylase acting in the synthetic direction (Figure 2) (*5,6,11*). Since levels of IMPD are higher in tumors as compared to normal tissues (*12-16*), inhibition of this enzyme might account for the observed antitumor activity of tiazofurin commensurate with the formation of TAD. Human IMPD exists as two isoforms (I and II) of which type I is predominant in normal cells and type II is upregulated in tumor cells (*17-19*). Unfortunately TAD and structurally similar analogues showed equal effective inhibitory activity against both isotypes (*20,21*). The recently solved X-ray structures of IMPD, particularly that of IMPD type II complexed with 6-chloropurine riboside 5'-monophosphate and with the selenium analogue of TAD (selenazole-4-carboxamide adenine dinucleotide, SAD) (*22*) suggest important strategies for the design of isoform-specific inhibitors (*23*).

The respective formation of TAD and SAD from tiazofurin and selenazofurin (**1b**, X = Se) is unique since other structurally related compounds such as ribavirin (**3**,R) and the natural purine precursor 5-amino-4-imidazolecarboxamide ribonucleoside (**4**, Z) do not form the corresponding dinucleotides (RAD and ZAD) *in vivo* (Table 3). Furthermore, synthetically made nicotinamide adenine dinucleotide analogues of **3** and **4**, prepared in the same manner as TAD and SAD, failed to show any inhibition of IMPD (Table 4) (*24*).

3, X = N (Ribavirin)
4, X = C-NH2

Table 3. Biosynthesis of NAD dinucleotide analogues by P388 cells and isolated NAD pyrophosphorylase

Cell Culture			NAD pyrophosphorylase			
Drug	Di-nucleotide	Specific activiy[a]	Starting nucleotide	Di-nucleotide	Specific activity[b]	% NAD[c]
1a	TAD	99.7±2	1a-MP	TAD	49.0	68
1b	SAD	270.1±4	2a-MP	SAD	54.0	75
3	(RAD)	0	3-MP	(RAD)	0	0
4	(ZAD)	0	4-MP	(SAD)	0	0

[a]nanomoles of dinucleotide/g of cell ± SD after 2 h. [b]nanomoles of dinucleotide h^{-1} (mg of protein)$^{-1}$. [c]Specific activity for NAD (100%) was 72.0

Table 4. Cytotoxicity and IMPD inhibitory activity of dinucleotides

Di-nucleotide	P388 cytotoxicity ID$_{50}$ (μM)	K$_i$ (μM)	Varying substrate	Type of inhibition
TAD	7.5	0.133	IMP	Uncompetitive
			NAD	Non-competitive
SAD	9.8	0.046	IMP	Uncompetitive
			NAD	Non-competitive
RAD	40	235	IMP	Non-competitive
			NAD	Non-competitive
ZAD	78	Inactive	IMP	Inactive
			NAD	Inactive

The synthetic pathway leading to TAD, which is catalyzed by NAD pyrophosphorylase, is countered by a degradative pathway catalyzed by a unique phosphodiesterase (TADase), and the balance between these two processes (Figure 3) can explain the sensitivity and natural resistance of some important tumor cell lines to tiazofurin (Table 5) (*11*).

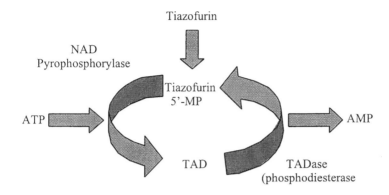

Figure 3. Mechanism of formation and hydrolysis of TAD

Table 5. TAD biosynthetic and degradative enzyme activities in murine tumors

		Specific activity (nmoles/h/mg protein)			
Tumor	*Sensitivity to tiazofurin*	*Tiazofurin kinase*	*NAD pyro-phosphorylase (N)*	*TADase (T)*	*N/T*
P388/S	S	2.6±0.6	75.0±5.0	24.0±6.7	3.13
Lewis lung	S	2.9±0.9	70.0±8.0	37.0±18	1.90
L1210	S	0.4±0.1	90.0±8.0	25.0±5.0	3.60
B16	R	1.0±0.2	47.0±2.0	100.0±19	0.47
Colon38	R	0.4±0.1	37.0±0.8	140.0±25	0.26
M5076	R	0.9±0.2	17.0±3.3	67.0±14	0.25
P388/R	R	2.4±0.4	12.6±0.9	25.0±3.7	0.46

The data in Table 5 also suggested that the degradative pathway is probably dominant and responsible for the spontaneous and acquired resistance to tiazofurin (*11*). A structure-activity investigation that sought to overcome the rapid cleavage of TAD led to the synthesis of the first hydrolytically resistant analogue of TAD, β-methylene-TAD, where the middle oxygen of the pyrophosphate unit was replaced by a methylene group (Figure 4) (*25*).

The three TAD phosphonate analogues synthesized were active as IMPD inhibitors with the β–methylene-TAD analogue being virtually indistinguishable from TAD (Table 6). Using partially purified TADase and the more general phosphodiesterase from snake venom, the control TAD produced equimolar amounts of AMP and tiazofurin-5'-monophosphate as demonstrated by HPLC, confirming that the enzymes are cleaving at the site of the middle oxygen of the pyrophosphate bridge of TAD. Both α- and γ-methylene-TAD analogues, which have this oxygen intact, were susceptible to cleavage, but to a lesser degree than TAD (Table 6).

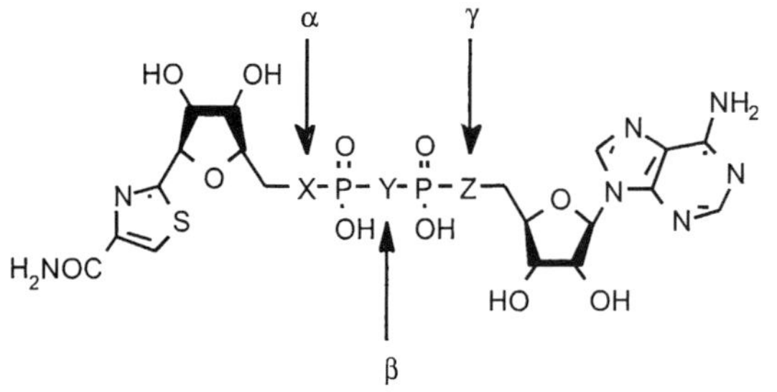

2, X = Y = Z = O (TAD)
5, X = CH_2, Y = Z = O (α)
6, Y = CH_2, X = Z = O (β)
7, Z = CH_2, X = Y = O (γ)

Figure 4. Structurally modified TAD analogues

Table 6. Biological activities of TAD phosphonates

		IMPD Inhibition		Degradation (% TAD)	
Di-Nucleot.	*Cytotoxicity P388*	K_i^a (μM)	*Type of Inhibition*	*TADase*	*Phosphodiesterase (snake venom)*
TAD	2.1	0.24	Not competitive	100	100
α-CH_2-TAD	100	1.97	Not competitive	10.4	27.2
β-CH_2-TAD	46	0.23	Not competitive	0	0
γ-CH_2-TAD	2.8	17[b]	Not determined	13.0	37.6

[a]with NAD as variable substrate. [b]ID_{50}

The interpretation of the toxicity results shown in Table 6 requires taking into account the hydrolytic stability of the dinucleotides during the 48-h experiment. TAD, which is equitoxic to tiazofurin on a molar basis, rapidly hydrolyzes to tiazofurin-5'-monophosphate, which in turn hydrolyzes to tiazofurin. The nucleoside can then be recycled back to TAD, thus accounting for its cytotoxicity. With the α- and γ-methylene-TAD analogues, the eventual breakdown of the molecules after 48-h would generate tiazofurin-5'-monophosphate —and eventually tiazofurin— only from γ-methylene-TAD, which would explain its similar cytotoxicity to tiazofurin and TAD and the much lower cytotoxicity displayed by the α-methylene-TAD analogue. On the other hand, the β-methylene-TAD analogue was completely stable after 48 h and its ID$_{50}$ value of 45 μM suggested that the intact molecule was even capable of entering the cell intact. The key experiment that confirmed the cell permeability of β-methylene-TAD was performed with an experimentally induced tiazofurin resistant (P388/R) line displaying extremely low levels of the enzyme responsible for the formation of TAD (NAD-pyrophosphorylase). If β-methylene-TAD were to be cleaved by some unknowm mechanism leading to the formation of tiazofurin, the cells would have remained insensitive to the drug as they already were to tiazofurin. The results showed that β-methylene-TAD was equally effective against both P388/S sensitive and P388/R resistant cell lines with a similar ID$_{50}$ of approximately 50 μM (25). At this dose level, the compound appeared to have blocked IMPD significantly, showing even more effective reduction of guanine nucleotide pools against the resistant cell line than against the sensitive line (Table 7). Against the resistant cell line, tiazofurin began to show effective reduction of nucleotide pools at doses 2000 times higher than those inhibiting growth of the sensitive strain.

Table 7. Guanine nucleotide pools in sensitive and resistant cells after treatment with β-methylene-TAD

P388/S (18 h after treatment)	Guanine nucleotide conc. (% of control)		
Treatment	GMP	GDP	GTP
β-CH$_2$-TAD, 50 μM	14	24	24
β-CH$_2$-TAD, 100 μM	6	19	39
Tiazofurin, 2.5 μM	12	41	51
Tiazofurin, 5.0 μM	8	36	34
P388/R (18 h after treatment)			
β-CH$_2$-TAD, 50 μM	3	24	5
β-CH$_2$-TAD, 100 μM	3	34	8
Tiazofurin, 25 μM	98	88	111
Tiazofurin 5.0 mM	0	74	52

Methods of Synthesis

The initial method for the synthesis of TAD was based on the synthesis of NAD itself where AMP and tiazofurin-5'-monophosphate were reacted in the presence of dicyclohexylcarbodiimide (9). However, only small amounts of TAD were obtained by this method, and better results were achieved by selectively activating one of the nucleotides before coupling (24). Hence, the tri-n-octylamine salt of tiazofurin-5'-monophosphate (8) was reacted with the activated phosphoromorpholidate form of AMP (9) to give TAD (Scheme 1).

Alternatively, tiazofurin-5'-monophosphate could be activated with carbonyldiimidazole and the resulting imidazolidate **10** coupled with AMP in a DMF solution containing tri-*n*-butylamine. The reactions were complete after 48 h, and, in the latter case, treatment with triethylamine was required to hydrolyze the 2',3'-cyclic carbonate formed during the reaction with carbonyldiimidazole. Isolation of TAD was performed by ion exchange chromatography, and the yields obtained were in the 30-50% range (*24*).

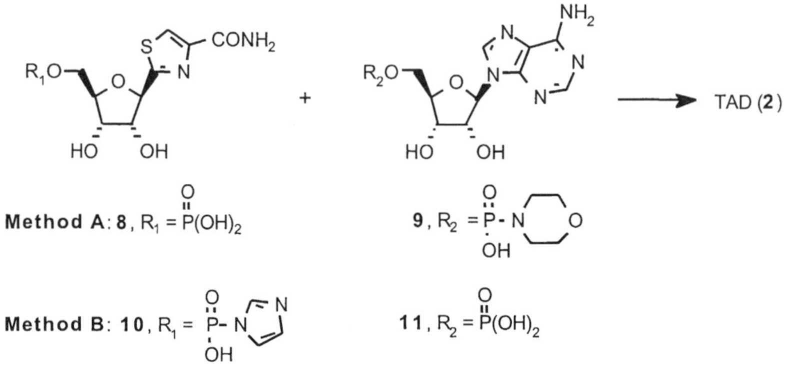

Scheme 1

Syntheses of α-methylene-TAD and γ-methylene-TAD proceeded essentially in the same manner starting with one the isosteric phosphonic acid analogues of AMP or tiazofurin-5'-monophosphate and the activated phosphoromorpholidate or imidazolidate of the other mononucleotide (Scheme 2) (*25*).

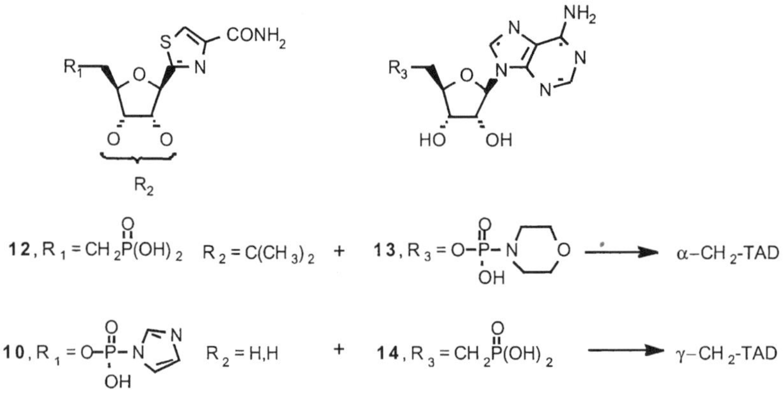

Scheme 2

Synthesis of the symmetric β-methylene-TAD required a different strategy which started with the ADP phosphonate analogue [5'-(α,β-methylene)diphosphate] (16) which was coupled with 2',3'-*O*-isopropylidenetiazofurine (17) in the presence of dicyclohexylcarbodiimide followed by a final treatment with acid to remove the acetonide group (25). β-Methylene-TAD and the other phosphonate analogues were purified using the same chromatograpic system that was developed for the isolation of TAD itself (5,6).

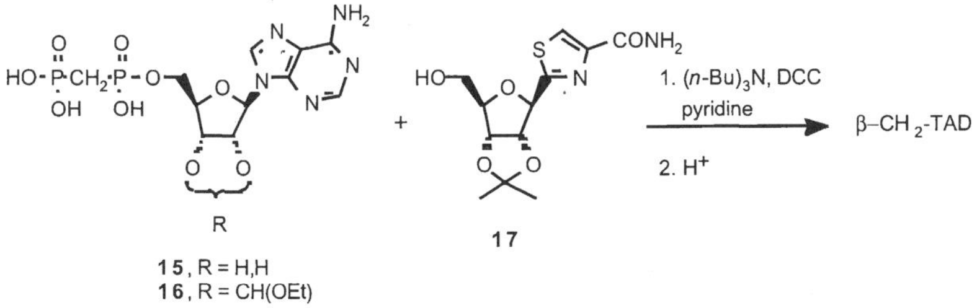

Scheme 3

One major drawback of the syntheses described above, including that of the biologically important β-methylene-TAD, is that all intermediates generated and the final products are charged molecules. Even the more elegant and efficient method developed later by Pankiewicz et al. still requires isolation

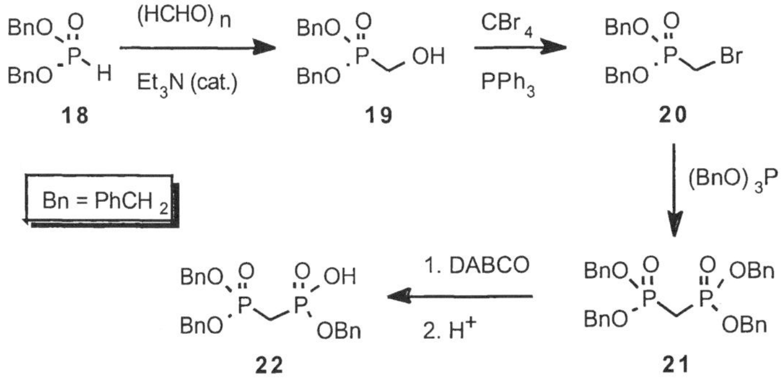

Scheme 4

of all key intermediates by reversed-phase HPLC (26). Thus, the capacity of these methods is limited to small-scale syntheses. To overcome some of these problems, a different approach that utilizes two rounds of Mitsunobu esterification reactions starting from the partially protected [[bis(benzyloxy)-phosphoryl]methyl]phosphonic acid monobenzyl ester (22) was developed (27). The key elements of this approach are based on the significant work of Mioskowski et al. who described the selective monodeprotection of phosphonate benzyl esters and the ensuing successful Mitsunobu esterifications of these monodeprotected substrates with conveniently protected nucleosides (28-30).

Critical to the success of this approach was the availability of adequate amounts of tetrabenzyl methylenebis(phosphonate) (21). A convenient method to obtain this compound started with dibenzyl bromomethylphosphonate (20), which was readily prepared by the Pudovik reaction starting with dibenzyl phosphite (18) and paraformaldehyde (31). The resulting dibenzyl hydroxymethylphosphonate (19) was then converted to 20 under standard bromination conditions, and following the Michaelis-Arbuzov reaction of 20 with tribenzyl phosphite, tetrabenzyl methylenebis(phosphonate) (21) could be obtained in 23 % overall yield (Scheme 4) (27).

The assembly of β-methylene-TAD relied on the selective monodeprotection of tetrabenzyl methylenebis(phosphonate) which was accomplished in the presence of equimolar amounts of 1,4 diazabicyclo[2,2,2]octane (DABCO) in refluxing toluene to give 22. This allowed the ensuing Mitsunobu esterification of 22 with 2',3'-O-isopropylidenetiazofurin (17) (Scheme 5) (27).

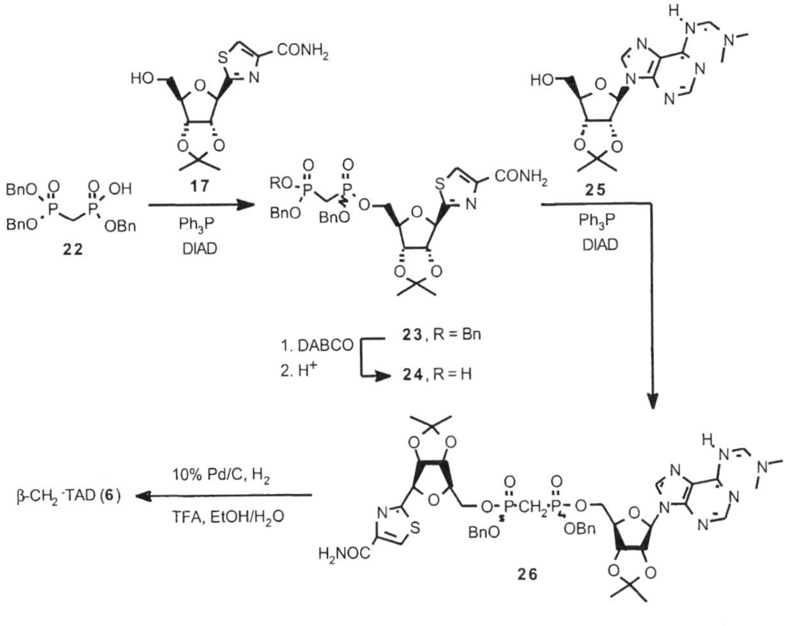

Scheme 5

Due to the chirality of the proximal phosphorous to tiazofurin, **23** was obtained as a mixture of diastereoisomers. The selective deprotection methodology again worked extremely well with **23** as a substrate to give **24** quantitatively. It is reasonable to speculate that the steric hindrance caused by the tiazofurin moiety directs the selective debenzylation to occur at the distal phosphonate benzyl ester. The second Mitsunobu esterification with 2',3'-O-isopropylidene-N^6-dimethylaminomethyleneadenosine proceeded even in better yield than the first Mitsunobu coupling giving the fully protected β-methylene-TAD precursor **26** in 72% yield after column chromatography (**27**). Final hydrogenolysis of **26** over Pd/C in EtOH performed under acidic conditions simultaneously removed all protective groups to give the final polar product β-methylene-TAD, which was obtained in 52% yield after purification by reversed-phase HPLC. With this process, all intermediates prior to the final deprotection step can be purified by conventional chromatography using organic solvents, and only a final HPLC purification is necessary at the end of the synthesis.

This process represents a new approach to β-methylene-TAD and other cell permeable analogues that offer promise as new therapeutic agents targeting IMPD.

Acknowledgment: I wish to acknowledge my former colleagues, Drs. David A. Cooney and David G. Johns, for inviting me to collaborate with them in their TAD adventure. I also wish to thank Dr. James A. Kelley for his help in mass spectrometry, which proved critical in the early studies of the structure of TAD. Finally, I would like to recognize my former postdoctoral fellows Drs. Gulilat Gebeyehu, Christopher K.-H. Tseng and Hisafumi Ikeda for their dedication and synthetic skills.

References

1. Fuertes M.; Garcia-Lopez, T.; Garcia-Muñoz, G.; Stud, M. *J. Org. Chem.* **1976**, *41*, 4076-4077.
2. Srivastava, P C.; Pickering, M. V.; Allen, L. B.; Streeter, D. G.; Campbell, M. T.; Witkowski, J. T.; Sidwell, R. W.; Robins, R. K. *J. Med. Chem.* **1977**, *20*, 256-262.
3. Huggins, J. W.; Robins, R. K.; Canonico, P. G. *Antimicrob. Agents Chemother.* **1984**, *26*, 476-480.
4. Robins, R. K.; Srivastava, P. C.; Narayanan, V. L.; Plowman, J.; Paull K. D. *J. Med. Chem.* **1982**, *25*, 107-108.
5. Cooney, D. A.; Jayaram, H. N.; Glazer, R. I.; Kelley, J. A.; Marquez, V. E.; Gebeyehu, G.; Van Cott, A. C.; Zwelling, L. A.; Johns, D. G. *Adv. Enzyme Regul.* **1983**, *21*, 271-303.
6. Ahluwalia, G. S.; Jayaram, H. N.; Cooney, D. A. In *Concepts, Clinical Developments, and Therapeutic Advances in Cancer Chemotherapy*, Muggia, F. M. Ed., Martinus Nijhoff Publishers: Boston, 1987, pp 63-102.
7. Cooney, D. A.; Jayaram, H. N.; Gebeyehu, G.; Betts, C. R.; Kelley, J. A.; Marquez, V. E.; Johns, D. G. *Biochem. Pharmacol.* **1982**, *31*, 2133-2136.

210

8. Kuttan, R.; Robins, R. K.; Saunders, P. P. *Biochem. Biophys. Res. Commun.* **1982**, *107*, 862-868.
9. Gebeyehu, G.; Marquez, V. E.; Kelley, J. A. *J. Med. Chem.* **1983**, *26*, 922-925.
10. Fridland, A; Connelly, M. C.; Robbins, T. J. *Cancer Res.* **1986**, *46*, 532-537.
11. Ahluwalia, G. S.; Jayaram, H. N.; Plowman, J. P.; Cooney, D. A.; Johns, D. G. *Biochem. Pharmacol.* **1984**, *33*, 1195-1203.
12. Yamada, Y.; Natsumeda, Y.; Weber, G. *Biochemistry* **1988**, *27*, 2193-2196.
13. Jackson, R. C.; Weber, G.; Morris, H.P. *Nature(London)* **1975**, *256*, 331-333.
14. Weber, G.; Prajda, N.; Jackson, R. C. *Adv. Enzyme Regul.* **1986**, *14*, 3-23.
15. Becher, H. J.; Lohr, G. W. *Klin. Wochenschr.* **1979**, *57*, 1109-1115.
16. Robins, R. K. *Nucleosides Nucleotides* **1982**, *1*, 35-44.
17. Carr, S. F.; Papp, E.; Wu, J. C.; Natsumeda, Y. *J. Biol. Chem.* **1993**, *268*, 27285-27290.
18. Nagai, M.; Natsumeda, Y.; Konno, Y.; Hoffman, R.; Irino, S.; Weber, G. *Cancer Res.* **1991**, *51*, 3886-3890.
19. Nagai, M.; Natsumeda, Y.; Weber, G. *Cancer Res.* **1992**, *52*, 258-261.
20. Zatorski, A.; Goldstein, B. M.; Colby, T. D.; Jones, J. P.; Pankiewicz, K. W. *J. Med. Chem.* **1995**, *38*, 1098-1105.
21. Zatorski, A.; Watanabe, K. A.; Carr, S. F.; Goldstein, B. M.; Pankiewicz, K. W. *J. Med. Chem*. **1996**, *39*, 2422-2426.

22. Colby, T. D.; Vanderveen, K.; Strickler, M. D.; Markham, G. D.; Goldstein, B. M. *Proc. Nat. Acad. USA* **1999**, *96*, 3531-3536.
23. Sintchak, M. D.; Nimmesgern, E. *Immunopharmacology* **2000**, *47*, 163-184.
24. Gebeyehu, G.; Marquez, V. E.; Van Cott, A.; Cooney, D. A.; Kelley, J. A.; Jayaram, H. N.; Ahluwalia, G. S.; Dion, R. L.; Wilson, Y. A.; Johns, D. G. *J. Med. Chem.* **1985**, *28*, 99-105.
25. Marquez, V. E.; Tseng, C. K.-H.; Gebeyehu, G.; Cooney, D. A.; Ahluwalia, G. S.; Kelley, J. A.; Dalal, M.; Fuller, R. W.; Wilson, Y. A; Johns, D. G. *J. Med. Chem.* **1986**, *29*, 1726-1731.
26. Pankiewicz, K. W.; Lesiak, K.; Watanabe, K. A. *J. Am. Chem. Soc.* **1997**, *119*, 3691-3695.
27. Ikeda, H.; Abushanab, E.; Marquez, V. E. *Bioorg. Med. Chem. Lett.* **1999**, *9*, 3069-3074.
28. Saady, M.; Lebeau, L.; Mioskowski, C. *J. Org. Chem.* **1995**, *60*, 2946-2947.
29. Saady, M.; Lebeau, L.; Mioskowski, C. *Synlett.* **1995**, 643-644.
30. Saady, M.; Lebeau, L.; Mioskowski, C. *Tetrahedron Lett.* **1995**, *36*, 2239-2242.
31. Kirby, A. J.; Warren, S. G. *The Organic Chemistry of Phosphorous*; Elsevier: New York, 1967.
32. Pudovik, A. N.; Konovalova, I. V. *Synthesis* **1979**, 81-

Inhibitor Design and Clinical Applications

Chapter 11

C-Nucleoside Analogs of Tiazofurin and Selenazofurin as Inosine 5′-Monophosphate Dehydrogenase Inhibitors

P. Franchetti[1], L. Cappellacci[1], M. Grifantini[1,*], H. N. Jayaram[2], and B. M. Goldstein[3]

[1]Dipartimento di Scienze Chimiche, Università di Camerino, Via S. Agostino, 1. 62032 Camerino, Italy
[2]Department of Biochemistry and Molecular Biology, Indiana University School of Medicine, 635 Barnhill Drive, MS 4053, Indianapolis, IN 46202–5122
[3]Department of Biochemistry and Biophysics, University of Rochester Medical Center, 601 Elmwood Avenue, Rochester, NY 14642

Inosine monophosphate dehydrogenase (IMPDH), the rate limiting enzyme in the de novo synthesis of guanine nucleotides, catalyzes the oxidation of inosine monophosphate (IMP) to xanthosine monophosphate (XMP). Because of its critical role in purine biosynthesis, IMPDH is a drug design target for antineoplastic, antiinfective, and immunosuppressive chemotherapy. Two type of IMPDH inhibitors are currently in clinical use or under development: nucleoside inhibitors, such as ribavirin, tiazofurin and mizoribine, and non-nucleoside, as mycophenolic acid. Tiazofurin, and its selenium analog selenazofurin, are *C*-glycosyl nucleosides endowed with antitumor activity which have to be converted in sensitive cells to the active forms, the dinucleotides NAD$^+$ analogs TAD and SAD, respectively. It was hypothesized that the inhibitory activity of both tiazofurin and selenazofurin is due to an attractive electrostatic interaction between the heterocyclic sulfur or selenium atom and the furanose oxygen 1'. This interaction constrains rotation about the *C*-glycosidic bond in tiazofurin and selenazofurin and in their active anabolites TAD and SAD. This hypothesis was confirmed by the investigation of several *C*-nucleosides related to these compounds. Structure-activity relationships studies revealed that S or Se in position 2 in thiazole or selenazole moiety with respect to the glycosidic bond is essential for cytotoxicity and IMPDH inhibitory activity of tiazofurin and selenazofurin, while the N atom is not. Computational methods suggested that rotational constraint around the *C*-glycosidic bond is determined both by favorable intramolecular (1-4) electrostatic interaction between the partial positive sulfur or selenium and the negative oxygen of the ribose and unfavorable van der Waals contacts between the heteroatoms and the ribose C2'-H and O4'.

Inosine monophosphate dehydrogenase (IMPDH; EC 1.1.1.205), which catalyzes the oxidation of inosine 5'-monophosphate (IMP) to xanthosine 5'-monophosphate (XMP) with concomitant reduction of NAD to NADH at the metabolic branch point in the *de novo* purine nucleotide synthetic pathway, is the rate-limiting enzyme for the guanine nucleotides biosynthesis. Blocking the conversion of IMP to XMP, IMPDH inhibitors lead to depletion of the guanylate (GMP, GDP, GTP and dGTP) pools, and, since GTP is a cofactor in the conversion of IMP to AMP (via succinyl AMP), ATP and dATP pools are depleted as well (*1-3*). IMPDH was shown to be increased significantly in cancer cells and therefore considered to be a sensitive target for cancer chemotherapy (*4-8*). It has proven to be a target not only for anticancer drugs (*8*), but also for antiviral, anti-infective, and immunosuppressive chemotherapy (*9-18*). Because IMPDH inhibitors lead to an accumulation of the intracellular pool levels of IMP, which can serve as phosphate donors for the phosphorylation by 5'-nucleotidases of 2',3'-dideoxynucleosides, these inhibitors are able to potentiate the anti-HIV activity of this type of nucleosides such as 2',3'-dideoxyinosine (ddI) (*3, 19*).

Inhibition of IMPDH is associated with changes in nucleic acid synthesis, gene expression, signaling and, ultimately, cell proliferation and differentiation (*20-22*). This inhibition also appears to compromise the ability of G proteins to fuction as transducers of intracellular signal (*22-26*). As a consequence *c-myc* and/or *Ki-ras* oncogenes are down-regulated in a variety of human tumor cell lines and in blast cells of leukemic patients treated with an IMPDH inhibitor (*27, 28*). IMPDH inhibitors have also been successfully employed in immunosuppressive therapy. Thus, mycophenolate mofetil (CellCept(R)), a prodrug of mycophenolic acid (MPA), and mizoribine (bredinin), two potent IMPDH inhibitors, are currently used as immunosuppressive agents in renal and heart transplant recipients.

The discovery of two IMPDH isoforms, labeled type I and type II, and of their differential regulation during neoplastic transformation, cancer cell differentiation and lymphocyte activation is attractive for possible novel development for isozyme-selective chemotherapy (*29-32*). Using specific cDNA probes for type I and type II IMPDHs, it was demonstrated that type I mRNA was the prevalent species in human normal leukocytes and lymphocytes, whereas type II predominated in human ovarian tumor cells, leukemic cell lines K562 and HL-60, and leukemic cells from patients with different types of leukemia (*29-32*). Enhanced expression of type II IMPDH was also demonstrated in solid tumors and cancer cell lines at mRNA and protein levels (*30, 33*).

Human type I and type II IMPDHs share 84% amino acid identity and show similar kinetic properties (*34*). The remarkable differences in type I and type II IMPDH gene expression suggest that the two isoforms may play different roles in *de novo* guanine nucleotide synthesis and have different regulatory mechanisms. Furthermore, the enhanced expression of the type II isoform in proliferating cells suggests that development of type II specific inhibitors may provide improved selectivity against target cells in anticancer and immunosuppressive chemotherapy. Several classes of IMPDH inhibitors are in clinical use or under development. These include agents that bind at either the substrate site (e.g. the nucleosides ribavirin and mizoribine) or at the NAD site (e.g. mycophenolic acid and other non-nucleoside compounds, and NAD analogs derived from *C*-nucleosides such as tiazofurin and selenazofurin). In this report we review some of our recent studies on structure-activity relationships of tiazofurin and its analogs.

Nucleoside IMPDH inhibitors and their mode of action

Several purine analogs, including 3-deazaguanine (**1**), the corresponding nucleoside (**2**), and 5'-nucleotide (**3**) (Figure 1) have been designed and synthesized by Robins and co-workers, starting from the observation that guanylic acid is known to be a natural inhibitor of IMPDH (*35*). These compounds showed broad-spectrum antiviral activity against a variety of DNA and RNA viruses and inhibitory action against several tumor cells. The biological activity of **1-3** has been associated with their ability to act as reversible IMPDH inhibitors (*35*). The active form of **1** is believed to be the 5'-monophosphate, since this was the most potent IMPDH inhibitor, acting at the substrate site. Several GMP analogs, structurally similar to the product of the IMPDH-catalyzed reaction, were found to be competitive IMPDH inibitors and to be active as antitumor agents (*35*).

Nucleoside analogs containing a five-membered heterocycle such as ribavirin, tiazofurin and mizoribine have been developped as reversible IMPDH inhibitors, and most of the research effort has been directed toward these compounds. Ribavirin, 1-β-D-ribofuranosyl-1,2,4-triazole-3-carboxamide (virazole, **4**), is a broad spectrum antiviral agent inhibiting the replication of a wide range of DNA and RNA viruses both *in vitro* and *in vivo* (*35*). Ribavirin is also active against various retroviruses, HIV included, and is in use for treatment of AIDS. Moreover, the combination of interferon-α and ribavirin was proved to be successful therapy for patients with chronic hepatitis C infection (*36*).

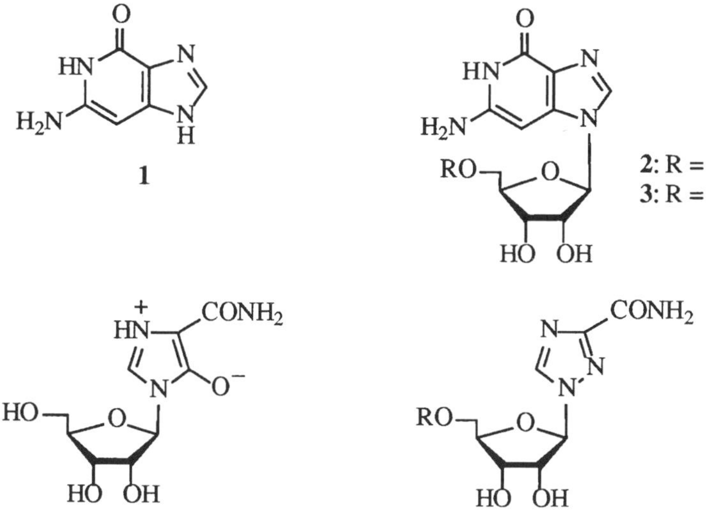

1

2: R = H
3: R = PO_3H_2

5: Mizoribine (Bredinin)

4: R = H (Ribavirin)
R = PO_3H_2 (Ribavirin-MP)

Figure 1. Structures of nucleoside IMPDH inhibitors.

Mizoribine (MZR, **5**), a nucleoside of the imidazole class, is a naturally occurring antibiotic isolated from the culture medium of the mold *Eupenicillium brefeldianum* M-2166. Although MZR proved to be active as an antitumor agent *in vitro*, it is used in Japan for the prevention of rejection in renal transplantation, and for the treatment of lupus nephritis, rheumatoid arthritis and the primary nephrotic syndrome (*37*). Ribavirin inhibits IMPDH after conversion to its 5'-monophosphate by cellular adenosine kinase and/or 5'-nucleotidase. This metabolite acts as substrate mimic of IMP (*9, 38*). Also mizoribine is metabolically activated by conversion into the 5'-monophosphate (MZR-5'P) which acts as a competitive transition state analog, binding IMPDH with a Ki of 0.5 nM respect to IMP (*37*).

Tiazofurin, 1-β-D-ribofuranosylthiazole-4-carboxamide (NSC 286193, **6**) (Figure 2), a C-glycosyl thiazole nucleoside synthesized by Robins and coworkers in 1977 (*39*), proved to be an antitumor agent active against a large number of tumor systems. It shows both *in vitro* and *in vivo* activity against human lymphoid, colon, liver, lung, renal, breast and ovarian tumor cells (*35*). Several phase I clinical trials have showed the efficacy of tiazofurin (Tr) for the treatment of patients with acute myelogenic leukemia or myeloid blast crisis and poor prognosis (*40*). In phase II trials tiazofurin has induced complete hematologic remission in patients with end-stage acute nonlymphocytic leukemia or in myeloblastic crisis of chronic myeloid leukemia (*41, 42*). Recently, the drug has been shown to induce apoptosis as well as differentiation in both the human leukemia K-562 and lymphoblastic MOLT-4 cell lines (*43*).

6: Tiazofurin	X = S,	Y = N, Z = O
7: Selenazofurin	X = Se,	Y = N, Z = O
8: Oxazofurin	X = O,	Y = N, Z = O
9: Thiophenfurin	X = S,	Y = CH, Z = O
10: Furanfurin	X = O,	Y = CH, Z = O
11: Selenophenfurin	X = Se,	Y = CH, Z = O
12: Furanthiofurin	X = O,	Y = CH, Z = S
13: Thiophenthiofurin	X = S,	Y = CH, Z = S
14: Imidazofurin	X = NH,	Y = N, Z = O

Figure 2. C-nucleoside analogs of tiazofurin and selenazofurin.

The efficacy of tiazofurin in the treatment of human leukemia appears to be related to its ability to kill tumor cells and to induce surviving cells to undergo maturation (*44*). Both the antiproliferative and maturation-inducing effects of the drug results from inhibition of IMPDH by the tiazofurin anabolite, thiazole-4-carboxamide adenine dinucleotide (TAD). This active anabolite is an analog of the cofactor NAD, in which the nicotinamde ring is replaced by a thiazole-4-carboxamide moiety, formed inside the cells from tiazofurin via its 5'-phosphate (TrMP) by the action of NMN adenylyltransferase (*44*). TAD is a noncompetitive inhibitor of IMPDH with respect to NAD, binding with a Ki of ~0.5 μM in both enzyme isoforms.

The selenium analog of tiazofurin, selenazofurin (2-β-D-ribofuranosyl-selenazole-4-carboxamide, 7) was found to be more cytotoxic than tiazofurin (44). Also selenazofurin is anabolized in the sensitive cells to its NAD analog (SAD). Interestingly, SAD binds IMPDH ~10 fold more tight than TAD, with Ki's of 0.06 and 0.03 μM with respect to NAD in the type I and II isoenzymes, respectively (45).

Structure-activity relationships

Several specific modifications of tiazofurin structure have been reported, but very few are tolerated (46). The amide function has been replaced by thioamide, amidine and imidate. The isomeric 2-β-D-ribofuranosylthiazole-5-carboxamide has also been synthesized. Structural modifications within the ribofuranosyl moiety have been reported including deoxygenated, ara- and xylofuranosyl, acyclic, pyranosyl, carbocyclic, and 5'-substituted analogs. Only carboxamidine, carboximidate and carbocyclic analogs of tiazofurin retain some biological activities; none of the other derivatives were significantly active against animal tumors.

It was found that certain conformational features are conserved in the thiazole and selenazole nucleosides. Crystal structures of tiazofurin and selenazofurin demonstrate a close intramolecular contact between the heterocyclic sulfur atom or selenium atom and the furanose ring oxygen 4' (47). Observed S/Se···O4' distances are significantly less than the sum of the sulfur or selenium and oxygen van der Waals radii. Quantum mechanical-based computational studies suggest that the close S/Se···O contacts observed in the thiazole and selenazole nucleosides are the result of an attractive electrostatic interaction between a negatively charged furanose oxygen and positively charged sulfur or selenium (47). The biological implications of this interaction are important. This interaction would be expected to constrain rotation about the C-glycosidic bond in the active anabolites TAD and SAD. Computational studies suggest that this constraint could be significant, creating a ~4 kcal/mol barrier to rotation about the C-glycosidic bond and favoring the conformation in which the base S or Se and furanose oxygen are adjacent (47). Thus, the S/Se···O interaction might influence the binding of the dinucleotide inhibitors to IMPDH.

In 1990 we started a research program aimed at the study of the structure-activity relationships of IMPDH inhibitors analogs of tiazofurin and selenazofurin. We first synthesized oxazofurin (8), an oxazole analog of tiazofurin and selenazofurin in which the S and Se heteroatoms are substituted by oxygen (48, 49). Oxazofurin proved to be inactive as antitumor and antiviral agent. So, it provided an ideal control for examining the significance of the S/Se···O interaction. The inactivity of oxazofurin might be due to the O---O1' repulsion which doesn't allow oxazole and furanose moieties in the putative anabolite oxazole-4-carboxamide adenine dinucleotide (OAD) to assume the right conformation to bind the enzyme. This hypothesis was supported by crystallographic data and quantum-mechanical-based computations (*50*).

To further investigate the structure-activity relationships of this type of *C*-nucleosides, we synthesized tiazofurin, oxazofurin and selenazofurin analogs in which the base was replaced by other five-membered ring heterocycles such as thiophene (thiophenfurin, **9**), furan (furanfurin, **10**) and selenophene (selenophenfurin, **11**) (*51, 52*). We found that thiophene and selenophene nucleosides in the solid state share several features observed in the structures of related thiazole and selenazole nucleosides. In fact, although the *C*-glycosidic torsion angle is somewhat higher than that observed in the referred nucleosides, the thiophene sulfur and selenophene selenium remains *cis* to the furanose oxygen, with a marginally close non-bonded S···O and Se···O contact of 3.04 Å and 3.12 Å, respectively. *Ab initio* molecular orbital studies suggested that this conformation was stabilized by an electrostatic interaction between the positively charged thiophene sulfur or selenophene selenium and the negatively charged furanose oxygen; this interaction is enhanced by an increase in positive charge on the heteroatom. This effect is similar to that noted for tiazofurin and selenazofurin.

Thiophenfurin was found to be cytotoxic *in vitro* toward murine lymphocytic leukemia P388 and L1210, human myelogenous leukemia K562, human promyelocytic leukemia HL-60, human colon adenocarcinoma LoVo, and B16 melanoma at concentrations similar to that of tiazofurin. In the same test, furanfurin proved to be inactive. Thiophenfurin was found active *in vivo* in BD_2F_1 mice inoculated with L1210 cells with a % T/C of 168 at 25 mg/kg. K562 cells incubation with thiophenfurin and selenophenfurin resulted in inhibition of IMPDH (52% and 76 %, respectively at 10 μM) and an increase in IMP pools (14.6 and 7-fold, respectively) with a concurrent decrease in GTP levels (40 and 58%, respectively). We found that thiophenfurin was more easily converted to the NAD analog (thiophene-3-carboxamide adenine dinucleotide, TFAD) than tiazofurin in myelogenous leukemia K562 cells, whereas furanfurin was converted to the NAD analog (FFAD) with only 10% efficiency. Thus, the inactivity of furanfurin as antitumor agent and IMPDH inhibitor may be due to its poor conversion to the dinucleotide in target cells and/or to the failure of the dinucleotide to inhibit the enzyme.

We hypothesized that the inactivity of furanfurin is due to an unfavorable glycosidic bond conformation by repulsive intramolecular O···O' interaction. This unfavorable conformation should be retained in FFAD. To check these hypotheses, we synthesized TFAD, FFAD and the selenophene NAD analog (SFAD) (Figure 3), and examined their ability to inhibit type I and type II mammalian IMPDH in comparison with TAD and SAD (*45*). These dinucleotides exhibited uncompetitive type of inhibition toward IMP and NAD substrates of both enzyme's isoforms. SAD was the most potent inhibitor showing a slight selectivity for type II. SFAD demonstrated similar inhibitory potency towards both types of IMPDH isoforms. TFAD proved to be slightly less potent than SFAD, and at the same time more potent than TAD exhibiting similar affinity toward IMPDH type I and type II. FFAD proved to be a poor inhibitor of IMPDH. The ranking of the inhibitory activity against IMPDH of these dinucleotides was: SAD > SFAD = TFAD = TAD >> FFAD, similar to the rank of the potency of corresponding C-nucleosides as IMPDH inhibitors in human myelogenous leukemia K562 cells. The data of the enzymes' inhibition confirm that the inactivity of furanfurin as antitumor agent is due, not only to its poor ability to be converted to the anabolite FFAD in target cells, but also to the low potency of this anabolite as an IMPDH inhibitor.

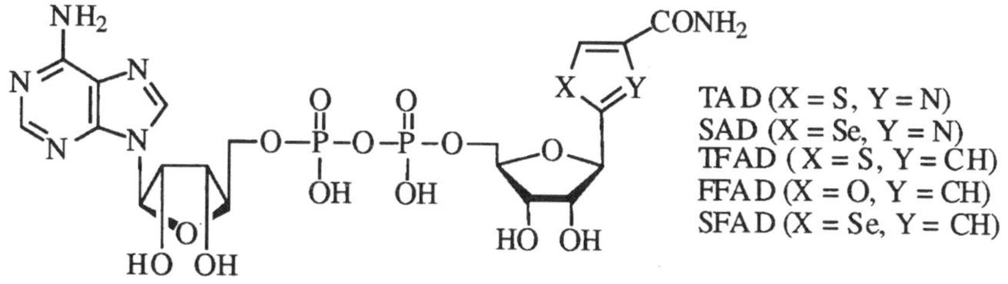

Figure 3. NAD analogs inhibitors of IMPDH.

To gain further information about the structure-activity relationships of this class of nucleosides, we have recently investigated furanfurin and thiophenfurin analogs in which the furanose ring oxygen O-4' was replaced by a sulfur atom (furanthiofurin, **12**, and thiophenthiofurin, **13**) (*53*). Furanthiofurin was synthesized by direct glycosylation of ethyl furan-3-carboxylate (**15**) with 1-*O*-acetyl-2,3,5-tri-*O*-benzyl-4-thio-D-ribofuranose (**16**) (Scheme 1). The glycosylation reaction performed in the presence of trifluoroacetic acid in methylene chloride afforded 2- and 5-glycosylated regioisomers as a mixture of α and β anomers. After removal of the benzyl groups using boron tribromide in methylene chloride and chromatographic separation, the 5-*β*-glycosylated ethyl ester **20β** was obtained. Treatment of **20β** with ammonium hydroxide gave furanthiofurin.

Thiophenthiofurin was synthesized in a similar way starting from ethyl thiophene-3-carboxylate (**21**) and using 1,2,3,5-tetra-*O*-acetyl-4-thio-D-ribofuranose (**22**) as glycosylation reagent in the presence of SnCl$_4$ in 1,2-dichloroethane. Also in this case a mixture of α and β anomers of 2- and 5-glycosylated regioisomers was obtained. After deprotection with methanolic

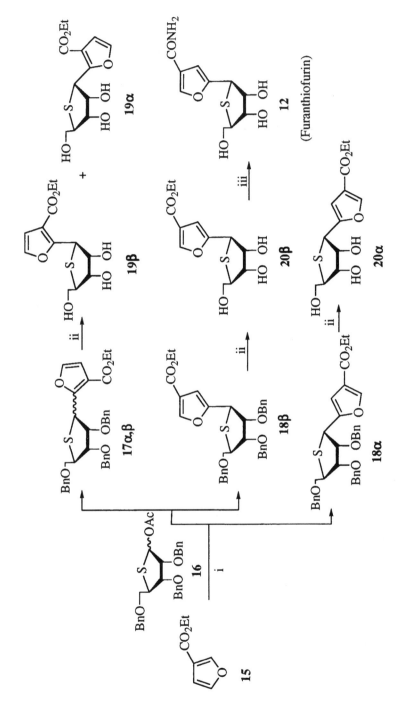

Scheme 1. Synthesis of furanthiofurin. (i) TFA, CH$_2$Cl$_2$; (ii) BBr$_3$/CH$_2$Cl$_2$, -78 °C; (iii) 30% NH$_4$OH.

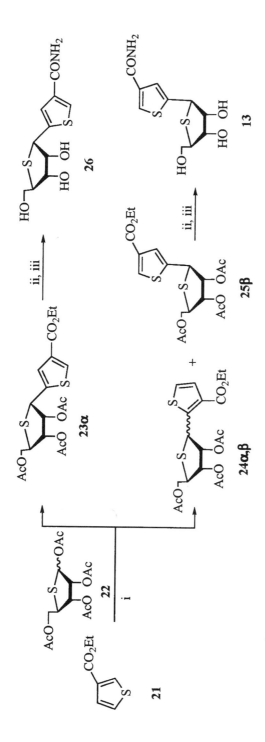

Scheme 2. Synthesis of thiophenthiofurin. (i) SnCl$_4$, ClCH$_2$CH$_2$Cl; (ii) NH$_3$/CH$_3$OH; (iii) 30% NH$_4$OH.

222

ammonia and treatment with ammonium hydroxide the ethyl ester **25β** was converted to thiophenthiofurin.

Furanthiofurin and thiophenthiofurin were evaluated for their ability to inhibit the growth of human myelogenous leukemia K562 cells using thiophenfurin as a reference compound. The IC_{50} (μM) values for these nucleosides are summarized in Table 1.

Furanthiofurin proved to be almost inactive until a maximum tested concentration of 100 μM. On the contrary, thiophenthiofurin showed cytotoxic

Table 1. Cytotoxicity and effect of compounds on IMPDH activity and on purine nucleotide concentration in human myelogenous leukemia K562 cells in culture

compd	Cytotoxicity IC_{50} (μM)[a]	IMPDH inhibition[b,c]	GTP pools[b,c]	IMP pools[b,c]
None	-	100.0	100.0	100.0
Thiophenfurin	1.7	76.0	96.6	88.2
Furanthiofurin	4700.0	75.2	66.1	105.0
Thiophenthiofurin	67.0	83.2	83.9	89.1

[a] Concentration required to inhibit 50% of cell proliferation at 48 h. [b] For studies related to IMPDH inhibition and nucleotide pools, cells were incubated with 2 μM thiophenfurin, or 100 μM of furanthiofurin or of thiophenthiofurin. [c] (% of control). Source: Reproduced from reference 53. Copyright 2000 American Chemical Society.

activity against K562 cells, albeit 39-fold lower than that of thiophenfurin. The inhibitory ability of these nucleosides against IMPDH from K562 cells in culture was also evaluated (Table 1). IMPDH activity was inhibited (25%) by furanthiofurin with a potency comparable to that of thiophenthiofurin (17% inhibition). This inhibition was ~50-fold lower than that exhibited by thiophenfurin.

Computational studies

A computational study on furanthiofurin and thiophenthiofurin in comparison with tiazofurin, selenazofurin, oxazofurin, furanfurin and thiophenfurin, was carried out (53). Heteroatom charges for these C-nucleosides are reported in Table 2. Charges were obtained at the DFT /631G**// HF/ 321G* level of theory using natural bond order partitioning as implemented in Gaussian 98 (*53*). For those compounds having a C-H group adjacent to the *C*-glycosidic carbon, the combined charge of the carbon and its attached proton is indicated.

Table 2. Comparison of heteroatom charges for tiazofurin, thiophenfurin, thiophenthiofurin, oxazofurin, furanfurin, and furanthiofurin

compd	X (charge)	Y (charge)	Z (charge)
Tiazofurin	S (+0.58)	N (-0.60)	O (-0.51)
Thiophenfurin	S (+0.52)	CH (-0.03)	O (-0.06)
Thiophenthiofurin	S (+0.52)	CH (-0.02)	S (+0.21)
Oxazofurin	0 (-0.48)	N (-0.50)	O (-0.58)
Furanfurin	0 (-0.46)	CH (-0.06)	O (-0.59)
Furanthiofurin	0 (-0.45)	CH (-0.06)	S (+0.22)

[a] Where Y = CH, the combined charge for the carbon, and its attached proton is listed.
Source: Adapted from reference 53. Copyright 2000 American Chemical Society.

General observations may be made based on the values shown in Table 2. Tiazofurin, thiophenfurin and thiophenthiofurin all contain a positively charged sulfur in the unsaturated five-membered heterocycle. Each of these compounds demonstrates activity, albeit to varying degrees. In oxazofurin, furanfurin and furanthiofurin, the positively charged sulfur is replaced by a negatively charged oxygen. These compounds show minimal or no activity. These observations suggest that an electrophilic atom adjacent to the *C*-glycosidic carbon is required for activity. This requirement may result from one or a combination of several factors (*53*).

The crystal structure of human IMPDH complexed with SAD, the selenium analog of the active tiazofurin anabolite, has been determined (*54*). The structure indicates an interaction between the SAD electrophilic heteroatom and the side chain of Gln 334 on the active site loop. Although the position of the loop is distorted in this complex by the presence of the 6-Cl analog of the IMP substrate, comparison with other structures suggests that a similar interaction may be maintained between the electrophilic heteroatom and Gln 441 when complexed with native substrate (54). The presence of a nucleophilic oxygen atom in place of the sulfur or selenium would destabilize this interaction. The presence of an electrophilic heteroatom would also be expected to influence binding of the parent compounds to anabolic enzymes. This may account for the variations in levels of NAD analogs observed between agents. The structure of the IMPH-SAD complex also suggests that the conformation about the C-glycosidic bond in the bound ligand is such that the Se atom and furanose oxygen remain *cis* to each other. Thus, conformational studies were carried out by *ab initio* computations and results compared in Figure 4 for tiazofurin, thiophenfurin, thiophenthiofurin, oxazofurin, furanfurin and furanthiofurin (53).

Tiazofurin shows a well defined minimum at χ = 0-20°, close to the conformation maintained in the enzyme-bound ligand. The barrier to rotation in tiazofurin is significantly higher than that found in the other agents, suggesting that tiazofurin anabolite may gain an entropic advantage in binding the target. Among the active agents, tiazofurin remains the most potent, and the most highly constrained. Thiophenfurin also shows an energetic miminum close to the favorable *cis* conformer (χ = 20°). However, replacement of the negatively charged thiazole nitrogen with the ~neutral C-H group also produces a local minimum at the unfavorable *trans* conformation (χ = 180°). Replacement of the oxolane oxygen with a sulfur in thiophenthiofurin results in a broader, shallower minimum in the favored region. Thus, any entropic advantage becomes succesively less significant in thiophenfurin and thiophenthiofurin respectively, as conformational constraints diminish. In the inactive compounds, replacement of the positively charged thiazole or thiophene sulfur with a negatively charged oxygen results in generally lower barriers to rotation. What constraints do exist are associated with global or local minimia shifted to higher, more unfavorable values of χ.

It is worth noting that the thiolane sulfur is electronically unique, mimicking neither the oxolane oxygen nor the thiazole or thiophene sulfur. Unlike the oxolane oxygen, the thiolane sulfur has a net positive charge. However, the magnitude of this charge is reduced to about half that of the more delocalized thiophene or thiazole sulfur. Thus, the conformational energy profiles for furan- and thiophenthiofurin are more heavily modulated by intramolecular steric and charge-transfer interactions. Nevertheless, the substitution of an electrophilic thiolane sulfur in place of the oxolane oxygen may be also be expected to influence binding of either the parent compounds or the NAD analogs to their targets. The net effect of this substitution is that neither furan- nor thiophenthiofurin appears to result in particularly effective IMPDH inhibition, despite wide variation in activity. These computational findings remain qualitatively consistent with biological results. Active compounds contain an

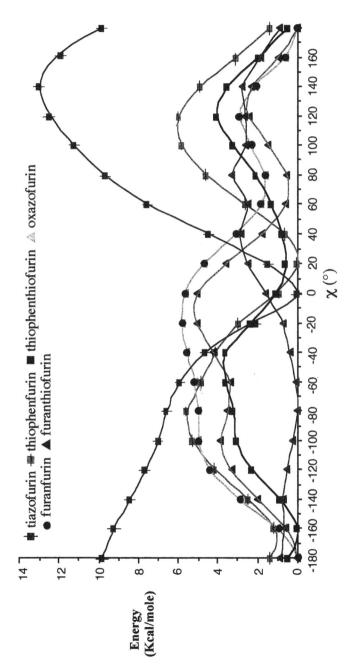

Figure 4. Energy vs C-glycosidic torsion angle χ for model fragments of tiazofurin, thiophenfurin, thiophenthiofurin, oxazofurin, furanfurin, and furanthiofurin. Each point on each curve was obtained at the RHF/3-21G*//3-21G* level. (Reproduced from reference 53. Copyright 2000 American Chemical Society.)

electrophilic sulfur in a delocalized environment adjacent to the C-glycosidic bond, and have an energetically favorable conformer around $\chi = 0°$. Among these, the more constrained (least flexible) compounds (tiazofurin and thiophenfurin) are more active than the less constrained thiophenthiofurin. Those compounds which contain a nucleophilic oxygen in place of the thiazole or thiophene sulfur appear less likely to adopt small angles of χ. These compounds (oxazofurin, furanfurin and furanthiofurin) show the least activity.

In 1997, Makara and Keserû have questioned the electrostatic hypothesis and suggested that the flexibility of the C-glycosidic bond in this type of C-nucleosides is ultimately determined by steric interactions of the heteroatoms with the C2'-H and O4' of the ribose (55). Application of this theory led these authors to design 2-β-D-ribofuranosylimidazole-4-carboxamide (imidazofurin, 14) as an analog that exhibits an almost identical behavior to tiazofurin in *ab initio* computations. This hypothesis prompted us to synthesize imidazofurin and to test its antitumor activity. Imidazofurin was synthesized as summarized in Scheme 3 (56). The reaction of 2,3,5-tri-O-benzyl-β-D-ribofuranosyl cyanide (27) with sodium methoxide in methanol gave the methyl imidate 28 which was reacted with 29 in methanol, to give intermediate 30. Treatment of 30 with methanolic ammonia produced amide 31, which was debenzylated by ammonium formate and Pd/C (10%) in methanol to afford imidazofurin in 35% yield.

Imidazofurin was evaluated for its ability to inhibit the growth of human myelogenous leukemia K562 cells. Tumor cell proliferation was evaluated by incubating the cells continuously with either the compound or saline for 48 h. Thiophenfurin was used as a reference compound. Contrary to what Makara and Keserû have predicted, imidazofurin proved to be non toxic to cell growth (no growth inhibition observed at 100 μM) as compared with thiophenfurin (IC_{50} = 4.6 μM). The poor activity of imidazofurin might be due to its inability to be phosphorylated by cellular kinases and nucleotidases, inability to be converted to the dinucleotide analog of NAD, or failure of the dinucleotide to bind to the target. We believe that the presence of imidazole annular prototropic tautomerism could destabilize the conformation of imidazofurin suitable to bind the enzyme. Makara and Keserû carried out their *ab initio* computations taking into consideration only the tautomeric form 14b. Thus, the contribution of tautomer 14a was neglected.

Conclusions

In conclusion, crystallographic and computation studies of a number of tiazofurin analogs and some of their dinucleotide anabolites have pointed out that

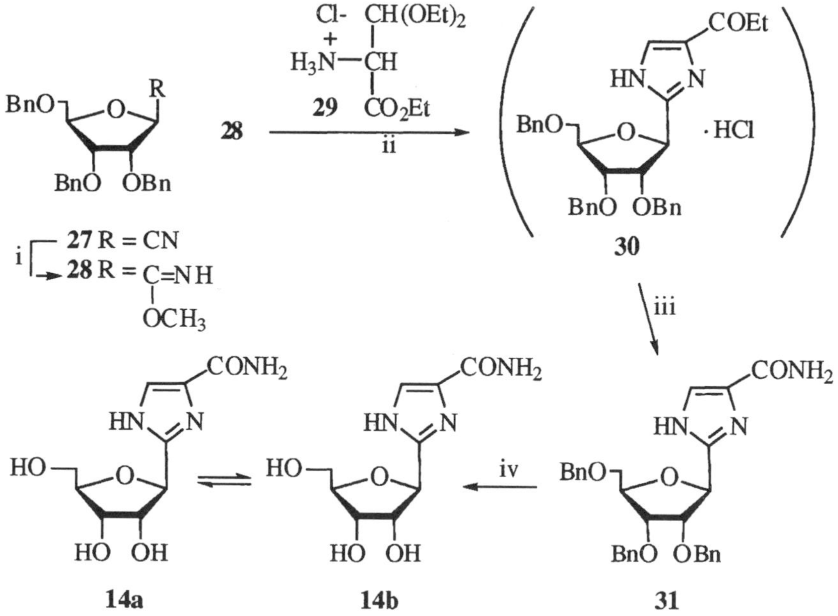

Scheme 3. Synthesis of imidazofurin. (i) CH_3ONa/CH_3OH (5 h, rt); (ii) CH_3OH (24 h, rt); (iii) NH_3/CH_3OH (2 h, rt); (iv) Pd/C, $HCOONH_4$, CH_3OH (1.5 h, reflux).

the ability to inhibit IMPDH of this type of C-nucleosides and their antitumor activity is conditioned by the intramolecular interaction between the heterocyclic atom (S, Se or O) and the furanose oxygen. This interaction, which is enhanced by an increase in positive charge on the heteroatom and compromised by an increase in negative charge, constrains rotation about the C-glycosidic bond. The constrain may be also attributed to other factors such as charge transfer interactions between antibonding orbitals on the heteroatom and the lone pair on the furanose oxygen (57, 58), and steric interactions between the heteroatom and ribose atom (55).

References

1. Lui, M. S.; Faderan, M. A.; Liepnieks, J. J.; Natsumeda, Y.; Olah, E.; Jayaram, H. N.; Weber, G. J. Biol. Chem. 1984, 259, 5078-5082.
2. Sokoloski, J. A.; Blair, O. C.; Sartorelli, A. C. Cancer Res. 1986, 46, 2314-2319.

228

3. Balzarini, J.; Lee, C.-K.; Herdewijn, P.; De Clercq, E. *J. Biol. Chem.* **1991**, *266*, 21509-21514.
4. Jackson, R. C.; Weber, G.; Morris, H. P. *Nature* **1975**, *256*, 331-333.
5. Weber, G. *N. Engl. J. Med.* **1977**, *296*, 541-551.
6. Weber, G. *Cancer Res.* **1983**, *43*, 3466-3492.
7. Natsumeda, Y.; Ikegami, T.; Murayama, K. *Cancer Res.* **1988**, *48*, 507-511.
8. Tricot, G.; Jayaram, H. N.; Lapis, E.; Natsumeda, Y.; Nichols, C. R.; Kneebone, P.; Heerema, N.; Weber, G.; Hoffman, R. *Cancer Res.* **1989**, *49*, 3696-3701.
9. Streeter, D. G.; Witkowski,J. T.; Khare, G. P.; Sidwell, R. W.; Bauer, R. J.; Robins, R. K.; Simon, L. N. *Proc. Natl. Acad. Sci. U.S.A.* **1973**, *70*, 1174-1178.
10. Wang, C. C.; Verham, R.; Chen, H.-W.; Rice, A.; Wang, A. L. *Biochem. Pharmacol.* **1984**, *33*, 1323-1329.
11. Nelson, P. H.; Eugui, E.; Wang, C. C.; Allison, A. C. *J. Med. Chem.* **1990**, *33*, 833-838.
12. Webster, H. K.; Whaun, J. M. *J. Clin. Invest.* **1982**, *70*, 461-469.
13. Quinn, C.; Bugeja, V.; Gallagher, J.; Whittaker, P. *Mycopathologia* **1990**, *111*, 165-168.
14. Wilson, K.; Berens, R. L.; Sifri, C. D.; Ullman, B. *J. Biol. Chem.* **1994**, *269*, 28979-28987.
15. Mizuno, K.; Tsujino, M.; Takada, M.; Hayashi, M.; Atsumi, K.; Asano, K.; Matsuda, T. *J. Antibiot.* **1974**, *27*, 775-782.
16. Abraham, E. P. *Biochem. J.* **1945**, *39*, 398-408.
17. O'Gara, M. J.; Lee, C.-H.; Weinberg, G. A.; Nott, J. M.; Queener, S. F. *Antimicrob. Agents Chemother.* **1997**, *41*, 40-48.
18. Hupe, D.; Azzolina, B.; Behrens, N. *J. Biol. Chem.* **1986**, *261*, 8363-8369.
19. Balzarini, J.; Lee, C.-K; Schols, D.; De Clercq, E. *Biochem. Biophys. Res. Commun.* **1991**, *178*, 563-569.
20. Weber, G.; Prajda, N.; Abonyi, M.; Look, K. Y.; Tricot, G. *Anticancer Res.* **1996**, *16*, 3313-3322.
21. Boritzki, T. J.; Berry, D. A.; Besserer, J. A.; Cook, P. D.; Fry, D. W.; Leopold, W. R.; Jackson, R. C. *Biochem. Pharmacol.* **1985**, *34*, 1109-1114.
22. Mandanas, R. A.; Leibowitz, D. S.; Gharehbaghi, K.; Tauchi, T.; Burgess, G. S.; Miyazawa, K.; Jayaram, H. N.; Boswell H. S. *Blood* **1993**, *82*, 1838-1847.
23. Manzoli, L.; Billi, A. M.; Gilmour, R. S.; Martelli, A. M.; Matteucci, A.; Rubini, S.; Weber, G.; Cocco, L. *Cancer Res.* **1995**, *55*, 2978-2980.
24. Kharbanda, S. M.; Sherman, M. L.; Kufe, D. W. *Blood* **1990**, *75*, 583-588.

25. Parandoosh, Z.; Robins, R. K.; Belei, M.; Rubalcava, B. *Biochem. Biophys. Res. Commun.* **1989**, *164*, 869-874.

26. Parandoosh, Z.; Rubalcava, B.; Matsumoto, S. S.; Jolley, W. B.; Robins, R. K. *Life, Sci.*, **1990**, *46*, 315-320.

27. Vitale, M.; Zamai, L.; Falcieri, E.; Zauli, G.; Gobbi, P.; Santi, S.; Cinti, C.; Weber, G. *Cytometry*, **1997**, *30*, 61-66.

28. Olah, E.; Csokay, B.; Prajda, N.; Kote-Jarai, Z.; Yeh, Y. A.; Weber, G. *Anticancer Res.*, **1996**, *16*, 2469-2477.

29. (a) Natsumeda, Y.; Ohno, S.; Kawasaki, H.; Konno, Y.; Weber, G.; Suzuki, K. *J. Biol. Chem.* **1990**, *265*, 5292-5295. (b) Carr, S. F.; Papp, E.; Wu, J. C.; Natsumeda, Y. *J. Biol. Chem.* **1993**, *268*, 27286-27290.

30. Konno, Y.; Natsumeda, Y.; Nagai, M.; Yamaji, Y.; Ohno, S.; Suzuki, K.; Weber, G. *J. Biol. Chem.* **1991**, *266*, 506-509.

31. Nagai, M.; Natsumeda, Y.; Konno, Y.; Hoffman, R.; Irino, S.; Weber, G. *Cancer Res.*, **1991**, *51*, 3886-3890.

32. Nagai, M.; Natsumeda, Y.; Weber, G. *Cancer Res.*, **1992**, *52*, 258-261.

33. Collart, F. R.; Chubb, C. B.; Mirkin B. L.; Huberman, E. *Cancer Res.* **1992**, *52*, 5826-5828.

34. (a) Holmes, E. W.; Pehlke, D. M.; Kelley, W. N. *Biochim. Biophys. Acta* **1974**, *364*, 209-217. (b) Xiang, B.; Taylor, J. C.; Markham, G. D. *J. Biol. Chem.* **1996**, *271*, 1435-1440.

35. Franchetti, P.; Cappellacci, L.; Grifantini, M. *Il Farmaco*, **1996**, 51, 457-469, and references cited therein.

36. Main, J.; McCarron, B.; Thomas, H. C. *Antiviral Chem. Chemother.* **1998**, 9, 449-460.

37. Ishikawa, H. *Current Med. Chem.* **1999**, 6, 575-597, and references cited therein.

38. Koyama, H.; Tsuji, M. *Biochem. Pharmac.*, **1983**, *32*, 3547-3351.

39. Srivastava, P. C.; Pickering, M.V.; Allen, L. B.; Streeter, D. G.; Campbell, M.T.; Witkowski, J. T.; Sidwell, R. W.; Robins, R. K. *J. Med. Chem.* **1977**, *20*, 256-62.

40. Maroun, J. A.; Stewart, D. J. *Invest. New Drugs* **1990**, *8*, S33-39.

41. Wright, D. G.; Boosalis, M. S.; Waraska, K.; Oshry, L. J.; Weintraub, L. R.; Vosburgh, E. *Anticancer Res.* **1996**, *16*, 3349-3351.

42. Tricot, G.; Weber, G. *Anticancer Res.* **1996**, *16*, 6A 3341-3347.

43. Jayaram, H. N.; Grusch, M.; Cooney, D. A.; Krupitza, G. *Current Med. Chem.* **1999**, *6*, 561-574, and references cited therein.

44. Jayaram, H. N.; Ahluwalia, G. S.; Dion, R. L.; Gebeyehu, G.; Marquez, V. E.; Kelley, J. A.; Robins, R. K.; Cooney, D. A.; Johns, D. G. *Biochem. Pharmacol.* **1983**, *32*, 2633-2636.

45. Franchetti, P.; Cappellacci, L.; Perlini, P.; Jayaram, H. N.; Butler, A.; Schneider, B. P.; Collart, F. R.; Huberman, E.; Grifantini, M. *J. Med. Chem.* **1998**, *41*, 1702-1707.

46. Franchetti, P.; Grifantini, M. *Current Med. Chem.* **1999**, *6*, 599-614, and references cited therein.

47. Burling, F. T.; Goldstein, B. M. *J. Am. Chem. Soc.*, **1991**, *114*, 2313-2320, and references cited therein.

48. Franchetti, P.; Cristalli, G.; Grifantini, M.; Cappellacci, L.; Vittori, S.; Nocentini, G. *J. Med. Chem.* **1990**, *33*, 2849-2852.

49. Franchetti, P.; Messini, L.; Cappellacci, L.; Grifantini, M.; Guarracino, P.; Marongiu, M. E.; Piras, G.; La Colla, P. *Nucleosides Nucleotides* **1993**, *12*, 359-368.

50. Goldstein, B. M.; Li, H.; Hallows, W. H.; Langs, D. A.; Franchetti, P.; Cappellacci, L.; Grifantini, M. *J. Med. Chem.*, **1994**, *37*, 1684-1688.

51. Franchetti, P.; Cappellacci, L.; Grifantini, M.; Barzi, A.; Nocentini, G.; Yang, H.; O'Connor, A.; Jayaram, H. N.; Carrell, C.; Goldstein, B. M. *J. Med. Chem.* **1995**, *38*, 3829-3837.

52. Franchetti, P.; Cappellacci, L.; Abu Sheikha, G.; Jayaram, H. N.; Gurudutt, V. V.; Sint, T.; Schneider, B. P.; Jones, W. D.; Goldstein, B. M.; Perra, G.; De Montis, A.; Loi, A. G.; La Colla, P.; Grifantini, M. *J. Med. Chem.* **1997**, *40*, 1731-1737.

53. Franchetti, P.; Marchetti, S.; Cappellacci, L.; Jayaram, H. N.; Yalowitz, J. A.; Goldstein, B. M.; Barascut, J-L.; Dukhan, D.; Imbach, J-L.; Grifantini, M. *J. Med. Chem.* **2000**, *43*, 1264-1270, and references cited therein.

54. (a) Colby, T. D.; Vanderveen, K.; Strickler, M. D.; Markham, G. D.; Goldstein, B. M. *Proc. Natl. Acad. Sci. USA* **1999**, *96*, 3531-3536; (b) Goldstein, B. M.; Colby, T. D. *Current Med. Chem.* **1999**, *6*, 519-536.

55. Makara, G. M.; Keserû, G. M. *J. Med. Chem.* **1997**, *40*, 4154-4159.

56. Franchetti, P.; Marchetti, S.; Cappellacci, L.; Yalowitz, J. A.; Jayaram, H. N.; Goldstein, B. M.; Grifantini, M. *Bioorg & Med. Chem. Lett.* **2001**, *11*, 67-69.

57. Barton, D. H. R.; Hall, M. B.; Lin, Z.; Parekh, S. I. *J. Am. Chem. Soc.* **1993**, *115*, 5056-5059.

58. Komatsu, H.; Iwaoka, M.; Tomoda, S. *Chem. Commun.* **1999**, 205-206.

Chapter 12

Studies with Benzamide Riboside, a Recent Inhibitor of Inosine 5′-Monophosphate Dehydrogenase

Hiremagalur N. Jayaram[1], Joel A. Yalowitz[1], Georg Krupitza[2], Thomas Szekeres[2], Karsten Krohn[3], and Krzysztof W. Pankiewicz[4]

[1]Department of Biochemistry and Molecular Biology, Indiana University School of Medicine, 635 Barnhill Drive, MS 4053, Indianapolis, IN 46202–5122
[2]University of Vienna, Wahringer Gurtel 18–20, A–1090 Vienna, Austria
[3]Division of Organic Chemistry, Universitat Gesamthochschule Paderborn, Paderborn 1621–4790, Germany
[4]Pharmasset, Inc., 1860 Montreal Road, Tucker, GA 30084

Benzamide riboside was synthesized as a potential inhibitor of poly(ADP-ribose)polymerase (PARP). However, it was a weak inhibitor of PARP, but instead an extremely potent inhibitor of cell proliferation. We demonstrated that it was a selective inhibitor of inosine 5'-monophosphate dehydrogenase (IMPDH). Benzamide riboside is converted intracellularly to an analogue of NAD, BAD (benzamide adenine dinucleotide) that is a potent inhibitor of IMPDH. We characterized the formation of BAD in cells, and it was followed by the chemical synthesis of BAD and its non-hydrolyzable methylene bis(phosphonate) analogues. On a molar basis, benzamide riboside exhibits more potent antitumor activity than tiazofurin, the first inhibitor of IMPDH that we had previously demonstrated to be acting through its NAD analogue, TAD (thiazole-4-carboxamide adenine dinucleotide). In the National Cancer Institute's panel of 60 tumors, benzamide riboside showed selective antitumor activity against human central nervous system tumors. In addition, benzamide riboside induces apoptosis in human ovarian carcinoma N.1 cells, a property not shared by tiazofurin. Induction of apoptosis by benzamide riboside is associated with a down-regulation of Cdc25A, a cell cycle specific phosphatase. New and potent analogues of benzamide riboside and its active metabolites are being synthesized to find specific inhibitors for IMPDH Type II, expression of which is up-regulated in tumor cells. IMPDH inhibitors are not only valuable as chemotherapeutic agents but also as probes for investigating biochemical functions of guanylates in intact cells.

Poly(ADP-ribose) polymerase [PARP] utilizes NAD as a substrate for the formation of poly(ADP-ribose)polymers that are utilized as a substrate for chromatin modification. In this process, protein-bound homopolymers are rapidly formed in response to DNA damage[1,2]. Therefore, action of agents (such as UV, alkylating agents, glucocorticoids, etc.) that induces DNA fragmentation stimulates the formation of protein bound poly(ADP-ribose) chains[2-5]. Nicotinamide and its analogues are known to potentiate cytotoxicity by inhibiting PARP, thereby increasing the strand breaks induced by DNA alkylation[6]. Although benzamide and related compounds are good *in vitro* inhibitors of PARP, they are known to be neurotoxic *in vivo* and thus difficult to use for cancer chemotherapy[7]. To overcome toxicity problem, Krohn and his associates[8] designed BR as an inhibitor of PARP. However, their studies suggested that BR was a poor inhibitor of PARP.

Inosine 5'-monophosphate dehydrogenase (IMPDH) oxidizes IMP to form XMP and in this process NAD accepts the electron. Thus, NAD and IMP are substrates for IMPDH. Our studies had demonstrated that tiazofurin by being converted into an analogue of NAD, TAD (thiazole-4-carboxamide adenine dinucleotide) was mimicking NAD and thus inhibited IMPDH utilization of NAD for the reaction[9-11]. Inhibition of IMPDH resulted in decreased guanylate levels including GTP and dGTP. TAD can also be hydrolyzed by a phosphodiesterase in resistant cells to tiazofurin 5'-monophosphate and AMP[12]. Further studies showed that the tiazofurin analogue selenazofurin, wherein selenium replaces sulfur on the thiazole ring, also acts as a potent IMPDH inhibitor by forming an analogue of NAD, SAD (selenazole-4-carboxamide adenine dinucleotide)[13]. Selenazofurin was 5-fold more cytotoxic to leukemic cells than tiazofurin, but was also more toxic to mice. Marquez and his associates chemically synthesized TAD and SAD[14].

TABLE I. COMPARE ANALYSIS OF TIAZOFURIN, BENZAMIDERIBOSIDE, AND OTHER IMPDH INHIBITORS

	TR	BR	SR	6-MP	3-DAG	ATH	PYF
TR	1.000						
BR	0.656	1.000					
SR	0.832	0.614	1.000				
6-MP	0.273	0.264	0.268	1.000			
3-DAG	0.323	0.178	0.310	0.618	1.000		
ATH	0.254	0.150	0.218	0.163	0.288	1.000	
PYF	0.503	0.259	0.510	0.382	0.489	0.165	1.000

TR, Tiazofurin; BR, Benzamide riboside; SR, Selenazofurin; 6-MP, 6-Mercaptopurine; 3-DAG, 3-Deazaguanine; ATH, Aminothiadiazole; and PYF, Pyrazofurin

COMPARE analysis to assign mechanism of action for benzamide riboside

As part of the screening program at the National Cancer Institute, compounds were tested for cytotoxicity against 60 different human tumor cell lines[15]. The data resulting from these tests were evaluated by utilizing a computer program, COMPARE developed by Dr. Kenneth Paull and his associates[16]. This program creates pair-wise correlations of the cell line data for a probe or "seed" compound with the corresponding cell line information for other compounds, permitting the estimation of the *in vitro* biochemical mechanisms of action of the "seed" compound. When tiazofurin (TR) was used as "seed" and the COMPARE analysis was performed, the correlation coefficients with BR (benzamide riboside) and SR (selenazofurin) were close to TR, suggesting that the two compounds might act through the same mechanism as TR[17]. As provided in Table I, it is evident that although other agents (6-mercaptopurine, 3-deazaguanine and aminothiadiazole) exhibit IMPDH inhibitory activity, they do not have close correlation coefficients. Pyrazofurin,

TABLE II. RELATIONSHIP OF INHIBITION OF IMPDH WITH BENZAMIDE RIBOSIDE CONCENTRATION

BR Concentration (μM)	IMPDH Activity (Percent of control)
0	100
10	90
25	63*
50	48*
100	7*

*Significantly different compared to saline controls ($p < 0.05$)

Human myelogenous leukemia K562 cells (1×10^7 cells/10 ml) were incubated with saline or indicated concentrations of BR for 2 hr at 37°C, and cells were harvested and IMPDH activity determined as cited[18]. Under the conditions of the assay, 17 ± 1.9 nmol of XMP/h/mg protein was formed in the cells incubated with saline.

an unrelated compound (an inhibitor of pyrimidine nucleotide biosynthesis) is included to demonstrate that it does not have any close resemblance in its correlation coefficient with IMPDH inhibitors. Therefore, it was of interest to conduct comparative examination of biochemical actions of TR, SR and BR in cancer cells.

Relationship of benzamide riboside and guanylate metabolism

We found that benzamide riboside was also metabolized in sensitive tumor cells to its 5'-monophosphate and then converted to its active metabolite, BAD (benzamide adenine dinucleotide)[18]. BAD inhibits IMPDH resulting in reduced guanylate synthesis. Relationship of the inhibition of IMPDH in human myelogenous leukemia K562 cells following incubation with graded concentrations of benzamide riboside is shown in Table II. This table demonstrates that there is a relationship between benzamide riboside dose and IMPDH inhibition in these human leukemic cells.

IMPDH inhibition should be associated with a decrease in guanylate nucleotide concentration in these cells. To ascertain this fact, we incubated K562 cells with benzamide riboside and examined the nucleotide pool sizes by high pressure liquid chromatography (HPLC). The results presented in Table III shows that guanylate (GMP and GTP) concentration was decreased without affecting adenylate concentration. Concurrently, IMP pools were elevated, suggesting that guanylate depletion was due to non-utilization of IMP for guanylate synthesis. We have always noticed that at the initial time points up to 2 hr, GDP levels were not influenced by IMPDH inhibitor action.

TABLE III. INFLUENCE OF BENZAMIDE RIBOSIDE TREATMENT ON THE LEVELS OF RIBONUCLEOTIDES IN HUMAN MYELOGENOUS LEUKEMIA K562 CELLS

Ribonucleotide	Control Concentration (nmol/g cells)	BR Treated (% of Control)
NAD	702.3 ± 24.6	87
IMP	10.5 ± 0.5	545*
GMP	7.8 ± 7.8	0*
GDP	62.0 ± 3.4	112
GTP	1199.4 ± 42.9	49*
AMP	52.8 ± 2.2	124
ADP	298.8 ± 9.8	157*
ATP	6023.1 ± 201.1	99

*Significantly different compared to saline controls ($p < 0.05$)

Human myelogenous leukemia K562 cells (1 x 10^7 cells/10 ml) were incubated with saline or 10 μM BR for 2 hr at 37°C. Cells were harvested and nucleotide levels were determined as cited[17].

The biochemical effects exerted by benzamide riboside on ribonucleotide pools were abrogated by replenishing guanylate concentrations with the addition of guanosine[18]. We examined the influence of benzamide riboside on deoxynucleotide concentration in K562 cells. This study provided in Table IV (taken from Reference 18) demonstrated a dose-related decrease in dGTP levels[18]. The concentrations of dCTP and dTTP were not altered due to treatment. The level of dATP was not affected when treated with 1 μM BR, whereas at higher concentrations of BR there was a decrease in dATP concentration[18].

TABLE IV. EFFECT OF BENZAMIDE RIBOSIDE ON 2'-DEOXYRIBONUCLEOTIDE POOLS IN HUMAN MYELOGENOUS LEUKEMIA K562 CELLS[1]

Nucleotides	Control Concentration (pmol/10^6 cells)	Percent of control		
		1 μM	2.5 μM	5 μM
DCTP	5.3 ± 0.4	95	82	96
DTTP	12.3 ± 1.2	91	107	116
DATP	47.4 ± 1.9	84	77	71*
DGTP	8.2 ± 0.1	82	60*	49*

[1]Data reproduced from Ref. 18.
*Significantly different from Control values (p <0.05)
K562 cells (1 x 10^8 cells) were incubated with saline or indicated concentrations of BR for 4 hr at 37°C. The cellular levels of deoxyribonucleotides were determined . Values of means ± SD of duplicate samples.

To validate the COMPARE analysis with the actual mechanism of action studies, we compared the biochemical actions of tiazofurin (TR), benzamide riboside (BR) and selenazofurin (SR) since all three compounds have very close correlation coefficients.

Mechanism of benzamide riboside resistance

Further indication that TR, BR and SR might share similar mechanism of activation came from the cells selected for resistance to TR and BR (Table V). We examined the sensitivity of TR, BR and SR in the cells selected for resistance to either TR or BR. These cells grew at the same rate as the sensitive cells in the presence of 2mM TR[19] (K562/TR) or in the presence of 0.1 mM BR[20] (K562/BR). It is evident from the results that there was cross resistance to TR and SR in the K562/BR cells and cross resistance to BR and SR in K562/TR cells. Earlier studies[19,20] have shown that in both tiazofurin-resistant and

TABLE V. CYTOTOXICITY OF TR, BR AND SR TO SENSITIVE AND RESISTANT CELLS

| | IC_{50} $[\mu M]$ | | |
Agent	K562/S	K562/TR	K562/BR
TR	9.1	12000	225.0
BR	2.5	1200	148.0
SR	0.5	1400	33.5

TR resistant K562 cells were selected by subculturing sensitive K562 cells with sublethal concentrations of TR. After about 60 generations, cells were selected which grew in the presence of 2 mM TR[19] at the same rate as the sensitive cells. BR resistant cells were similarly selected for resistance to BR and the cells grew in the presence of 100 μM BR[20] at a rate similar to that of sensitive cells.

benzamide riboside-resistant variants, the enzyme responsible for the conversion of their respective monophosphates to NAD analogues, NMN adenylyltransferase (NAD pyrophosphorylase) was decreased to almost undetectable levels. Therefore, these resistant cells formed very little of active metabolite, TAD or BAD leading to very little inhibition of cellular IMPDH and hence exhibited resistance. NMN adenylyltransferase activity in the hepatoma 3924A cells resistant to TR was only decreased to 53% of sensitive cells and yet the hepatoma 3924A/TR cells exhibited resistance to TR[21]. In the K562/TR cells, the activity of the target enzyme IMPDH was unaltered, whereas in the K562/BR cells, IMPDH activity was 3-fold amplified. In a line of hepatoma 3924A cells rendered resistant to TR, the activity of IMPDH was increased 2.6-fold[21].

Synthesis of NAD analogues by IMPDH inhibitors and influence on IMPDH inhibition

Further studies were conducted to validate COMPARE analyses, K562 cells were labeled with adenosine that would label all adenylates including ATP. Since ATP is utilized by NMN adenylyltransferase along with the respective 5'-monophosphates of TR, SR, or BR to synthesize the respective radiolabeled TAD, SAD and BAD that could be analyzed on HPLC. The results presented in Table VI indicate that BR formed the most of the NAD analogue, followed by SR and TR. K562 cells converted BR avidly and in 2 hr about 10% of the agent was converted to its active metabolite, BAD. The cells synthesized BAD about 2-3-fold more than that of SAD or TAD.

TABLE VI. SYNTHESIS OF NAD ANALOGUES OF BR (BAD), SR (SAD) , AND TR (TAD)

| Agent | NAD Analogue concentration | |
	(nmol/10^7 cells)	μM
BR	10.91	1.091
SR	5.66	0.566
TR	4.04	0.404

K562 cells (1 x 10^7) in culture were preincubated with [2,8-^{3}H]adenosine (10 μCi/flask; 200 mCi/mmol) for 2 hr at 37°C and then exposed to 10μM each of the agents for 2 hr. Drug metabolites was analyzed on HPLC[18].

All three NAD analogues were examined for their interaction with the NAD site of human IMPDH, Type I and Type II. The results are compiled in Table VII and they indicate that SAD was 5-10-fold more potent than TAD and BAD in inhibiting both IMPDH Type I and Type II. None of these analogues showed absolute specificity towards either Type I or Type II IMPDH. However, all three NAD analogues were very potent in inhibiting IMPDH activity.

TABLE VII. INHIBITION OF PURIFIED IMPDH BY TAD, SAD AND BAD

| NAD Analogues | IMPDH Inhibition [K_i, μM] | |
	Type I	Type II
TAD[22]	0.47	0.44
SAD[23]	0.06	0.03
BAD[24]	0.78	0.88

Cytotoxicity of IMPDH inhibitors to human tumor cells

We next compared the patterns of toxicity of BR, SR and TR to the human tumor cell lines. Screening program at the National Cancer Institute has developed a panel of 60 different human tumor cell lines representing leukemia, non-small cell lung cancer, small cell lung cancer, colon cancer, melanoma ovarian cancer and renal cancer, against which new compounds are examined for their sensitivity[15]. The data generated from the cytotoxicity studies were evaluated by a computer COMPARE analyses[16]. The results of the sensitivity of TR, BR and SR, are presented in Table VIII and parts of them are published[18,21]. SR and BR demonstrated 3- to 10-fold greater cytotoxicity than tiazofurin to human leukemic cell lines. With few exceptions, lung and renal cancer cell lines were exceptionally sensitive to BR than SR or TR. Colon and melanoma cell lines were equally sensitive to BR and SR but less sensitive to TR. CNS tumors were selectively sensitive to BR. Of the three agents, SR showed greater cytotoxicity to ovarian cancer cell lines than BR or TR. Overall, BR and SR exhibited 3-10-fold greater potency against the sixty human tumor cell line systems compared to TR.

Effect of IMPDH inhibitors on nucleotide synthesis

The effect of BR, TR and SR on IMPDH activity of K562 cells was examined (Figure 1). Following 2 hr incubation with equimolar concentration (10 μM) of each of the agents, IMPDH activity was decreased by 71% in the case of SR, 49% with BR and 26% with TR. This is in accordance with the degree of sensitivity of leukemic cells to these agents.

When the influence of BR, SR and TR on the ribonucleotide concentration was examined (Table IX), none of the three agents influenced ATP concentration. However, they did significantly reduce the pool sizes of GMP and GTP with a concurrent increase in IMP level.

To examine differences between the three agents further, the effect on deoxyribonucleoside triphosphates was next examined[24]. The results are provided in Table X and they indicate that treatment with TR, SR or BR significantly decreases only dGTP pools.

Summary

In summary, tiazofurin, selenazofurin and benzamide riboside exhibit close correlation coefficients in the COMPARE algorithm program demonstrating that they share similar mechanism of drug metabolism leading to the formation of NAD analogues that are potent inhibitors of IMPDH. Tiazofurin was the first in

TABLE VIII. CYTOTOXICITY OF TR, BR AND SR TO HUMAN TUMOR CELL LINES[1]

Disease category	Cell line	IC_{50} [μM]					
		TR	D^2	BR	D^2	SR	D^2
Leukemia	CCRF-CEM	9.5	S	0.6	S	0.8	S
	HL-60	8.7	S	1.3	S	0.9	S
	K562	4.5	S	1.6	S	0.6	S
	MOLT-4	5.5	S	0.4	S	0.5	S
Non-Small Lung	A549/ATCC	97.1	R	15.2	S	8.6	S
	HOP-18	1170.6	R	26.0	M	>100	R
	HOP-62	130.7	R	5.4	S	10.2	S
	HOP-92	113.3	R	4.3	S	>100	R
	NCI-H226	662.6	R	16.5	M	58.6	R
	NCI-H23	62.4	R	7.8	S	11.0	S
	NCI-H322M	39.0	M	38.1	M	6.0	S
	NCI-H460	211.0	R	2.2	S	17.4	M
	NCI-H522	48.1	R	15.1	S	6.2	S
Small Lung	DMS 114	61.1	R	15.3	S	25.4	M
	DMS 273	30.8	M	7.6	S	3.5	S
Colon	COLO 205	7.0	S	6.4	S	1.2	S
	DLD-1	8.5	S	5.4	S	1.8	S
	HCC-2998	315.6	R	26.4	S	32.4	S
	HCT-116	19.0	M	11.1	S	3.4	S
	HCT-15	4.2	S	2.0	S	0.5	S
	HT-29	44.0	M	10.4	S	3.5	S
	KM-12	43.1	M	4.1	S	0.5	S
	SW-620	13.6	S	4.2	S	1.4	S
CNS	SF-268	66.8	R	3.1	S	6.8	S
	SF-295	12.8	S	4.9	S	2.3	S
	SF-539	17.9	M	2.0	S	2.4	S
	SNB-19	35.4	M	6.8	S	6.3	S
	SNB-75	31.4	M	2.6	S	59.0	S
	U-251	8.7	S	5.6	S	0.9	S
	XE 498	63.0	R	2.0	S :	8.9	S
Melanoma	LOX IMVI	266.5	R	26.1	S	32.1	M
	M19-MEL	29.5	M	21.0	S	5.2	S
	SK-MEL-5	82.4	R	18.2	M	10.3	S
	SK-MEL-28	13.0	S	3.3	S	2.6	S
	UACC-62	40.0	M	17.9	S	12.2	S
Ovarian	IGROVI	40.9	M	23.3	M	4.9	S
	OVCAR-3	15.6	M	13.9	S	0.7	S
	OVCAR-5	440.7	R	38.4	M	31.0	M
	OVCAR-8	27.5	M	2.4	S	4.3	S
	SV-OV-3	121.6	R	4.0	S	28.9	M
Renal	CAKI-1	57.9	R	11.8	S	6.9	S
	SN12C	478.0	R	25.6	S	84.6	R
	UO-31	50.4	R	3.2	S	6.5	S

[1]Data compiled from References 18 and 21

D^2Degree of sensitivity of human tumors towards TR, BR and SR. Symbols S, M and R represent Sensitive (<15 μM), Moderately sensitive (15-40 μM) and Resistant (>40 μM) to cell lines, respectively.

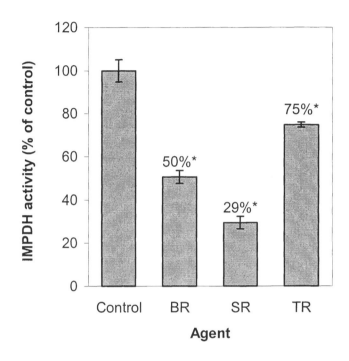

Figure 1: Effect of benzamide riboside (BR), selenazofurin (SR) and tiazofurin (TR) on IMPDH activity. Logarithmically growing K562 cells (1 x 10[7] cells) were incubated with saline or 10 μM each of BR, SR or TR for 2 hr and IMPDH activity was assayed[17]

TABLE IX. EFFECT OF BENZAMIDE RIBOSIDE, SELENAZOFURIN AND TIAZOFURIN ON RIBONUCLEOTIDE CONCENTRATION IN HUMAN MYELOGENOUS LEUKEMIA K562 CELLS[1]

| Nucleotide | Control Concentration (pmol/10^6 cells) | Percent of Control | | |
		BR	SR	TR
ATP	3966 ± 16	100	96	102
IMP	20 ± 1	691*	934*	805*
GMP	16 ± 3	27*	15*	22*
GDP	142 ± 2	90	85	81*
ATP	836 ± 16	54*	44*	63*

[1] Source: Reproduced with permission from reference 24. Copyright 1994 Elsevier.

*Significantly different compared to Control (p <0.05)

Human myelogenous leukemia K562 cells (1 x 10^7 cells/10 ml) were incubated with saline or 10 µM BR, SR or TR for 2 hr at 37°C. Cells were harvested and nucleotide levels were determined as cited[17,24].

TABLE X. EFFECT OF BENZAMIDE RIBOSIDE, SELENAZOFURIN AND TIAZOFURIN ON THE LEVELS OF 2'-DEOXYNUCLEOSIDE TRIPHOSPHATES IN HUMAN MYELOGENOUS LEUKEMIA K562 CELLS[1]

| dNTP | Control Concentration (pmol/10^6 cells) | Percent of Control | | |
		BR	SR	TR
dCTP	60 ± 4	183	130	153
dTTP	13 ± 1	161	153	146
dATP	30 ± 2	100	100	83
dGTP	39 ± 1	43*	41*	38*

[1] Source: Reproduced with permission from reference 24. Copyright 1994 Elsevier.

K562 cells (5 x 10^7 cells) were incubated with saline or 10 µM each of BR, SR or TR for 4 hr at 37°C. Concentration of deoxynucleoside triphosphates was analyzed by HPLC[24].

this series and has shown good activity in the treatment of acute myelogenous leukemia in humans[25-27]. All three compounds are C-nucleosides, and as IMPDH inhibitors they will find new applications. Selenazofurin, although more active than TR, is also more toxic to mice. Benzamide riboside is the most recent IMPDH inhibitor belonging to this class. Our studies have demonstrated that it is more active than tiazofurin. In addition, benzamide riboside, unlike tiazofurin, induces apoptosis without down-regulating *c-myc* expression[28,29]. Benzamide riboside induces apoptosis in human ovarian carcinoma N.1 cells, whereas tiazofurin does not. Induction of apoptosis with benzamide riboside is associated with down-regulation of Cdc25A expression[28,29]. Future studies with benzamide riboside to determine its *in vivo* activity, therapeutic index, toxicity and bioavailability should determine its therapeutic use. Synthesis of a new and potent methylenebis(phosphonate) analogue of BAD, the active metabolite of BR, has been reported[30]. Synthesis of novel analogues as specific inhibitors of IMPDH, Type II that is preferentially expressed in proliferating cells, such as cancer, is in progress.

References

1. Ueda, K. and Hayaishi, O. ADP-ribosylation. *Annu. Rev. Biochem.* **1985,** *54,* 73-100.

2. Juerez-Salinas, H., Sims, J.L. and Jacobsen, M.K. Poly(ADP-ribose) in carcinogen-treated cells. *Nature (London)* **1979,** *282,* 740-741.

3. Wielckens, K., Schmidt, A, George, E., Bredehorst, R. and Hilz, H. DNA fragmentation and NAD depletion. *J. Biol. Chem.* **1982,** *257,* 12872-12877.

4. Jacobsen, E.L., Antol, K.A., Juarez-Salinas, H. and Jacobson, M.K. Poly(ADP-ribose) metabolism in ultraviolet irradiated human fibroblasts. *J. Biol. Chem.* **1983,** *258,* 103-107.

5. Wielckens, K. and Delfs, T. Glucocorticoid-induced cell death and poly[adenosine diphosphate (ADP)-ribosylation]: Increased toxicity of dexamethasone on mouse S49.1 lymphoma cells with poly(ADP-ribosyl)ation inhibitor benzamide. *Endocrinology* **1986,** *119,* 2383-2392.

6. Durkacz, B.W., Omidiji, O., Gray, D.A. and Shall, S. (ADP-ribose)$_n$ participates in DNA excision repair. *Nature (London)* **1980,** *283,* 593-596.

7. Cosi. C., Chopin, P. and Marien M. Benzamide, an inhibitor of poly(ADP-ribose) polymerase, attenuates methamphetamine-induced dopamine neurotoxicity in the C57B1/6N mouse. *Brain Research* **1996,** *735,* 343-8.

8. Krohn, K., Heins, H. and Wielckens, K. Synthesis and cytotoxic activity of C-glycosidic nicotinamide riboside analogues. *J. Med. Chem.* **1992,** *35,* 511-517.

244

9. Jayaram, H.N., Dion, R.L., Glazer, R.I., Johns, D.G., Robins, R.K., Srivastava, P.C. and Cooney, D.A. Initial studies on the mechanism of action of a new oncolytic thiazole nucleoside, 2-β-D-ribofuranosylthiazole-4-carboxamide. *Biochem. Pharmacol.* **1982**, *31*, 2371-2380.

10. Jayaram, H.N., Dion, R.L., Glazer, R.I., Johns, D.G., Robins, R.K., Srivastava, P.C. and Cooney, D.A. Studies on the mechanism of action of 2-β-D-ribofuranosylthiazole-4-carboxamide (NSC 286193). II. Relationship between dose level and biochemical effects in P388 leukemia *in vivo*. *Biochem. Pharmacol.* **1982**, *31*, 2557-2560.

11. Cooney, D.A., Jayaram, H.N., Gebeyehu, G., Betts, C.R., Kelley, J.A., Marquez, V.E. and Johns, D.G. The conversion of 2-β-D-ribofuranosylthiazole-4-carboxamide to an analogue of NAD with potent IMP dehydrogenase-inhibitory properties. *Biochem. Pharmacol.* **1982**, *31*, 2133-2136.

12. Ahluwalia, G.S., Cooney, D.A., Marquez, V.E., Jayaram, H.N. and Johns, D.G. Studies on the mechanism of action of 2-β-D-ribofuranosylthiazole-4-carboxamide. VI. Biochemical and pharmacological studies on the degradation of thiazole-4-carboxamide adenine dinucleotide (TAD). *Biochem. Pharmacol.* **1986**, *35*, 3783-3790.

13. Jayaram, H.N., Ahluwalia, G.S., Dion, R.L., Gebeyehu, G., Marquez, V.E., Kelley, J.A., Robins, R.K., Cooney, D.A. and Johns, D.G. Conversion of 2-β-D-ribofuranosylselenazole-4-carboxamide to an analogue of NAD with potent IMP dehydrogenase inhibitory properties. *Biochem. Pharmacol.* **1983**, *32*, 2633-2636.

14. Gebeyehu, G., Marquez, V.E., Van Cott, A., Cooney, D.A., Kelley, J.A., Jayaram, H.N., Ahluwalia, G.S., Dion, R.L., Wilson, Y.A. and Johns, D.G. Ribavirin, tiazofurin, and selenazofurin: Mononucleotides and nicotinamide adenine dinucleotide analogues. Synthesis, structure and interactions with IMP dehydrogenase. *J. Med. Chem.* **1985**, *28*, 99-105.

15. Monks, A., Scudiero, D., Skehan, P., Shoemaker, R., Paull, K., Vistica, D., Hose, C., Langley, J., Cronise, P., Walgro-Wolff, A. Gray-Goodrich, M., Campbell, H., Mayo, J. and Boyd, M. Feasibility of a high-flux anticancer drugs screen utilizing a diverse panel of human tumor cell lines in culture. *J. Natl. Cancer Inst.*, **1991**, *83*, 757-766.

16. Paull, K.D., Shoemaker, R., Hodes, L., Monks, A., Scudiero, D., Rubinstein, L., Plowman, J. and Boyd, M.R. Display and analysis of patterns of differential activity of drugs against human tumor cell lines: Development of mean graph and COMPARE algorithm. *J. Natl. Cancer Inst.*, **1989**, *81*, 1088-1092.

17. Jayaram, H.N., Gharehbaghi, K., Jayaram, N.H., Rieser, J., Krohn, K. and Paull, K.D. Cytotoxicity of a new IMP dehydrogenase inhibitor, benzamide riboside, to human myelogenous leukemia K562 cells. *Biochem. Biophys. Res. Commun.* **1992**, *186*, 1600-1606.

18. Gharehbaghi, K., Paull, K.D., Kelley, J.A., Barchi, Jr. J.J., Marquez, V.E., Cooney, D.A., Monks, A., Scudiero, D., Krohn, K. and Jayaram, H.N. Cytotoxicity and characterization of an active metabolite of benzamide riboside, a novel inhibitor of IMP dehydrogenase. *Int. J. Cancer,* **1992**, *56*, 892-899.

19. Jayaram, H.N., Zhen, W. and Gharehbaghi, K. Biochemical consequences of resistance to tiazofurin in human myelogenous leukemia K562 cells. *Cancer Res.,* **1993**, *53*, 2344-2348.

20. Jayaram, H.N., O'Connor, A., Grant, M.R., Yang, H., Grieco, P.A. and Cooney, D.A. Biochemical consequences of resistance to a recently discovered IMP dehydrogenase inhibitor, benzamide riboside, in human myelogenous leukemia K562 cells. *J. Exp. Ther. Oncol.* **1996**, *1*, 278-285.

21. Jayaram, H.N., Pillwein, K., Lui, M.S., Faderan, M.A. and Weber, G. Mechanism of resistance to tiazofurin in hepatoma 3924A. *Biochem. Pharmacol.,* **1986**, *35*, 587-593.

22. Franchetti, P., Cappellacci, L., Perlini, P., Jayaram, H.N., Butler, A., Schneider, B.P., Collart, F.R., Huberman, E. and Grifantini, M. Isosteric analogues of nicotinamide adenine dinucleotide derived from furanfurin, thiophenfurin, and selenophenfurin as mammalian inosine monophosphate dehydrogenase (type I and II) inhibitors. *J. Med. Chem.* **1998**, *41*, 1702-1707.

23. Zatorski, A., Watanabe, K.A., Carr, S.F., Goldstein, B.M. and Pankiewicz, K.W. Chemical synthesis of benzamide adenine dinucleotide: Inhibition of inosine monophosphate dehydrogenase (Type I and II). *J. Med. Chem.* **1996**, *39*, 2422-2426.

24. Gharehbaghi, K., Shreenath, A., Hao, Z., Paull, K.D., Szekeres, T., Cooney, D.A., Krohn, K. and Jayaram, H.N. Comparison of biochemical parameters of benzamide riboside, a new inhibitor of IMP dehydrogenase, with tiazofurin and selenazofurin. *Biochem. Pharmacol.* **1994**, *48*, 1413-1416.

25. Tricot, G.J., Jayaram, H.N., Lapis, E., Natsumeda, Y., Nichols, C.R., Kneebone, P., Heerema, N., Weber, G. and Hoffman, R. Biochemically directed therapy of leukemia with tiazofurin, a selective blocker of inosine 5'-monophosphate dehydrogenase activity. *Cancer Res.,* **1989**, 3696-3701.

26. Jayaram, H.N., Lapis, E., Tricot, G., Kneebone, P., Paulik, E., Zhen, W., Engeler, G.P., Hoffman, R. and Weber, G. Clinical

pharmacokinetic study of tiazofurin administered as a 1-hour infusion. *Int. J. Cancer* **1992,** 182-188.

27. Wright, D.G., Boosalis, M.S., Waraska, K., Oshry, L.J., Weintraub, L.R. and Vosburgh, E. Tiazofurin effects on IMP-dehydrogenase activity and expression in the leukemia cells of patients with CML blast crisis. *Anticancer Res.,* **1996,** *16,* 3349-51.

28. Grusch, M., Rosenberger, G., Fuhrmann, G., Braun, K., Titscher, B., Szekeres, T., Fritzer-Szekeres, M., Oberhunber, G., Krohn, K., Hengstschlaeger, M., Krupitza, G. and Jayaram, H.N. Benzamide riboside induces apoptosis independent of Cdc25A expression in human ovarian carcinoma N.1 cells. *Cell Death Diff.,* **1999,** *6,* 736-744.

29. Jayaram, H.N., Grusch, M. Cooney, D.A. and Krupitza, G. Consequences of IMP dehydrogenase inhibition, and its relationship to cancer and apoptosis. *Curr. Med. Chem.,* **1999,** *6,* 561-574.

30. Pankiewicz, K.W., Lesiak, K., Zatorski, A., Goldstein, B., Carr, S.F., Sochacki, N., Majumdar, A., Seidman, M. and Watanabe, K.A. The practical synthesis of methylenebis(phsophonate) analogues of benzamide riboside adenine dinucleotide. Inhibition of human inosine monophosphate dehydrogenase (Type I and II). *J. Med. Chem.* **1997,** *40,* 1287-1291.

Chapter 13

Cofactor Analogues as Inhibitors of IMP Dehydrogenase: Design and New Synthetic Approaches

Krzysztof W. Pankiewicz[1], Steven Patterson[1],
Hiremagalur N. Jayaram[2], and Barry M. Goldstein[3]

[1]Pharmasset, Inc., 1860 Montreal Road, Tucker, GA 30084
(email: kpankiewicz@pharmasset.com or spatterson@pharmasset.com)
[2]Indiana University School of Medicine, Indianapolis, IN 46202 (email:
hjayaram@iupui.edu)
[3]University of Rochester Medical Center, 601 Elmwood Avenue,
Rochester, NY 14642 (email: barry_goldstein@urmc.rochester.edu)

Inosine monophosphate dehydrogenase (IMPDH, E.C.1.1.1.205) is a key enzyme in *de novo* synthesis of purine nucleotides. IMPDH uses nicotinamide adenine dinucleotide (NAD) as a cofactor, which abstracts hydride from the C2 of inosine-5'-monophosphate (IMP). A structural analogue of the cofactor that binds tightly the enzyme but cannot participate in hydride transfer would therefore inhibit IMPDH. The cofactor binding domain is not conserved among IMPDH enzymes from different sources (human, mammalian, bacterial, parasitic). In addition the two isoforms of the human enzyme; type I expressed in normal cells and type II, dominant in cancer cells and activated lymphocytes, differ at the cofactor binding site. This indicates that specific inhibition of a bacterial, parasitic IMPDH or even a human isoform is possible by synthetic cofactor mimics. Specific inhibition of IMPDH therefore has enormous potential in the development of novel antibacterial, antaparasitic, immunosuppressive, and anticancer agents. In this chapter the design, synthesis, IMPDH inhibition, and biological activity of cofactor analogues is described.

Introduction

IMP dehyrogenase (IMPDH), the key enzyme in *de novo* synthesis of purine nucleotides, emerged recently as a major therapeutic target. In the last few years several inhibitors of the enzyme have been developed. Mycophenolic mofetil (MMF, CellCept®, Figure 1, manufactured by Roche) is in clinical use as an immunosuppressant in the US and mizoribine (Asahi) is successfully used in Japan. Ribavirin (ICN) in combination with interferon-α has been approved for the treatment of Hepatitis C virus infections. The L-form of the drug has been synthesized and showed reduced toxicity.[2]

A novel non-nucleoside IMPDH inhibitor (VX-497, Figure 2, available from Vertex) is in Phase II clinical studies as an antiviral agent.[3] Tiazofurin reached clinical trials as an anti-leukemic agent but it was too toxic for a broad clinical application. Benzamide riboside showed an improved pharmacological profile and is studied as a potential replacement for tiazofurin. In an early stage of the development is EICAR[4] synthesized in Japan. Finally, a group of bis(phosphonate) analogues of nicotinamide adenine dinucleotide (NAD), the cofactor used by IMPDH, have been found to be potent inhibitors of IMPDH and are under development in Pharmasset.[5]

The majority of IMPDH inhibitors may be categorized in three groups. The first consists of nucleosides, such as ribavirin, EICAR, and mizoribine, which are phosphorylated in the cell to the 5'-monophosphates and bind IMPDH at the substrate-binding domain. The second are nucleosides, such as tiazofurin and benzamide riboside, that are uniquely converted in cells into an active metabolite, tiazofurin adenine dinucleotide (TAD, Figure 3) and benzamide adenine dinucleotide (BAD), respectively. TAD and BAD are analogues of NAD and inhibit IMPDH binding at the cofactor site. The third group includes compounds such as mycophenolic acid and VX-497, which do not require metabolic activation and bind at the cofactor-binding domain of IMPDH.

Of course, the mode of action of all these compounds is much more complicated than such categorization could indicate. For example, EICAR-5'-monophosphate not only forms a covalent adduct with Cys331 of IMPDH but it is also coupled in cells with AMP to give EICAR adenine dinucleotide[4], the cofactor type inhibitor. On the other hand, mizoribine 5'-monophosphate was proved[6] to be a potent transition-state inhibitor – the formal negative charge located on the 5-oxygen of the aglycone of mizoribine resembles the negative charge which develops on N3 of IMP during the reaction with Cys331 of IMPDH. It has been suggested that ribavirin broad antiviral activity is due to its diversified mode of action.[2] Indeed, inhibition of IMPDH by ribavirin 5'-monophosphate results in a depletion of the cellular guanosine triphosphate (GTP) pool. However, it is known that ribavirin is also metabolized to the triphosphate level. Thus, ribavirin-5-triphosphate could inhibit viral-specific RNA polymerases more effectively when the competition of GTP is severely reduced (by inhibition of IMPDH). In addition, it was reported that with certain

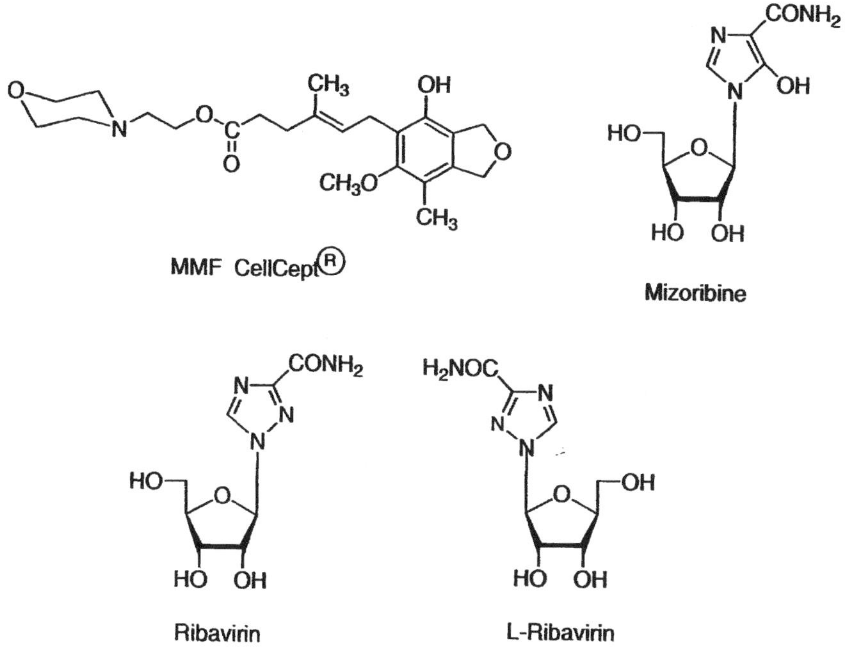

Figure 1. Structure of MMF, Mizoribine, Ribavirin and L-Ribavirin

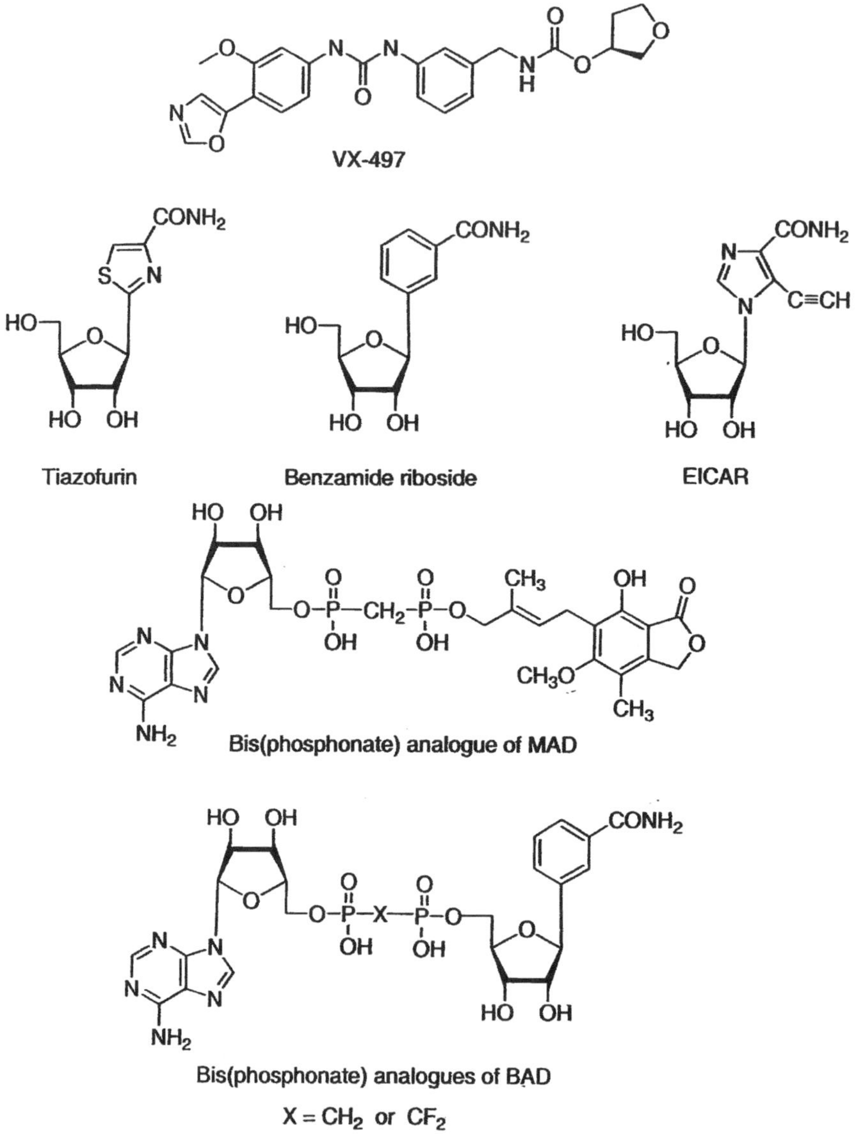

250

VX-497

Tiazofurin

Benzamide riboside

EICAR

Bis(phosphonate) analogue of MAD

Bis(phosphonate) analogues of BAD

X = CH₂ or CF₂

Figure 2. Structure of VX-497, Tiazofurin, Benzamide riboside and EICAR, Bis(phosphonate) analogues of MAD and BAD

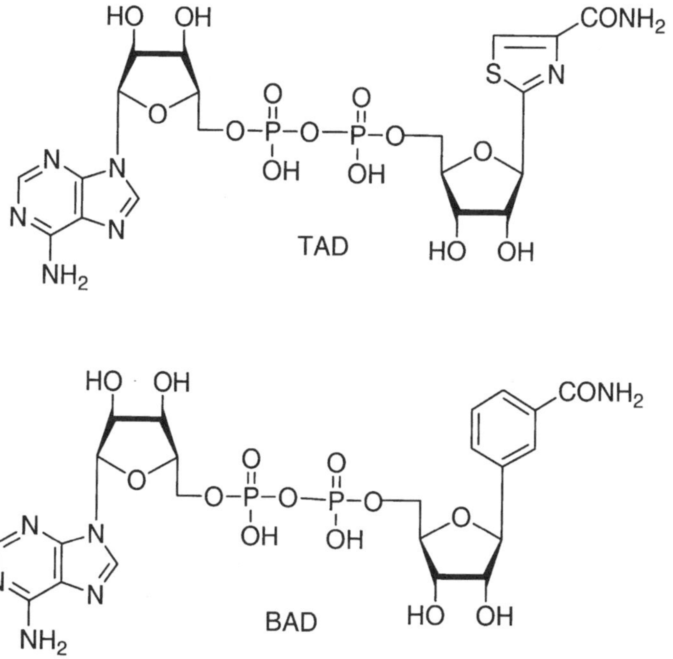

Figure 3. Structure of TAD and BAD

RNA and DNA viruses, ribavirin 5'-triphosphate inhibits the viral-specific mRNA capping enzymes.

Human IMPDH is well characterized and this, in addition to its importance to neoplasia and immunosuppression, makes it an exceptionally attractive target. Two isoforms (type I and type II) are known. Both contain 514 amino acids and have 84% sequence homology. IMPDH cDNA was first cloned and sequenced by Collart and Huberman.[7] The two isoforms are regulated differently. Type I is expressed in normal cells while type II is selectively up-regulated in cancer cells and activated lymphocytes and emerges as the dominant form.[8]. Thus, the specific inhibition of type II IMPDH would provide significant therapeutic advantage by eliminating or reducing potential toxicity caused by inhibition of type I isoform.

Recently solved[9] X-ray structure of human IMPDH type II in complex with the cofactor analogue SAD and the substrate analogue 6-chloropurine riboside showed clearly that the cofactor-binding domain is not conserved between the two isoforms. The cofactor-binding domain of this structure differs from that of type I isoform by three amino acids involved in NAD binding. These differences can be exploited in design of specific inhibitors that bind IMPDH at the cofactor site. In addition, IMPDH enzymes from nonhuman sources emerged as attractive targets. These are enzymes from bacterial sources such as *Streptococcus pyogenes*,[10] human pathogens such as *Pneumocystis carinii*[11](AIDS patients) or *Borelia burgdorferi*[12](Lime disease), and parasites such as *Tritrichomonas foetus*.[13] The cofactor-binding domain of these enzymes differs from that of the human enzyme since cofactor-type inhibitors show different activity against these enzymes. For example, mycophenolic acid (MPA), the most potent inhibitor of the human IMPDH, is not active against bacterial enzymes at al.[10] Consequently, inhibitors that bind bacterial IMPDH at the cofactor site are expected to exhibit potent antibacterial activity without affecting the host (human enzyme).

All the above observations indicate that inhibitors, which bind at the cofactor site of IMPDH, are expected to show some selectivity and therefore are of great therapeutic interest. On the other hand, IMPDH, along with a large number of cellular dehydrogenases, utilize NAD as the cofactor and it seems likely that NAD analogues designed as inhibitors of IMPDH would inhibit a number of NAD dependent enzymes. However, this is not the case. Human IMPDH follows different kinetics than all other dehydrogenases. Binding of IMP results in a conformational change of the enzyme.[14] IMPDH adopts an *open* conformation around its nucleotide binding site in the absence of IMP and binding of IMP stabilizes a *closed* conformation, *which has a higher affinity for NAD*.[14] Association of IMP and NAD with IMPDH is random and release of products is ordered.[15] NADH release precedes hydrolysis of the activated enzyme-xantosine monophosphate complex (E-XMP*).[16] This complex has a great affinity to NAD and a significant fraction of IMPDH exists as an E-XMP*-NAD complex when the concentration of NAD is sufficient. The formation of this complex may provide another mechanism for specific inhibition of IMPDH by NAD analogues.[16] Thus, both the conformational reorganization induced by

IMP and formation of E-XMP*-NAD complex may contribute to the improved and specific binding of NAD (and cofactor type inhibitors) to IMPDH. This may explain the exceptional sensitivity of IMPDH to inhibition from the cofactor site. In fact, mycophenolic acid, which binds at the cofactor-binding domain, is an extremely specific inhibitor of IMPDH – no inhibition of other NAD dependent enzymes by mycophenolic acid has been ever reported.[17] In contrast, inhibitors bound at the substrate site such as ribavirin 5'-monophosphate could also be recognized by the nucleotide-binding site in other enzymes and are therefore not specific.

The above observations indicate that synthetic modifications of NAD analogues may provide new inhibitors with improved IMPDH affinity and the desired specificity towards type II of the human isoform. In addition, NAD analogues may be poor inhibitors of the human enzyme but may show activity against enzymes from other than human sources such as bacterial enzymes. With this in mind we studied the chemistry of NAD analogues and their interactions with IMPDH enzymes.

Bis(phosphonate) analogues of TAD and BAD

It is well established that tiazofurin[18, 19] (TR) is converted in the cell into active metabolite thiazole-4-carboxamide adenine dinucleotide, TAD (Scheme 1). The first step in the biosynthesis of TAD is the phosphorylation of TR by adenosine kinase to give tiazofurin 5'-monophosphate (TRMP). TRMP is coupled with AMP by NMN-adenylyl transferase to give TAD.[20] TAD mimics the natural NAD, however, it cannot participate in hydride transfer and inhibits human IMPDH. The enzyme's binding affinity to TAD is 3 orders of magnitude higher than that of NAD. Recently developed benzamide riboside[21] (BR) is metabolized to BAD[22] in a similar manner. In spite of this TAD and BAD are useless as potential therapeutics. These NAD analogues are pyrophosphates by nature and therefore they are not able to penetrate cell membranes and are vulnerable to degradation by cellular enzymes. TAD is cleaved to TRMP and AMP by phosphodiesterases, including a specific "TADase." Resistance to tiazofurin is associated both with decreased anabolic activity by NMN-adenylyl transferase and with increased degradation of TAD by TADase.[23, 24] Very likely resistance to BR followes a similar mechanism.

Simple replacement of the pyrophosphate oxygen of TAD and BAD with a $-CH_2-$ or $-CF_2-$ group gave stable bis(phosphonate), which were found to be as active inhibitors of IMPDH as the parent dinucleotides.[25-27] Their shape and the size resemble the parent pyrophosphates and may be of therapeutic interest. In contrast to the pyrophosphates, bis(phosphonate) analogues are able to penetrate cell membranes. Thus, concentration of these analogues in the cell does not depend on the activity of the anabolic enzymes. In addition, bis(phosphonate)s have built-in resistance to degradation enzymes, and consequently are active in TR-resistant cell lines.

Scheme 1

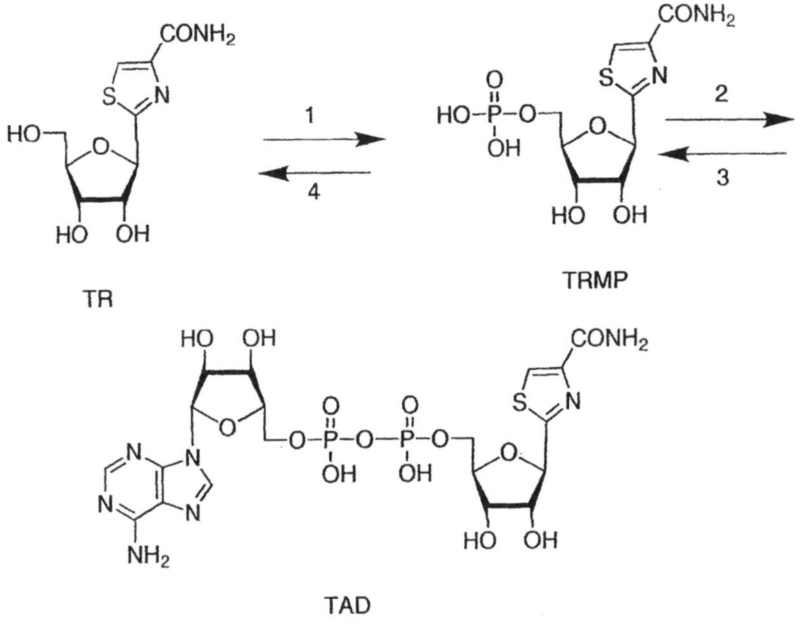

1. Adenosine kinase, 2. NMN - adenylyl transferase,
3. TADase, phosphodiesterase, 4. Phosphatases

The major restriction in examination of an ultimate therapeutic value of bis(phosphonate) analogues as potential drugs was lack of efficient methods of their synthesis. These compounds were usually prepared in milligram amounts by dicyclohexylcarbodiimide (DCC) coupling of 2',3'-O-isopropylidene-adenosine-5'-methylenebis(phosphonate) (**1**, A = 2'3'-O-isopropylidene-adenosin-5'-yl, Scheme 2) with 2',3'-O-isopropylidene-tiazofurin or –benzamide riboside (B-OH) followed by de-isopropylidenation. It had been generally accepted that the reaction went through the formation of an amidine intermedide (**2**, R = cyclohexyl), which was subsequently displaced by the nucleophilic 5'-hydroxyl group of B-OH to give the protected bis(phosphonate) **7** (A = 2'3'-O-isopropylidene-adenosin-5'-yl, B= 2'3'-O-isopropylidene-tiazofurin-5'-yl). However, this is not the case. We have discovered[28] that **1** as well as other mono-substituted methylenebis(phosphonate)s undergo reaction with DCC [or other coupling reagents such as diisopropyl-carbodiimide (DIC)] forming the corresponding tetra-phosphonate **3**, which is further dehydrated to give the monocyclic (**4**) and then bicyclic trisanhydride intermediate **5**. Since intermediate **5** is neutral, it readily reacts with B-OH or other nucleophilic reagents, preferentially at P^2 and P^3 due to steric hindrance provided by bulky adenosine groups at P^1 and P^4, to give the tetraphosphonate derivative **6**. Hydrolysis of **6** with water followed by deprotection afforded the desired methylenebis(phosphonate) analogue of TAD (**7**, A = adenosin-5'yl, B = tiazofurin-5'yl) in high yield.

Adenosine 5'-bis(phosphonate) can be prepared according to Poulter's precedure[29] by reaction of 5'-O-tosyl adenosine with tris(tetra-n-butylammonium salt of bis(phosphonic) acid in acetonitrile. This method, however, fails in the case of 2'3'-O-isopropylideneadenosine (protection of 2',3'-hydroxyl groups is essential for DCC coupling). It is known that acetonide protection brings N-3 of the adenine base and C-5' of the sugar moiety to a close proximity and therefore intramolecular displacement of the 5'-tosyl group with the formation of the corresponding 3,5'-cycloadenosine is prefered (Figure 4).[30]

We therefore developed[31] an alternate procedure based on the mechanism discussed above. 2-(4-Nitrophenyl)ethyl alcohol (**8**, Scheme 3) was treated with an equimolar amount of a commercially available methylenebis(phosphonyl) tetrachloride and tetrazole to give 2-(4-nitrophenyl)ethyl methylenebis(phosphonate) (**9**) as the major product, which was separated by preparative HPLC. Compound **9** was further converted into the corresponding intermediate **10** by dehydration with DIC. Reaction of **10** with 2',3'-O-isopropylideneadenosine afforded the desired protected derivative **11** from which the nitrophenylethyl group was removed by elimination with 1,8-diazabicyclo[5.4.0]undec-7-ene (DBU). Several *ribo*- and *deoxyribo*-nucleoside methylenebis(phosphonate)s have also been prepared in a similar manner.[31]

We found[28] that our methods might have a general application to the synthesis of isosteric methylenebis(phosphonate) analogues of biologically important P^1, P^2-disubstituted pyrophosphates. Indeed, the reaction of **5** (A = 2',3'-O-isopropylideneadenosin-5'-yl) with benzyl 2,3-isopropylidene-β-D-

Scheme 2

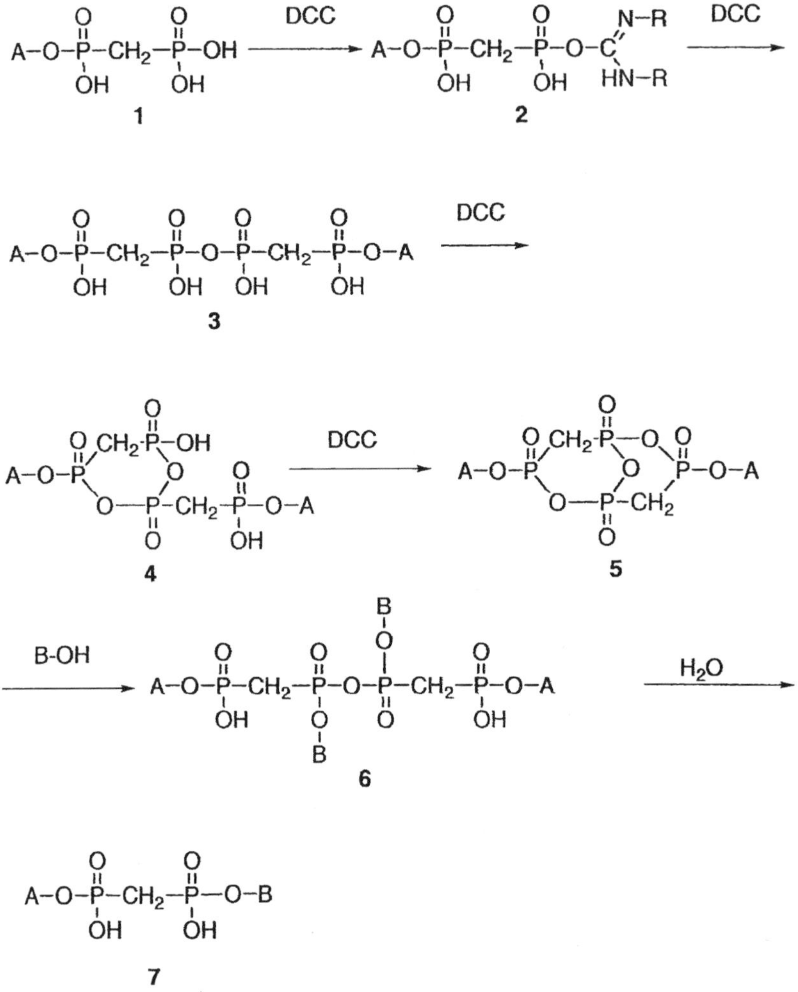

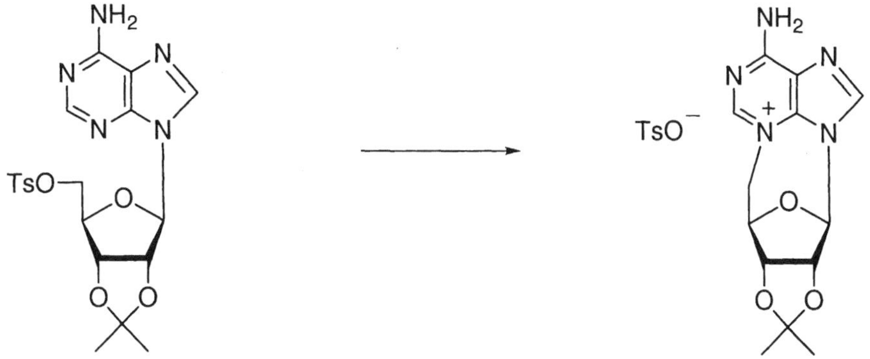

Figure 4. Formation of 3, -5'cycloadenosine

Table I. Growth Inhibition and Differentiation of K562 cells.

Compound	Inhibition of IMPDH IC$_{50}$ (μM)	Inhibition of cell growth IC$_{50}$ (μM	Conctn (μM)	Differen tiation - % of positive cells	% of positive cells / conctn.
TR	-	3.0	15.0	80	5.3
TAD	0.1	3.7	15.0	70	4.6
2'F-ara-TAD	2.6	3.8	13.0	56	4.3
3'F-TAD	0.7	3.2	2.5	26	10.4
β-CH$_2$-TAD	0.1	18.0	14.3	35	2.4
β–CF$_2$-TAD	0.3	11.0	14.0	48	3.4
2'F-ara-β-CH$_2$-TAD	6.0	70.0	19.0	9.5	0.7
2'F-ara-β-CF$_2$-TAD	0.8	13.0	13.0	40	3.1
β-CH$_2$-BAD	0.8	68.0	15.0	9.5	0.6
2'F-ara-β-CH$_2$-BAD	175	Not active	75.0	9.0	0.1

ribose followed by hydrolysis and deprotection afforded a novel ADP-ribose analogue (Figure 5). In a similar manner treatment of **5** (A = 2'3'-O-isopropylideneadenosin-5'-yl) with an unprotected riboflavin gave, after work-up, the FAD analogue. Also, treatment of **5** (A = cytidin-5'-yl) with N-acetylethanolamine or 1,2-dipalmitoyl-*sn*-glycerol afforded methylene-bis(phosphonate) analogues of CDP-ethanolamine and CDP-DAG, respectively.

Using this methodology we synthesized methylenebis(phosphonate) and difluoromethylenebis-(phosphonate) analogues of TAD and BAD (Figure 6) and examined their inhibitory activity against IMPDH as well as their influence on growth and differentiation of human erythroleukemia K562 cells.[32] Bis(phosphonate) derivative of TAD and its analogues substituted with a fluorine atom at the adenine sugar were of special interest since we found earlier[33] that TAD and its corresponding pyrophosphate analogues, 2'F-*ara*-TAD, (containing fluorine at the C2' in *arabino* configuration) and 3'F-TAD (fluorine in *ribo* configuration) showed a similar differentiation activity to that of tiazofurin (TR, Table I). Examination of bis(phosphonate) analogues revealed that methylenebis(phosphonate) analogues are weak inhibitors of growth and they did not induce differentiation. This has been attributed to the metabolic stability of methylenebis(phosphonate) derivatives and to their less efficient transport across the cell membrane as compared with nucleosides. In contrast, pyrophosphate TAD analogues could be hydrolyzed in cell culture to the corresponding nucleotides and nucleosides. Consequently, TR is released and then transported readily across the cell membrane to be reanabolized to the active TAD. Indeed, we found that during incubation in cell extracts both TAD and 2'-F-ara-TAD released significant amount of TR.[34]

Interestingly, both analogues containing the –CF$_2$- linkage, e.g. β-CF$_2$-TAD and 2'-F-ara-β-CH$_2$-TAD showed improved inhibitory and differentiation-inducing activity compared to methylenebis(phosphonate) analogues. It has been suggested[35] that the isopolar and isosteric –CF$_2$- group is a better mimic of the pyrophosphate oxygen than the –CH$_2$- linkage. Since we found that compounds containing –CF$_2$- bridge are as stable in cell extracts as methylenebis(phosphonate)s it is reasonable to conclude that difluoromethylene substitution of the methylene linkage *does* improve transport of the intact bis(phosphonate)s.[32]

Bis(phosphonate) analogues of mycophenolic adenine dinucleotide. MAD analogues.

Among IMPDH inhibitors only mycophenolic acid (Figure 7) showed more potent (4 fold) inhibitory activity against human IMPDH type II isoform than type I isoform.[8] In spite of this MPA is not active against cancer. Its therapeutic potential is limited by its undesirable metabolism. In humans, MPA is rapidly metabolized into the inactive glucuronide (*via* its C7-phenolic function) and as much as 90% of the drug circulates in this inactive form.[36] Protection of the C7 function or replacement of the phenol group with a fluorine atom, amino or nitrile group in order to prevent glucuronidation resulted in inactive compounds.[37, 38] Mycophenolic mofetil, a prodrug of MPA, is administered in high doses (2-3 grams per day) in order to maintain the therapeutic level of MPA in activated lymphocytes where there is little glucuronidation. However, since the level of glucuronyltransferase activity is rather high in many cancer cells it is not surprising that MPA lacks any significant anticancer activity.

In 1996 the first crystal structure of IMPDH was solved.[39] It was a Chinese hamster IMPDH (which differs only by 4 amino acids from the human enzyme) in complex with MPA. It was found that MPA mimics binding of the nicotinamide mononucleotide (NMN) leaving the adenosine monucleotide (AMP) subsite empty. Its carboxyl group is positioned at the space occupied by the phosphoryl group of NMN.[39] This observation inspired us to use MPA as the substitute for NMN and construct the whole NAD mimic by attachment of AMP to MPA. Such a mycophenolic adenine dinucleotide (MAD) analogue would have a much better binding capacity than MPA alone due to the additional binding at the adenosine subsite of IMPDH. Simple linking of the carboxyl group of MPA with AMP would provide a mycophenolic adenine mixed anhydride (Figure 8). This mixed anhydride, however, would not be stable in protic solvents. Therefore, we linked adenosine and mycophenolic alcohol through a methylenebis(phosphonate) bridge to form a novel mimic of the natural cofactor.[40] The carboxyl group of MPA was reduced to give C6-mycophenolic alcohol[41] (Scheme 4, C6-MPAlc) which was then coupled with 2',3'-O-isopropylideneadenosine 5'-methylenebis(phosphonate) (1) *via* the intermediate **5** to give the protected C6-MAD analogue (**12**) in 32% yield.[40] Deisopropylidenetion of **12** with a mixture of CF_3COOH/H_2O gave the final C6-MAD. C6 designates the number of carbon atoms in the linker between the aromatic ring of MPA and the β-phosphorus atom of the adenosine bis(phosphonate) moiety. The assignment of the structure of C6-MAD was confirmed by ^1H and ^{31}P NMR. The resonance signal of the 6'-methylene group of the MPAlc moiety in the proton NMR at 3.79 ppm was observed as a quartet ($J_{H-P} = 6.3$ Hz, $J_{H-H} = 6.3$ Hz) showing coupling with the β-phosphorus atom.

We found that C6-MAD was an equally potent inhibitor of both type I and type II isoforms of IMPDH with $K_i = 0.33$ μM and $K_i = 0.29$ μM (Table II), respectively. Contrary to expectations, C6-MAD showed a 10 to 30-fold less potent inhibitory activity against IMPDH than MPA indicating that the expected binding of the adenosine moiety of C6-MAD at the adenosine subsite of IMPDH is not achieved or is only partially accomplished. On the other hand, antiproliferative activity of C6-MAD in K562 cells was found to be high ($IC_{50} = 1.5$ μM) only 5 fold lower than that of MPA ($IC_{50} = 0.3$ μM). The most important observation was that C6-MAD was resistant to glucuronidation *in*

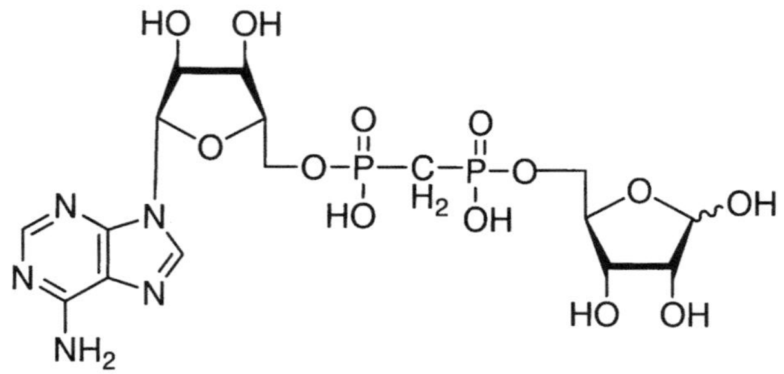

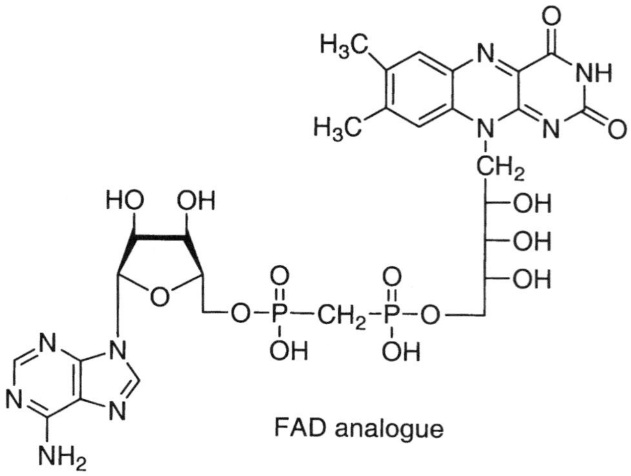

FAD analogue

ADP-ribose analogue

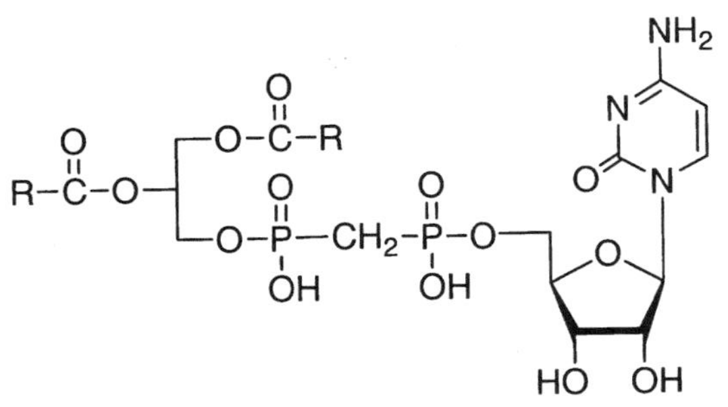

Figure 5. Structure of ADP-ribose, FAD and CDP analogues

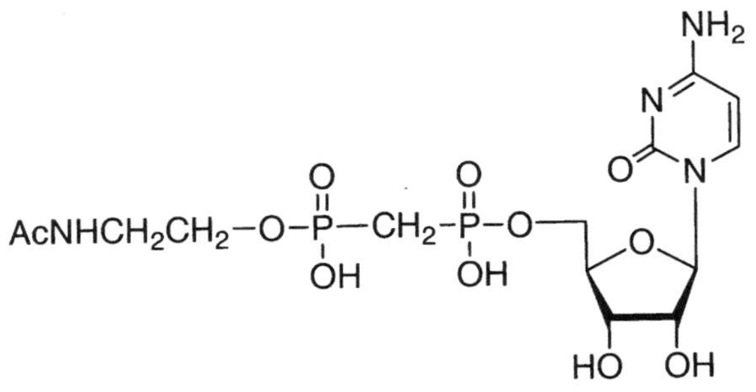

CDP-ethanolamine analogue

CDP-DAG analogue

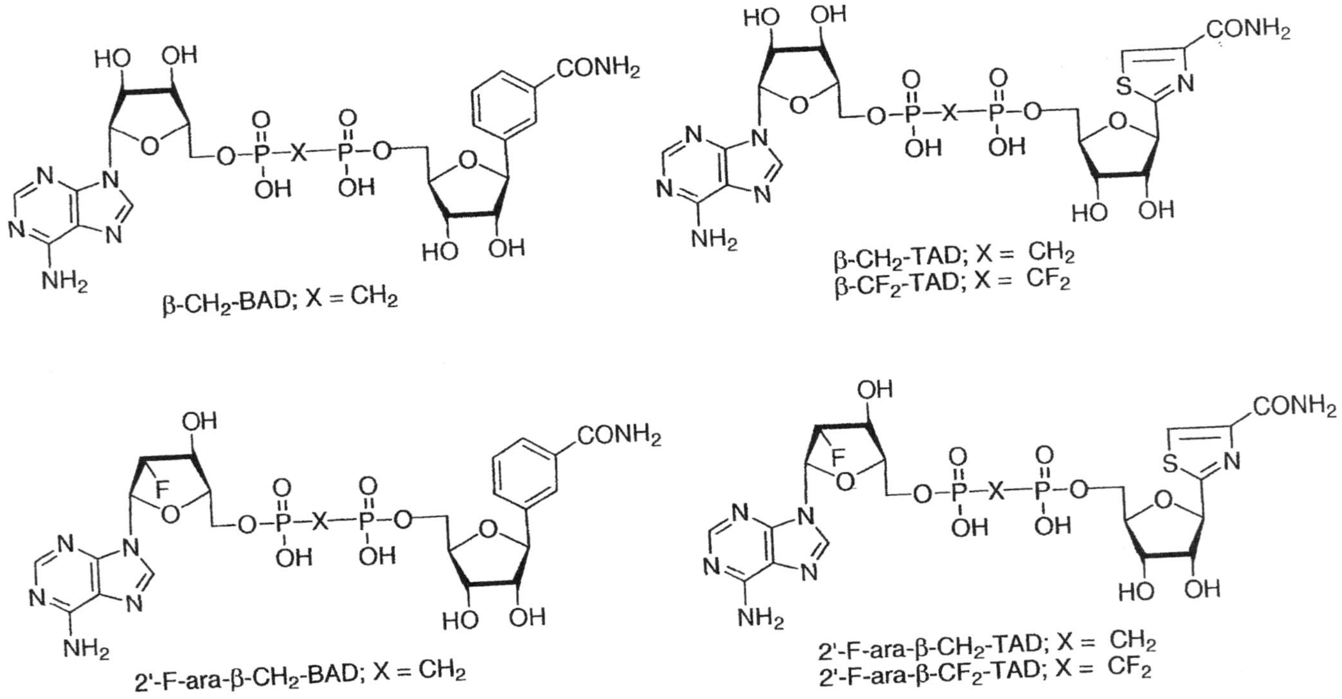

Figure 6. Structure of bis(phosphonate) analogues of TAD and BAD

β-CH₂-TAD; X = CH₂
β-CF₂-TAD; X = CF₂

2'-F-ara-β-CH₂-TAD; X = CH₂
2'-F-ara-β-CF₂-TAD; X = CF₂

β-CH₂-BAD; X = CH₂

2'-F-ara-β-CH₂-BAD; X = CH₂

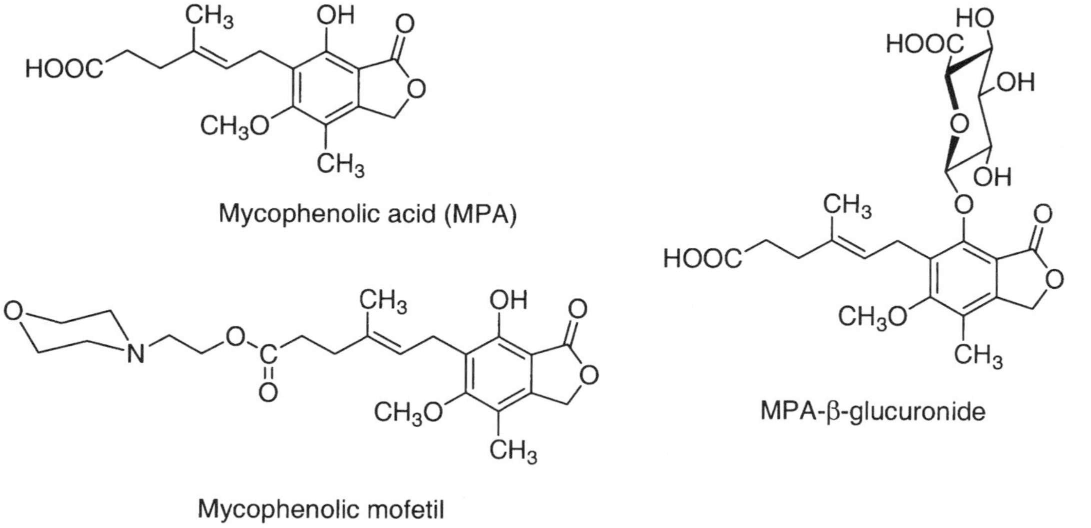

MPA-β-glucuronide

Mycophenolic acid (MPA)

Mycophenolic mofetil

Figure 7. Structure of MPA, mycophenolic mofetil, and MPA-β-gluconuride

264

Scheme 3

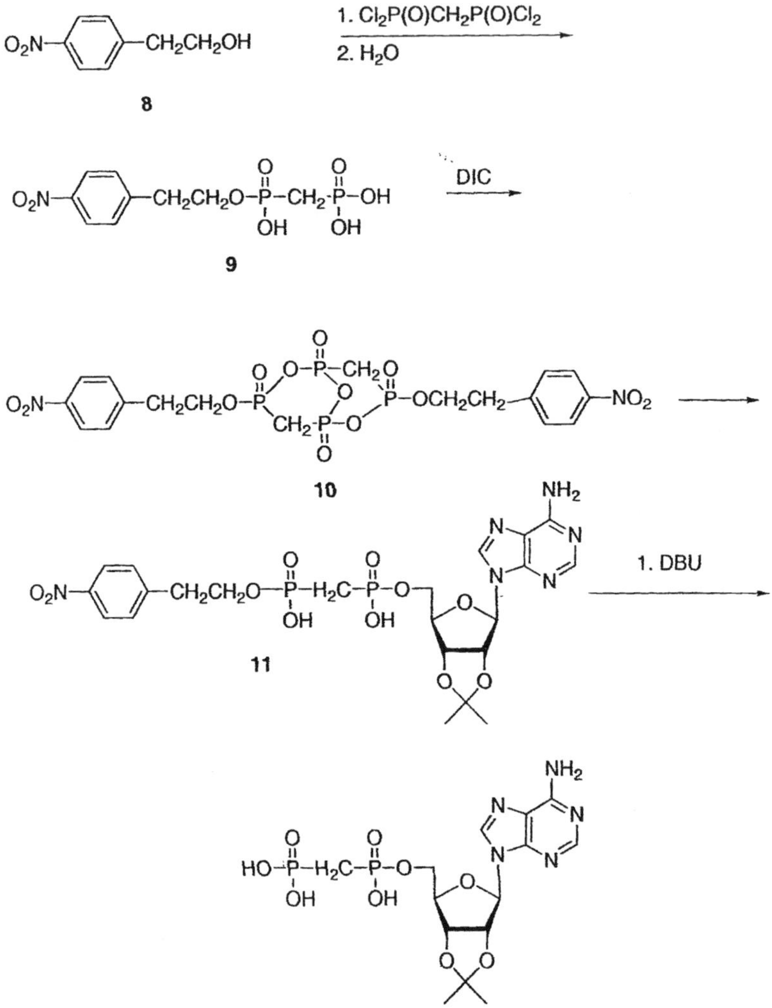

Scheme 4

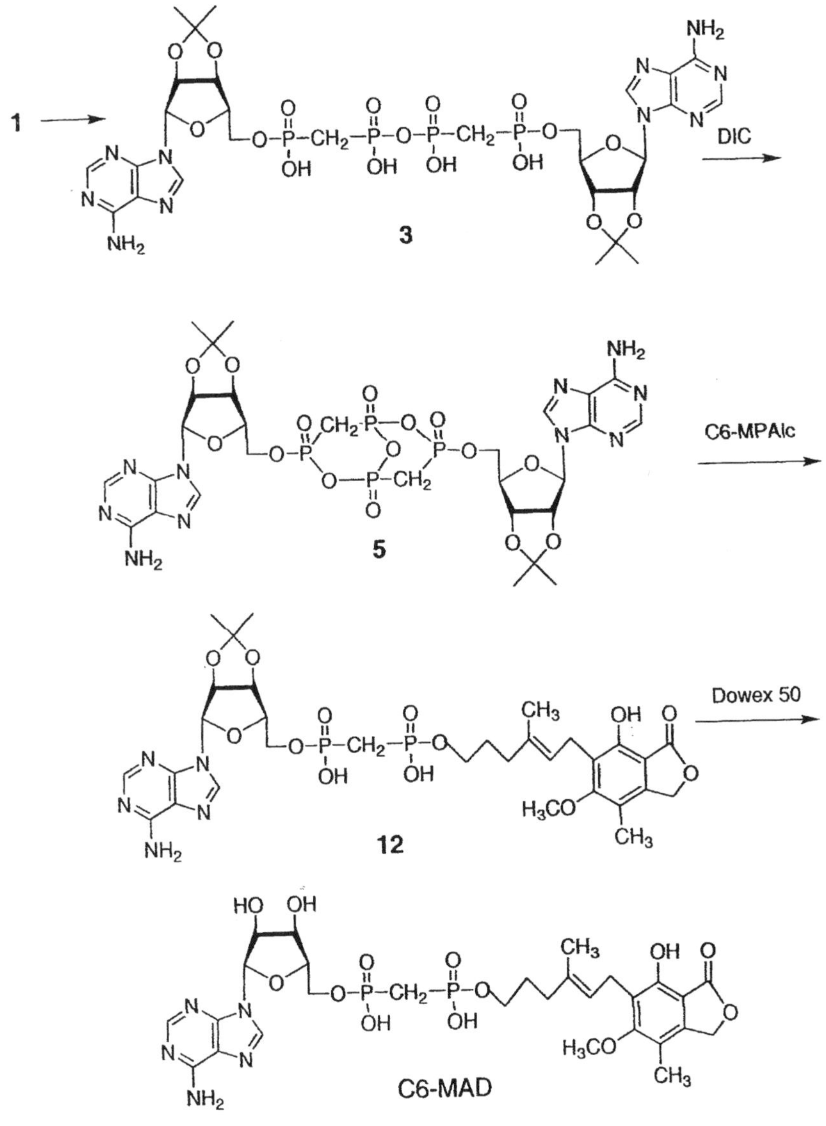

vitro. When C6-MAD and MPA were incubated with uridine 5'-diphosphoglucuronyltransferase (UDPGT) and uridine 5'-diphosphoglucuronic acid at 30 °C for 2 h extensive conversion of MPA to the 7-O-glucuronide was observed. This was further supported by the fact that MPA-glucuronide was converted back to the starting MPA by hydrolysis with β-glucuronidase. In contrast, no formation of the corresponding glucuronide of C6-MAD has been detected in similar conditions. Although, it is known that glucuronidation efficacy by UDPGT may differ depending on incubation conditions and that significant species differences exists in substrate specificity and the reaction rate *in vivo*, these results indicate, that contrary to MPA, C6-MAD may show resistance to glucuronidation *in vivo*. If this is the case C6-MAD may have some therapeutic potential in cancer treatment.

It is worthy to note that C6-MAD is not a perfect mimic of the mixed anhydride. It contains six carbon atoms and the oxygen atom in the linker between the mycophenolic aromatic ring and the β-phosphorus atom of the adenosine bis(phosphonate) moiety. This linker is two atoms longer than that in the mixed anhydride (five carbon atoms). It is likely, therefore, that the binding affinity of C6-MAD to IMPDH may be reduced. With this in mind we synthesized two new analogues with shorter linkers, C4-MAD (four carbon atoms) and C2-MAD (two carbons).

We used MPA as a starting material for the preparation of mycophenolic alcohols (Scheme 5). We modified the known procedure[41] of synthesis of the key aldehyde **13**. Instead of ozonolysis of MPA, which worked poorly in our hands, we used a one pot OsO_4 oxidation of the double bond of MPA followed by oxidative cleavage ($NaIO_4$). The desired aldehyde crystallized from the reaction mixture and was obtained in 95% yield . Reduction of **13** with $NaBH_4$ gave the C2-MPAlc whereas treatment of the aldehyde with 2-(triphenylphosphoranylidene)propionaldehyde followed by reduction with $NaBH_4$ afforded the desired C4-MPAlc.[30]

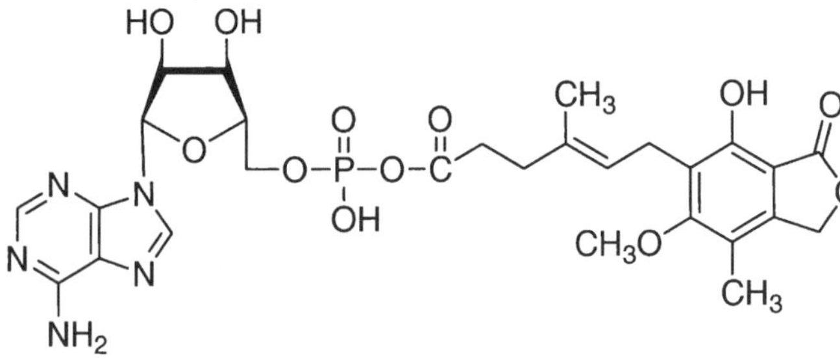

Mixed anhydride

Figure 8. mycophenolic AMP mixed anhydride

Scheme 5

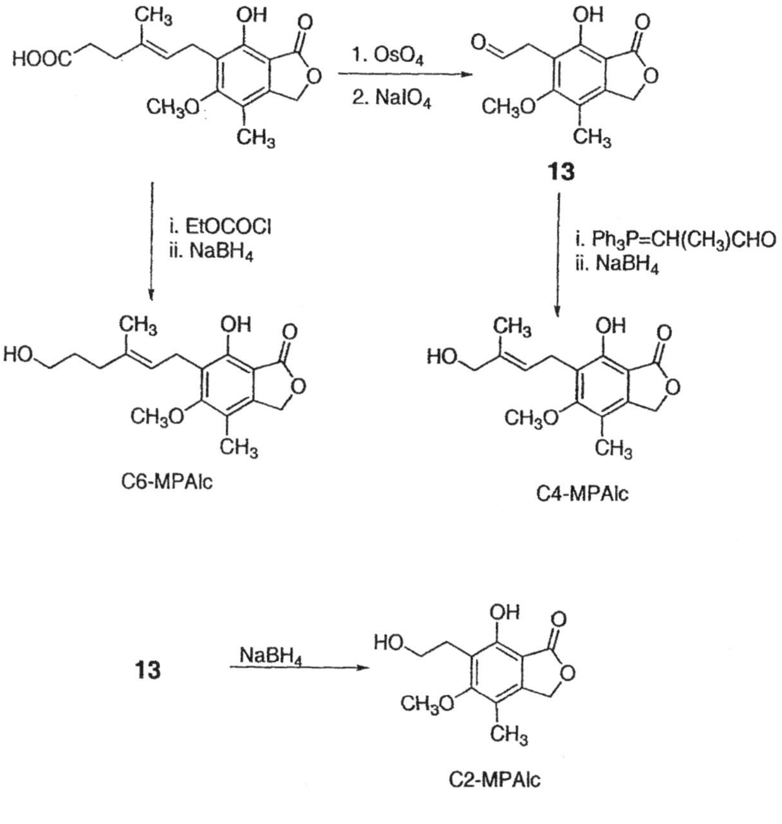

Coupling of C2-MPAlc with the 2',3'-O-isopropylideneadenosine 5'-methylenebis(phosphonate) (1) was performed in similar conditions as the coupling of C6-MPAlc. After hydrolysis the reaction mixture was treated with an aqueous CF_3COOH to remove the 2',3'-O-isopropylidene protecting group and then the mixture was purified on a preparative HPLC column. Three compounds were eluted from the column: 1) the desired C2-MAD was isolated in 60% yield as the major compound, 2) the minor component 14 (Figure 9) was isolated in 11 % yield, and 3) a small amount of unreacted tetraphosphonate derivative 3 (7 % yield) was also obtained. The minor compound 14 was found to contain C2-MPAlc linked to the β-phosphorus atom of the methylenebis(phosphonate) moiety through the 7-phenol function. The assignment of the structure of 14 was confirmed by ^{1}H and ^{31}P NMR. Bis(phosphonate) 14 did not show any inhibitory activity against human IMPDH.

The formation of the 7-O-linked derivative in the coupling reaction was not observed during the synthesis of C6-MAD, however, such reaction should be expected. Therefore the protection of the phenol function may be necessary in order to improve efficiency of the coupling procedure. We found that mycophenolic alcohols could be selectively protected with a benzyl group at the 7-OH by reaction with benzyl bromide in a 1M solution of tetrabutylammonium fluoride in THF. The desired 7-O-benzyl derivatives C2- and C4- mycophenolic alcohols were obtained in crystalline forms and high yields.

Finally, coupling of 7-O-benzyl-C2-MPAlc with 1 followed by deisopropylidenation with CF_3COOH/H_2O and hydrogenolytic debenzylation (Pd/C) afforded the desired C2-MAD in 55 % yield. The reaction conditions were not optimized. We found that the coupling of the benzyl protected C2-MPAlc with adenosine bis(phosphonate) 1 requires longer reaction time than the similar coupling of the unprotected C2-MPAlc.

We also examined an alternative synthetic strategy by using 7-O-benzyl-C2-MPAlc for the preparation of the methylenebis(phosphonate) analogue 17 (Scheme 6) and subsequent coupling of 17 with 2',3'-O-isopropylideneadenosine. Mesylation of the C2-MPAlc afforded a good yield of the mesylate 15, however, further reaction with tris(tetra-n-butylammonium)-methanebis(phosphonate) salt according to Poulter's procedure[31] gave exclusively the elimination product 16. Nevertheless, the desired mycophenolic bis(phosphonate) 17 was obtained in a high yield by reaction of 7-O-benzyl-C2-MPAlc with a commercially available methylenebis(phosphonic) dichloride in conditions of Yashikawa's phosphorylation in $(EtO)_3P(O)$. Although, a small amount of P^1, P^2-disubstituted derivative 19 was also formed in the reaction, compounds 17 and 19 were easily separated by preparative HPLC. The benzyl protecting group was than removed from 17 and 19 by hydrogenolysis on Pd/C to give bis(phosphonate) derivatives 18 and 20, respectivelt These compounds showed a potent inhibition of IMPDH (Table II)   The benzyl protected bis(phosphonate) 17 was coupled with 1 to give C2-MAD after hydrolysis and deprotection.

Finally, C4-MAD (Figure 10) was obtained by coupling of C4-MPAlc with adenosine bis(phosphonate) 1. This is the best mimic of the mixed anhydride as far as its linker's length is concerned.

All these compounds were assayed for inhibitory activity against human IMPDH type I and type II and their activity as antiproliferative agent and differentiation inducers were measured in human erythroleukemia K562 cells (Table II and Table III).

Table II. Inhibition of IMPDH and K562 cell growth by MPA and methylenebis(phosphonate) analogues of MAD.

INHIBITORS	K_I (type I) [μM]	K_I (type II) [μM]	IC_{50} [μM]
MPA	0.04	0.01	0.3
C2-MAD	0.33	0.25	0.7
C4-MAD	0.52	0.38	0.1
C6-MAD	0.33	0.29	1.5
Bis(phosphonate) 18	0.89	0.40	ND
Bis(phosphonate) 20	2.3	0.47	ND

Regardless of the linker size, MAD analogues inhibited IMPDH 10 fold less potently (approximately an order of magnitude) than mycophenolic acid itself. Contrary to expectations, the attachment of adenosine moiety to MPA did not improve enzyme-inhibitor binding. It is reasonable to assume that mycophenolic moiety of MAD binds strongly to the enzyme but the additional binding of the adenosine part of MAD is not accomplished as expected. However, examination of the inhibitory activity of the bis(phosphonate) derivatives 18 and 20 wihich do not contain adenosine moiety and binds less tightly to IMPDH than C2-MAD, suggests that adenosine contributes to some extend to the affinity of C2-MAD to the enzyme.   The data in Table II also indicate that MADs were as potent inhibitors of IMPDH as the most potent methylenebis(phosphonate) analogues of TAD and BAD (Table I). However MADs were found to be much superior than bis(phoshonate)s of TAD and BAD in terms of the inhibition of the growth of K562 leukemic cells (Table II and Table III). Bis(phosphonate) analogues of TAD and BAD were less effective than their parent nucleosides (TR or BR) in inhibition of K562 cells growth (IC_{50}'s) as well as in a stimulation of cell differentiation. In contrast, the most potent MAD analogue, *C4-MAD, shows more potent growth inhibition of K562 cells than MPA*, although it is less active than MPA as

270

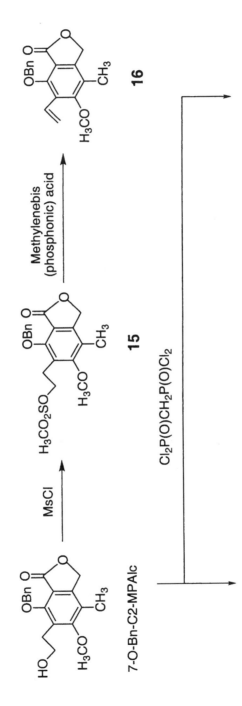

Scheme 6

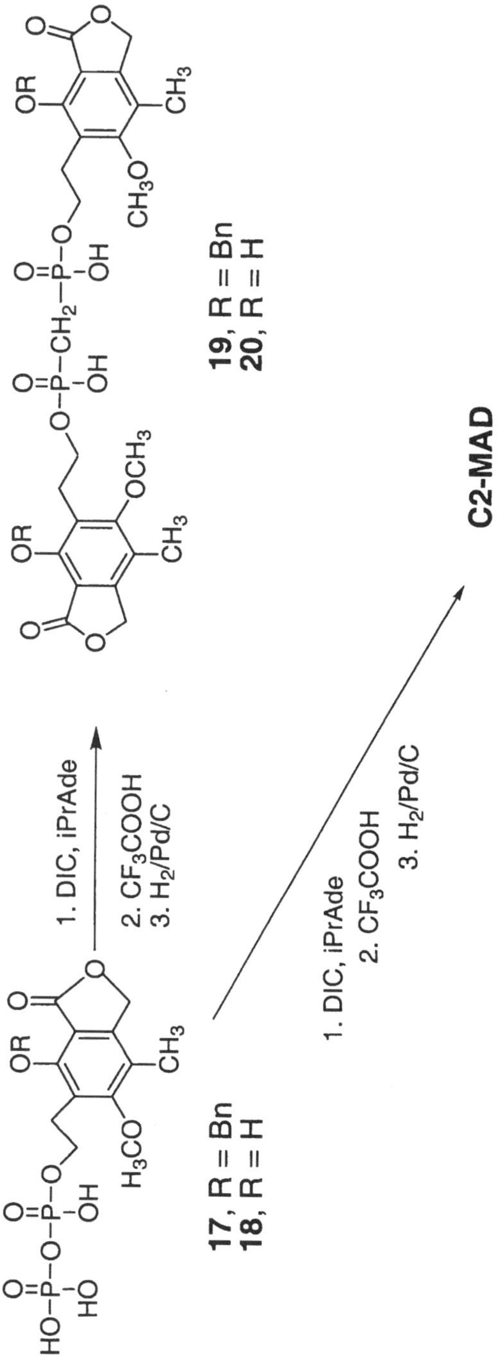

C2-MAD

1. DIC, iPrAde
2. CF₃COOH
3. H₂/Pd/C

17, R = Bn
18, R = H

19, R = Bn
20, R = H

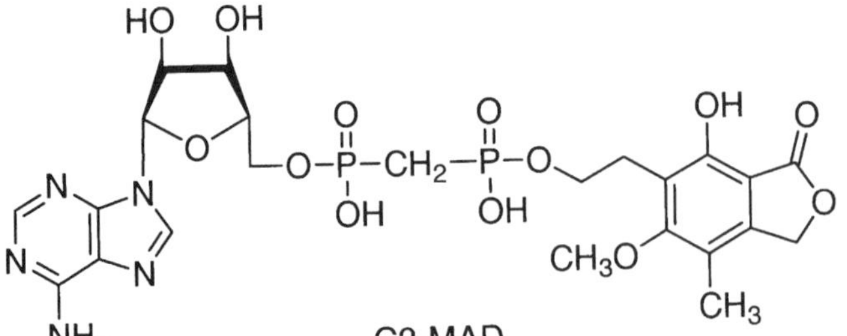

Figure 9. Structure of C2-MAD and 7-0-C2-MAD

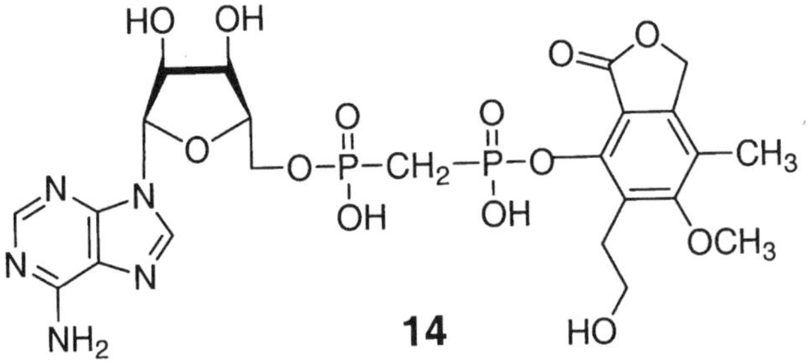

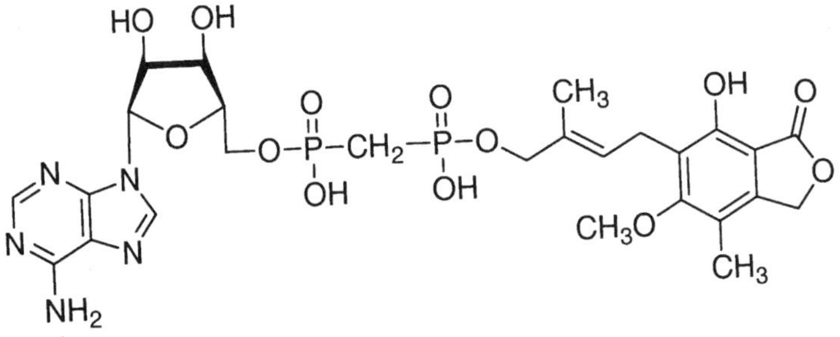

Figure 10. Structure of C4-MAD

inhibitor of IMPDH. C4-MAD and surprisingly C2-MAD analogue inhibit the cells' growth much more potently than analogues of TAD and BAD or even tiazofurin. That is probably due to the efficient (active) transport of MAD analogues into cells.

We also found that, in contrast to MPA, C2-MAD and C4-MAD showed similar resistance to glucuronidation as that of C6-MAD. Incubation of MADs or MPA with uridine 5'-diphospho-glucuronyltransferase and uridine 5'-diphosphoglucuronic acid resulted in extensive (70%) glucuronidation of MPA, whereas no formation of the corresponding MAD glucuronides was observed under similar conditions. These results indicate that if MAD analogues are not glucuronidated in cancer cells they would be potent anticancer agents.

In addition, MAD analogues, specifically C2-MAD and C4-MAD, were found to be highly potent inducers of K 562 cell differentiation (Table III), an order of magnitude more potent than TR. The potency of the C6-MAD analogue was 1.7 fold higher than that of TR. Taking together with their resistance to glucuronidation and potent inhibition of cell growth, MAD analogues may stimulate the differentiation of leukemic cells *in vivo*. If this is the case, MAD analogues may have good potential in treatment of human leukemias.

Table III. Effect of tiazofurin and methylenebis(phosphonate) analogues of TAD and MAD on differentiation of K562 cells.

INHIBITORS	Concentration [μM]	% of benzidine positive cells	% of benzidine positive cells / concentration[a]
Tiazofurin	15.0	80.0	5.3
TAD	14.0	35.0	2.5
C2-MAD	1.6	65.4	40.9
C4-MAD	1.5	95.0	63.3
C6-MAD	10.0	90.0	9.0

[a]The last column of the table shows the percentage of differentiated cells caused by 1.0 μM concentration of the compounds.

Synthesis of gram amounts of above methylenebis(phosphonate) analogues of MAD for pharmacokinetic and toxicological studies is now in progress.

NAD analogues with modifications in the pyrophosphate moiety.

Little is known about the importance of intact pyrophosphate moiety of NAD analogues on their biological activity. The analogues of NAD with an altered, non-ionic (or less-ionic) bridge replacing the pyrophosphate linker have not been examined as potential inhibitors of IMPDH yet. Certainly, such modifications should affect the ability of the analogues to penetrate cell membranes as well as their resistance to degradation by anabolic enzymes. If the replacement of phosphate group(s) of the pyrophosphate moiety of TAD with isosteric or non-isosteric groups would not affect much the binding affinity and consequently the inhibitory activity of modified TAD analogues then such compounds may have some therapeutic potential. Sulfone or sulfonate esters are non-ionic, achiral, isosteric analogues of phosphate esters whereas carboxylate esters geometry differs from that of phosphates. Thus, we synthesized TAD analogues containing –CO– and –SO_2 groups in order to determine the importance of steric factors on inhibitory activity of these analogues.

The 2',3'-O-isopropylideneadenosine under treatment with chloroacetic anhydride afforded the 5'-O-chloroacetyl ester (**21**, Scheme 7), which was converted into the corresponding phosphonoacetyl derivative **22** by Arbuzow reaction with tris(trimethylsilyl)phosphite followed by hydrolysis. Coupling of **22** with 2',3'-O-isopropylidenetiazofurin followed by hydrolysis and deprotection gave exclusively pyrophosphate derivative **23**. This result again supports our novel mechanism of DCC coupling of mono-substituted bis(phosphonate) analogues with nucleosides.[28] Since further DCC dehydration of **23** is not possible and the bicyclic intermediate **5** (Scheme 2) cannot be formed there is no coupling with the tiazofurin derivative. On the other hand MSTN catalyzed coupling of **22** with the tiazofurin derivative afforded the desired phosphonoacetyl analogue of TAD **24** along with a small amount of **23**.[43]

The similar approach for introduction of isosteric –SO_2– group instead of the phosphate group of methylenebis(phosphonate) analogue of TAD was unsuccessful. Treatment of 2',3'-O-isopropylideneadenosine with $ClCH_2SO_2Cl$ gave the cyclic N^3-C5' anhydro derivative as chloromethanesulfonyl chloride salt by a similar reaction to that of 5'-O-tosyladenosine derivative (*vide supra*). When, however, N^6-(dimethylamino)methylene-2',3'-O-isopropylidene-adenosine (**26**, Scheme 8) was treated with phosphonomethanesulfonyl trichloride in CH_2Cl_2 in the presence of 2,6-di-tert-butylpyridine as the acid acceptor the desired sulfonylmethanephosphonic acid derivative **28** was obtained. MSTN catalyzed coupling of **28** with 2',3'-O-isopropylidene-tiazofurin followed by deisopropylidenation with CF_3COOH/H_2O afforded a low yield of TAD anlogue **29**. Interestingly, reaction of adenosine derivative **25** with phosphonomethanesulfonyl trichloride in triethylphosphate, according to

Scheme 7

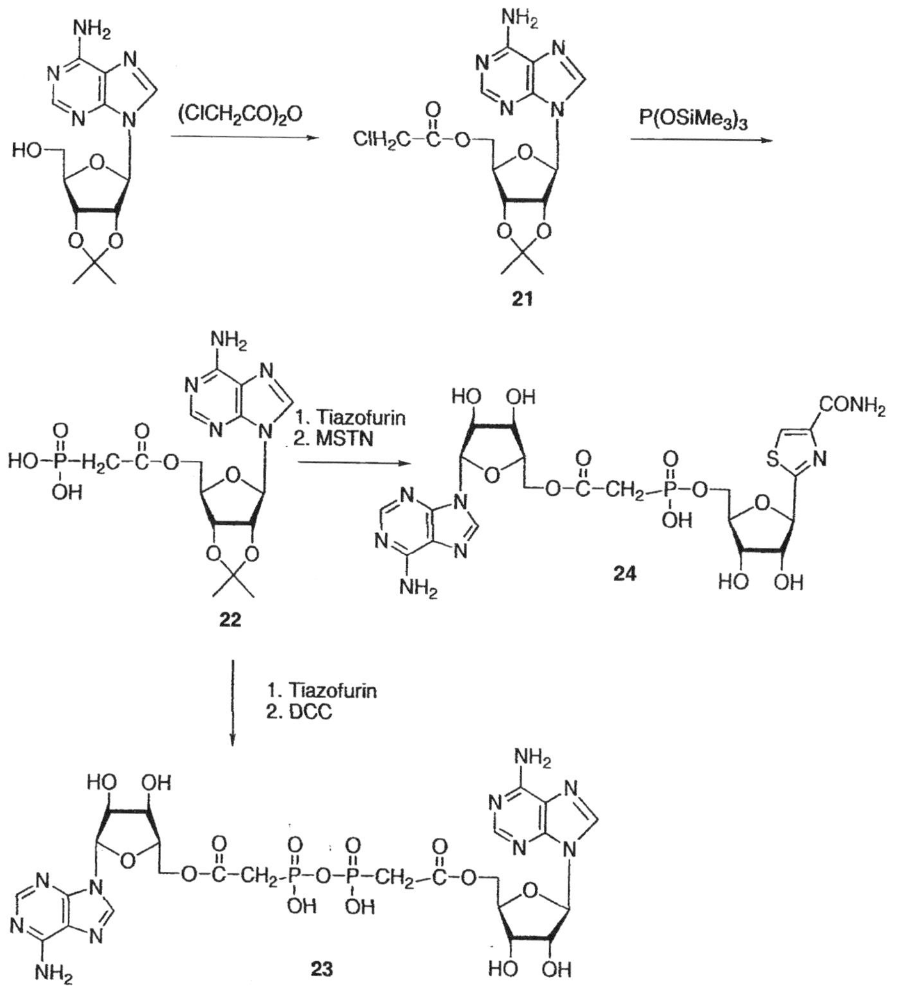

21

22

23

24

Scheme 8

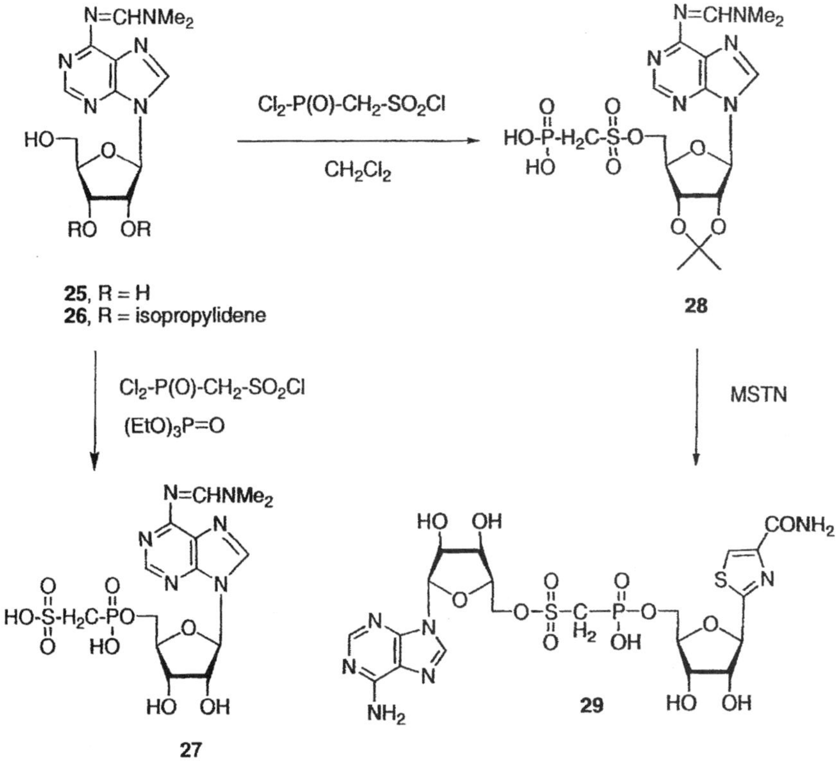

25, R = H
26, R = isopropylidene

27

28

29

Yashikawa's phosphorylation procedure afforded, after hydrolysis, adenosine 5'-O-phosphonyl-methanesulfonic acid (27)

Evaluation of biological activity of these compounds and further studies on chemical modification of the pyrophosphate moiety of these analogues are in progress.

Towards specific inhibition of human IMPDH type II.

Although, the appropriate potency of inhibitors of IMPDH is an important issue, the specificity of these compounds is crucial. The development of an inhibitor specific for type II isoform expressed in cancer cells, which does not affect the activity of isoform I, predominant in normal cells, could afford an agent with an unique cytotoxicity directed exclusively or primarily against cancer cells.

All known inhibitors, including nucleosides and their active metabolites [including their bis(phosphonate) analogues], did not show any specificity. They inhibit both type I and type II isoforms with equal potency. Only mycophenolic acid inhibits isoform II with slightly greater (4-fold) potency than type I. These results are not surprising in the light of new observations based on the crystal structure of a ternary complex of human type II IMPDH with 6-chloropurine and SAD, (selenazofurin adenine dinucleotide)[42] solved recently by Goldstain et al[9]. It was found that although the substrate site and nicotinamide subsite of IMPDH are conserved in the type I and type II isoforms, the adenosine subsite of the enzyme's cofactor binding domain is not. The adenosine moiety of the SAD analogue is bound in the previously unidentified cleft between two adjacent monomers, forming an unusual number of inter-monomer hydrogen bonds. This may explain, in part, the reason for requirement for the formation of tetramer, which is an active form of the enzyme. *This again shows the uniqueness of the cofactor-binding domain of IMPDH.* The adenine ring is stacked between the side chains of Phe-282 and His-253. Density is observed between the adenine amino group and the side chain of Thr-252. Also the hydrogen bonding is observed between adenine N3 and the side chain of Thr-45 on the adjacent monomer. His-253, Phe-282 and Thr-45 in the type II isoform are replaced by Arg, Tyr and Ile, respectively, in the type I enzyme. Thus, the structural differences of the catalytic site between the type I and type II isoform are expressed exclusively in the adenine subsite. Since tiazofurin and BR are metabolized to the cofactor analogues in such a manner that the active dinucleotides always contain adenine moiety, which fits equally well to both isoforms, no specificity should be expected. *The specific inhibition of type II isoform could be only attained by chemical construction of the bis(phosphonate) analogues of the active metabolites, which contained a modified purine base.*

The desired modification of the cofactor analogue should be such that a new inhibitor would fit much better to the type II than to the type I isoform.

The replacement of bulky isoleucine of the type I by smaller threonine in IMPDH type II can be exploited for the synthesis of new NAD analogues with improved specificity against the type II isoform. Thus, modication of the bis(phosphonate) analogue of BAD by replacement of the adenine moiety of BAD by isoguanine or 2, 6-diaminopurine (Figure 11) may lead to specificity. We expect that the introduction of a substituent at C2 of adenine ring, such as – OH or $-NH_2$, may result in formation of bifurcated hydrogen bonds between the hydroxyl group of Thr-45 of isoform II and both the adenine N3 and the oxo group of isoguanine or the amino group of diaminopurine. In addition, the hydroxyl or amino group at C2 of the purine ring is expected to be well accommodated in the adenine-binding pocket of the type II isoform (where the relatively small Thr-45 fills the cleft). In contrast, BAD analogues containing isoguanine (BID) or diaminopurine (BDD) would not fit well to IMPDH type I because the bulky isoleucine should prevent tight binding due to the steric hindrance caused by the presence of –OH or $-NH_2$ group at C2 of the purine ring.

This hypothesis is supported by modeling studies, based on the Goldstein's crystal structure.[9] The crystal structure indicated that the cofactor adenine pocket is tightly constrained by Thr-252 and Thr-45 (Plate 1). As noted above, these residues form edge-on contacts with the adenine moiety. Replacement of the hydrogen atom at C2 of the adenine with the amino group (as in 2,6-diaminopurine) appears favorable in the type II enzyme. Here, the extra NH_2 at the 2 position of the heterocyclic ring is well placed to form a hydrogen bond with the hydroxyl group of Thr-45 (Plate 1 [left]). However, in the type I enzyme, Thr-45 is replaced by Ile-45. In this case, the bulky Ile methyl groups form unfavorable steric contacts with both N3 and the 2-amino substituent of the ligand (Plate 1 [right]). A similar result is obtained by replacing the adenine ring with isoguanosine, which contains exocyclic oxygen at the 2 position.

Synthesis of NAD analogues with modification at the C2 of the adenine ring is now in progress in our laboratory.

References

1. NAD analogues 19. For part 18 see: Pankiewicz, K.W.; Goldstein, B.M. The chemistry of nucleoside and dinucleotide inhibitors of inosine monophosphate dehydrogenase (IMPDH) in *Recent Advances in Nucleosides: Chemistry and Chemotherapy*, Chu, C.K, Matsuda, A., ed., Elsevier 2002

2. Tam, R. C.; Ramasamy, K.; Bard, J.; Pai, B.; Lim, C.; Averett, D. R. *Antimicrob. Agents Chemother.* **2000**, *44*, 1276-1283.

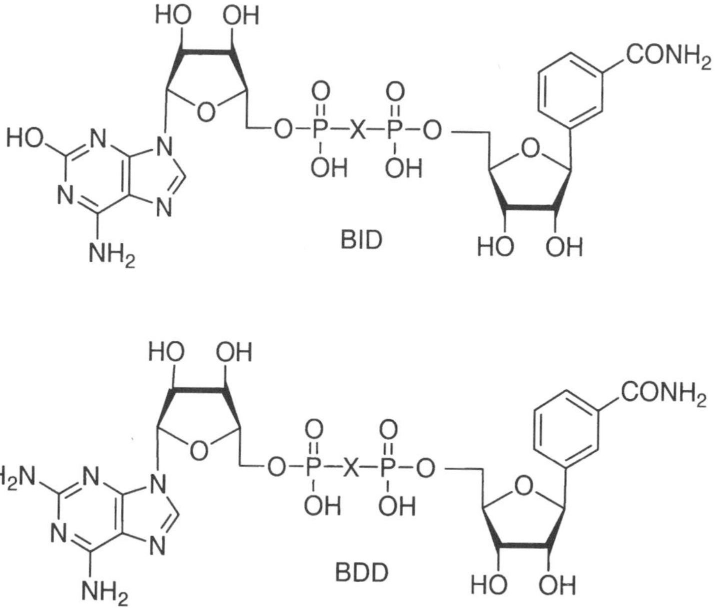

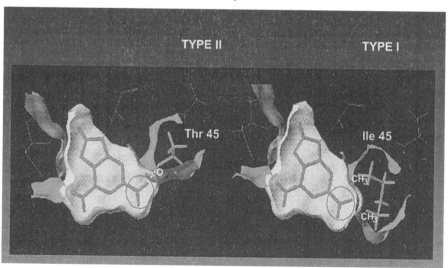

Figure 11. Structure of BID and BDD

Plate 1. Model of IMPDH adenine-binding site occupied by 2,6,-diaminopurine containing dinucleotide. Van der Waals surfaces of adjacent residues are illustrated. At left: replacement of adenine with 2,6,-diaminopurine allows formation of hydrogen bonds with Thr-45 in IMPDH type II. At right: 2,6,-diamino purine exocyclic nitrogen at the C2 forms unfavorable steric contacts with Ile-45 in IMPDH type I .

(This figure is also in color insert.)

3. Markland, W.; McQuaid, T.J.; Jain, J.; Kwong, A. D. *Antimicrob. Agents Chemother.* **2000**, *44*, 859-866.
4. Minakawa, N.; Matsuda, A. *Curr. Med. Chem.* **1999**, *6*, 615-628.
5. Pankiewicz, K. W. *Exp. Opin. Ther. Patents.* **1999**, *9*, 55-65.
6. Kerr, K. M; Hedstrom L. *Biochemistry* **1997**, *36*, 13365-13373.
7. Collart, F.; Huberman, E. *J. Biol. Chem.* **1988**, *263,*15769-15772.
8. Carr, S. F.; Papp, E.; Wu, J. C.; Natsumenda, Y. *J. Biol. Chem.* **1993**, *268*, 27286-27290.
9. Colby, T. D.; Vanderveen, K.; Strickler, M. D.; Markham, G. D.; Goldstein, B. M. *Proc. Natl. Acad. Sci. USA* **1999**, *96*, 3531-3536.
10. Zhang, R-G.; Evans, G.; Rotella, F. J.; Westbrook, E. M.; Beno, D.; Huberman, E.; Joachimiak, A.; Collart, F. R. *Biochemistry* **1999**, *38*, 4691-4700.
11. O'Gara, M. J.; Lee, C-H.; Weinberg, G. A.; Nott, J. M.; Queener, S. F. *Antimicrob. Agents Chemother.* **1997**, *41*, 40-48.
12. McMillan, F. M.; Cahoon, M.; White, A.; Hedstrom, L.; Petsko, G. A.; Ringe, D. *Biochemistry* **2000**, *39*, 4533-4542.
13. Whitby, F. G.; Luecke, H.; Kuhn, P.; Somoza, J. R.; Huete-Perez, J. A.; Phillips, J. D.; Hill, C. P.; Fletterick, R. J.; Wang, C.C. *Biochemistry* **1997**, *36*, 10666-10674.
14. Nimmesgern, E.; Fox, T.; Fleming, M. A.; Thomson, J. A. *J. Biol. Chem.* **1996**, *271*, 19421-19427.
15. Wang, W.; Hedstrom, L. *Biochemistry* **1997**, *36*, 8479-8483.
16. Fleming, M. A.; Chambers, S. P.; Conelly, P. R.; Nimmesgern, E.; Fox, T.; Bruzzese, F. J.; Hoe, S. T.; Fulghum, J. R.; Livingston, D. J.; Stuver, C. M.; Sintchak, M. D.; Wilson, K. P.; Thomson, J. A. *Biochemistry* **1996**, *35*, 6990-6997.
17. Wu, J. C. *Prespectives in Drug Discovery and Design;* Wyvratt, M. J., Sigal, N.H., Eds.: ESCOM Science Publisher: Leiden, **1994**, Vol. 2, pp. 185-204.
18. Fuertes, M.; Garzia-Lopez, T.; Garzia-Munoz, G.; Stud, M. *J. Org. Chem.* **1976**, *41*, 4074-4077.
19. Robins, R. K. *Nucleosides, Nucleotides,* **1982**, *1*, 35-44.
20. Cooney, , D. A.; Jayaram, H. N.; Glazer, R. A.; Kelly, J. A.; Marquez, V. E.; Gebeyehu, G.; Van Cott, A. C.; Zwelling, L. A.; Johns, D. G. *Adv. Enzyme Regul.* **1983**, *21*, 271-303.
21. Krohn, K.; Heins, H.; Wielckens, K. *J. Med. Chem.* **1992**, *35*, 511-517.
22. Gharehbaghi, K.; Paull, K. D.; Kelley, J. A.; Barhi Jr., J. J.; Marquez, V. E.; Cooney, D. A.; Monks, A.; Scudiero, D.; Krohn, K.; Jayaram, H. N. *Int. J. Cancer* **1994**, *56*, 892-899.
23. Ahluwalia, G. S.; Jayaram, H. N.; Plowman, J. P.; Cooney, D. A.; Johns, D. G. *Biochem. Pharm.* **1984**, *33*, 1195
24. Carney, D. N.; Ahluwalia, G. S.; Jayaram, H. N.; Cooney, D. A.; Johns, D. G. *J. Cin. Invest.* **1985**, *75*, 175

25. Marquez, V. E.; Tseng, C. K. H.; Gebeyehu, G.; Cooney, D. A.; Ahluwalia, G. S.; Kelley, J. A.; Dalal, M.; Fuller, R. W.; Wilson, Y. A.; Johns, D. G. *J. Med. Chem.* **1986**, *29*, 1726-1731.

26. Pankiewicz, K. W.; Lesiak, K.; Zatorski, A.; Goldstein, B. M.; Carr, S. F.; Sochacki, M.; Majumdar, A.; Seidman, M.; Watanabe, K. A. *J. Med. Chem.* **1997**, *40*, 1287-1291.

27. Lesiak, K.; Watanabe, K. A.; Majumdar, A.; Seidman, M.; Vanderveen, K.; Goldstein, B. M.; Pankiewicz, K. W. *J. Med. Chem.* **1997**, *40*, 2533-2538.

28. Pankiewicz, K. W.; Lesiak, K.; Watanabe, K. A. *J. Am. Chem. Soc.* **1997**, *119*, 3691-3695.

29. Davisson, V. J.; Davis, D. R.; Dixit, V. M.; Poulter, C. D. *J. Org. Chem.* **1987**, *52*, 1794-1801.

30. W. Jahn. *Chem. Ber.* **1965**, *98*, 1705-1708.

31. Lesiak, K.; Watanabe, K. A.; George, J.; Pankiewicz, K. W. *J. Org. Chem.* **1998**, *63*, 1906-1909.

32. Lesiak, K.; Watanabe, K. A.; Majumdar, A.; Seidman, M.; Vanderveen, K.; Golldstein B. M.; Pankiewicz, K. W. *J. Med. Chem.* **1997**, *40*, 2533-2538.

33. Zatorski, A.; Goldstein, B. M.; Colby, T. D.; Jones, J. P.; Pankiewicz, K. W. *J. Med. Chem.* **1995**, *38*, 1098-1105.

34. Pankiewicz, K. W.; Lesiak, K. *Nucleosides, Nucleotides* **1999**, *18*, 927-932.

35. Blackburn, M.; England, D. A.; Kolkman, F. *J. Chem. Soc. Chem. Commun.* **1981**, 930-932.

36. Anderson, W.K., Lee, J., Swann, R.T., Bohem, T.L. *Advances in New Drug Development*, Kim, B.K., Lee, E.B., Kim, C.H., Han, Y.N., Eds., The Pharmaceutical Society of Korea, **1991**, p-8-17.

37. Franklin, T.J., Jacobs, V.N., Jones, G., Ple, P. *Drug Metabolism Disposition* **1997**, *25*, 367.

38. Franklin, T.J., Jacobs, V., Jones, G., Ple, P., Bruneau, P. *Cancer Res.* **1996**, *56*, 984.

39. Sintchak, M. D.; Fleming, M. A.; Futer, O.; Raybuck, S. A.; Chambers, S. P.; Caron, P. R.; Murcko, M. A.; Wilson, K. P. *Cell* **1996**, *85*, 921-930.

40. Lesiak, K.; Watanabe, K. A.; Majumdar, A.; Powell, J.; Seidman, M.; Vanderveen, K.; Goldstein, B. M.; Pankiewicz, K. W. *J. Med. Chem.* **1998**, *41*, 618-622.

41. Jones, D. F.; Mils, S. D. *J. Med. Chem.* **1971**, *14*, 305-311.

42. SAD is an analogue of TAD in which the sulfur atom of the thiazole-4-carboxamide moiety of tiazofurin is replaced with a selenium atom.

43. Zatorski, A.; Lipka, P.; Pankiewicz, K.W. *Collect. Czech. Chem. Commun.* **1993**, *58*, 122-126

Chapter 14

Identification of Specific Inhibitors of IMP Dehydrogenase

Frank R. Collart[1],* and Eliezer Huberman[2]

[1]Biosciences Division and [2]Biochip Technology Center, Argonne National Laboratory, 9700 South Cass Avenue, Building G 202, Argonne, IL 60439
*Corresponding author: email: Fcollart@anl.gov;
telephone: 630–252–4859; fax: 630–252–5517

IMP dehydrogenase (IMPDH) is an important therapeutic target and IMPDH inhibitors are used in cancer chemotherapy and for immunosuppression. Although IMPDH inhibitors may have therapeutic potential as antimicrobial, antifungal or antiprotozoal agents, no specific IMPDH inhibitors have been identified for microbial organisms. The recent availability of crystal structures of IMPDH from different organisms will facilitate the identification of these agents. We have developed a screening method for identifying IMPDH inhibitors that is applicable to any class of organism. The system is amenable to high throughput systems for the screening of inhibitors generated by combinatorial chemistry or other methods and can be used to screen for inhibitors to IMPDH from any source for which a coding sequence is available. The addition of exogenous guanosine can be used as a method to identify inhibitory chemicals that specifically target IMPDH. This work was supported by the U.S. Department of Energy, Office of Health and Environmental Research, under Contract W-31-109-ENG-38.

IMPDH, an important therapeutic target and inhibitors of this enzyme have clinical importance in the treatment of neoplastic disease[1] and viral infections[2] and as immunosuppressive agents[3]. These inhibitors may also be useful antimicrobial, antifungal or antiparasitic therapeutic agents, however, this application has not been widely investigated. The basis for the therapeutic utility of this enzyme stems form several unique features. IMPDH is a required enzyme for the de novo synthesis of purine nucleotides. Because of this essential function, all free-living organisms contain the necessary genetic information to produce the IMPDH enzyme. From a therapeutic perspective, the only method to circumvent a block in IMPDH activity is via the salvage pathways. The utility of this enzyme as a therapeutic target is further enhanced by its unusual reaction mechanism that involves the sequential binding of IMP and NAD with the subsequent release of NADH and XMP[4]. It is likely that this distinctiveness relative to other dehydrogenase enzymes contributes to the specificity of IMPDH inhibitors.

A characteristic of IMPDH enzymes that has not been fully exploited from a therapeutic perspective is the biochemical diversity. Bacterial IMPDH enzymes show biochemical and kinetic characteristics that are different than the mammalian IMPDH enzymes (Table I). These differences include cofactor-binding affinity[5], sensitivity to inhibitors[6] and cation requirements[7]. These differences likely reflect an alteration in the reaction mechanism as a consequence of the variance of amino acid residues in the active site. Identification of these residues or combination of residues that impart this mammalian or microbial enzyme signature can be exploited for the development of agents that specifically target bacterial or protozoan IMPDH enzymes.

Basis for the development new IMPDH inhibitors

Progress in the identification IMPDH inhibitors has been leveraged by several developments in the scientific arena including:
- The availability of genomic sequence information from many different organisms
- The development of bacterial systems for expression of IMPDH enzymes
- The determination of the crystal structure for IMPDH from several sources including mammalian and bacterial enzymes

At the present time, IMPDH coding regions have been identified for greater than 50 different species with representatives from the eukaryotic, prokaryotic, and archaeal domains. The rapid accumulation of sequence information from various organisms provides a resource of new IMPDH enzymes to discern the full spectrum of diversity and will provide a basis for the development of specific IMPDH inhibitors that target selected organisms. The wealth of genomic information has also provided the incentive for the development of bacterial systems for expression of IMPDH enzymes. IMPDH enzymes from more that 20 different organisms have been expressed in *Escherichia coli*. A number of expression clones developed at the Robotic Molecular Biology Facility at Argonne National Laboratory are listed in Table II. Many of these clones are expressed as tagged fusion proteins to facilitate downstream protein assay and purification procedures. For most bacterially expressed IMPDH enzymes, addition of an N-terminal fusion tag does appear to significantly affect enzyme activity. These expression clone resources have contributed to characterization of the biochemical and kinetic properties and provided an enzyme resource for crystallization screens and the subsequent determination of the crystal structures for several IMPDH enzymes. The availability of IMPDH crystal structures from eukaryotic[8-9] and bacterial[5, 10] organisms provides detailed molecular information about the catalytic site and insight into the reaction mechanism. This information suggests there are a number of areas that can be attractive target regions (Table III) for therapeutic development.

IMPDH inhibitor screen

Developments in combinatorial chemistry allow the rapid and economical synthesis of hundreds to thousands of discrete compounds. These compounds are typically arrayed in moderate-sized libraries of small organic molecules designed to provide targeted or directed diversity. However, there is not a published method to systematically screen the various chemical agents for utility as IMPDH inhibitors. Development of such an assay would enhance and accelerate the discovery of therapeutically useful inhibitors.

Description of the method.

We have developed an efficient method to identify specific inhibitors of IMPDH. The method includes a prokaryotic or eukaryotic host organism that is a guanosine auxotroph. Our present system used the *E. coli* H712 strain originally described by Nijkamp, *et al.*[11] This strain does not produce a function IMPDH enzyme and requires the addition of a guanine nucleotide precursor(s) to the culture medium for cell growth. For the screening method, the H712 strain is transformed with an IMPDH expression vector capable of producing a functional IMPDH enzyme. The present system uses a broad host range vector pJF118EH, constructed by Fürste *et al.*[12] This expression system uses ampicillin resistance as a selectable marker and permits the regulated expression of foreign coding sequences in an *E. coli* host. The pJF118EH vector has the properties of inducible expression such that in the absence of an inducer the expression of the cloned foreign is low. Induction of the foreign gene is initiated by the addition of IPTG to the bacterial culture medium. H712 bacteria transformed with the

Table I. Differential inhibitor sensitivity of human and *S. pyogenes* IMPDH

Inhibitor	Human K_i (μM)	Streptococcus K_i/IC_{50} (μM)	Ratio Human/Strep
MPA	0.01	150	15,000
Mizoribine-P	0.01	1	50
Ribavirin-P	0.4	6	15
TAD	0.19	10-20	80

Table II. Expression of Recombinant IMPDH in *E. coli*

	Organism	Product	Enzyme Activity
Eukaryotic	Human	Enzyme	▸
	Human	Truncated enzyme	▸
	Arabidopsis thaliana	Enzyme	▸
	Trypanasoma bruceii	Enzyme	▸
Bacteria	*Streptococcus pyogenes*	Enzyme	▸
	Haemophilus influenzae	Enzyme	▸
	Pseudomonas aeruginosa	Enzyme	▸
	Bacillus stearothemophilus	PCR fragment	
	Mycobacterium tuberculosis	Enzyme	▸
Archaea	*Pyrococcus furiosus*	Enzyme	▸
	Halobacterium halobium	PCR fragment	

Table III. Molecular target regions amenable for IMPDH inhibitors.

Target region	Therapeutic relevance
IMP binding site	Several inhibitors of bacterial and mammalian enzymes are currently available (e.g. ribavirin and mizoribine)
Cofactor binding site	Inhibitors that occupy this site display a wide variance in effectiveness for mammalian and bacterial IMPDH enzymes.
Cation requirement	Some IMPDH exhibit a reduce requirement for monovalent cations.
CBS dimmer domain	Domain function is unknown but contains a cleft with binding potential.
Tetrameric contacts	The tetrameric form of the enzyme is essential for attainment of the catalytic pocket environment

pJF118EH vector containing the human or *S. pyogenes* IMPDH coding regions produce function enzyme and no longer require the addition of guanosine to the culture medium for growth. Furthermore, the IMPDH enzymes produced in the *E. coli* host retain the biochemical and kinetic characteristics of the source (i.e., either human or *S. pyogenes*). This system allows an evaluation of the effect of various agents on the growth of the recombinant bacteria and the identification of putative IMPDH inhibitors. However, there are chemicals that inhibit host cell proliferation regardless of whether or not the host is expressing a prokaryotic or eukaryotic IMPDH enzyme. To identify agents specifically targeting IMPDH, the inhibitor profile for a particular agent is compared to the same profile obtained when a guanine nucleotide precursor such as guanine or guanosine is added to the culture medium. Guanosine is the preferred agent due to its high solubility in aqueous solutions. These guanine nucleotide precursors are able to circumvent the block on IMPDH activity imposed by IMPDH inhibitors and can be used to exclude nonspecific growth inhibitory or toxic compounds. Agents specifically targeting IMPDH will exhibit a decrease ability to inhibit cell proliferation in the presence of guanine/guanosine while the growth inhibitory effects of nonspecific or toxic agents will not be ameliorated by guanine/guanosine.

Plate assay system for candidate inhibitors.

A plate assay system containing various amounts of several different IMPDH inhibitors were added to 7mm filter disks and the disks placed on a lawn of H712 bacteria containing either the human or *S. pyogenes* IMPDH expression plasmid and IPTG for induction of enzyme expression. Inhibitors were selected on the basis of reports in the literature regarding specificity for inhibition of human IMPDH. Both mycophenolic acid (MPA) and ribavirin are clinically useful and MPA is known to inhibit human, but not bacterial IMPDH[6]. Some inhibitors, ribavirin, tiazofurin, and mizoribine, require cellular activation for utility as IMPDH inhibitors. For several of the inhibitors, a clear area was observed around the filters corresponding to the degree of inhibition (Figure 1A). Various inhibition patterns were observed for bacteria containing the human or *S. pyogenes* IMPDH expression vectors that ranged from no inhibition, to inhibition of human IMPDH, to inhibition of both forms of IMPDH. Furthermore all of the growth inhibitory chemicals showed a differential between the low and high does. The inhibition profiles obtained with this panel of IMPDH inhibitors demonstrate the IMPDH enzyme produced in bacteria retains the biochemical and kinetic characteristics of the source (i.e. human or *S. pyogenes*). The results also illustrate the utility of this approach for identification of the inhibitory spectrum of IMPDH inhibitors.

This method also has the capability to identify clinically useful features of the potential IMPDH inhibitors. Clinically useful agents must be transported into the target organism and may require activation for therapeutic effectiveness. In this screening procedure, the use of a host organism that mimics the characteristics of the eventual therapeutic target can provide useful information regarding clinically useful properties of potential therapeutic agents. The results presented in Figure 1A demonstrate ribavirin and mizoribine are transported into the bacterial cells and are activated by the host system to a form, which is capable of inhibiting IMPDH. Thus, this method inherently excludes compounds that might be impermeable to the host organism. Alternatively, because many inhibitory compounds require activation by chemical modification, this method can be applied to determine the competency of a host organism for metabolic activation of IMPDH inhibitors.

To demonstrate that the observed growth inhibition was specific for IMPDH, some of the inhibitor filter disks were placed on a lawn of H712 bacteria on plated supplemented with guanosine. On these plates, the exogenous guanosine reduced the inhibitory effect of the mycophenolic acid (Figure 1B). The ability of guanosine to ameliorate the growth inhibition indicates the specificity of these chemicals for IMPDH.

High throughput screening.

The microbial system is amenable to high throughput systems for the screening of inhibitors generated by combinatorial chemistry or other methods. To examine this approach we used microwell plates and a Biomek workstation to implement automated protocols to assess the effect of various inhibitors on bacterial growth. Inhibitors were added to an arrayed culture of recombinant H712 bacteria and the effect on bacterial growth assessed after an overnight incubation by measurement of absorbance at 600 nm. The inhibition pattern observed for H712 bacteria containing the human IMPDH expression plasmid (Figure 2) is similar to that observed for the plate assay. All of the tested IMPDH inhibitors decrease the growth of the recombinant strain as a function of inhibitor concentration with MPA being the most effective inhibitor of bacterial growth. A parallel analysis using H712 bacteria containing the *S. pyogenes* IMPDH expression plasmid (Figure 3) illustrated the difference between the bacterial and mammalian IMPDH enzymes. The growth of the variant containing the *S. pyogenes* enzyme is not inhibited by MPA consistent with results observed with the purified bacterial enzyme[5]. Bacterial growth is inhibited by treatment with ribavirin and mizoribine consistent with results obtained using the purified enzyme (Table I). In contrast to the results obtained using purified IMPDH enzymes, the *in* vivo screening method shows ribavirin is a more effective inhibitor of the bacteria expression the human and *S, pyogenes* IMPDH enzymes than mizoribine. This effect is likely represents a differential ability related to transport or metabolic activation of these agents.

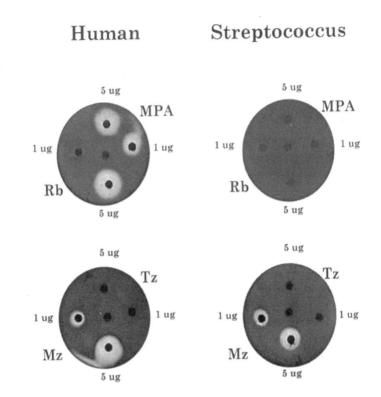

Figure 1. Assays of candidate inhibitors. (A) Different amounts (1 and 5 μg) of several different IMPDH inhibitors were added to 7mm filter disks and the disks placed on a lawn of H712 bacteria containing either the human or S. pyogenes IMPDH expression plasmid and IPTG for induction of enzyme expression. A control disk containing the solvent but no inhibitor was placed in the center of the dish. The inhibitors used were MPA, ribavirin (Rb), tiazofurin (Tz), and mizoribine (Mz). After incubation at 37 °C overnight, the plates were examined for inhibition of bacterial cell growth. (B) IMPDH inhibitors were added to 7mm filter disks placed on a lawn of H712 bacteria containing the human IMPDH expression plasmid. The growth medium of the indicated plate was supplemented with 50 μg/ml of guanosine. After incubation at 37 °C overnight, the plates were examined for inhibition of bacterial cell growth. The restoration of bacterial growth on the plates containing exogenous guanosine indicates the specificity of this chemical for IMPDH.

B

+guanosine

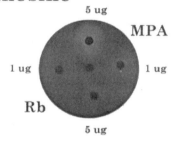
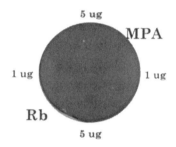

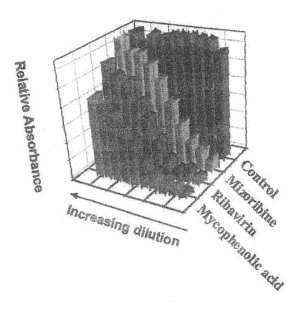

Figure 2. Inhibitor response of H712 bacteria containing the human IMPDH expression plasmid. Inhibitors were added to the initial well to a final concentration of 2.5 mM. The final well of the dilution series contained an inhibitor concentration of 5 μM. The absorbance (at 600 nm) of the plate cultures were determined after incubation at 37 °C overnight.

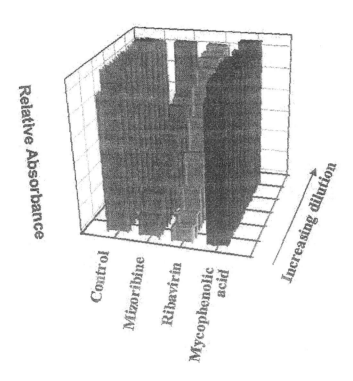

Figure 3. Inhibitor response of H712 bacteria containing the S. pyogenes
*IMPDH expression plasmid. Inhibitors were added to the initial well to a final
concentration of 2.5 mM. The final well of the dilution series contained an
inhibitor concentration of 5 μM. The absorbance (at 600 nm) of the plate
cultures were determined after incubation at 37 °C overnight*

292

Summary

The microbial screening system of the present invention has a number of advantages for the identification of specific IMPDH inhibitors:

1. The microbial system is amenable to high throughput systems for the screening of inhibitors generated by combinatorial chemistry or other methods
2. This method can be used to screen for inhibitors to IMPDH from any source for which a coding sequence is available. At the present time, more than 50 different IMPDH coding sequences from the eukaryotic, bacterial and archaeal domains are available in the DNA sequence databases
3. The addition of exogenous guanosine can be used as a method to verify inhibitory chemicals that specifically target IMPDH and to distinguish from other causes of host cell inhibition.
4. The method provides metabolic information such as the ability of the inhibitors to enter the cell and to be activated directly or after their cellular metabolism.
5. This method can be used to identify chemicals capable of differentially inhibiting IMPDH from organisms that are genetically dissimilar.

This method can be applied for the identification of specific IMPDH inhibitors, which could be used to as the basis for the design of pharmaceuticals. Additional human IMPDH inhibitors could be identified for use as chemotherapeutic agents for the treatment of neoplasm's and as immunosuppressive agent. This method is also useful for the identification of agents selective for various IMPDH isoforms (e.g. human Type I and II). This method also has the potential for identification of inhibitors of bacterial, fungal, protozoan or viral IMPDH that do not inhibit the human or other mammalian enzymes and may be effective therapeutic agents.

Acknowledgment

The submitted manuscript has been created by the University of Chicago as Operator of Argonne National Laboratory ("Argonne") under Contract No. W-31-109-ENG-38 with the U.S. Department of Energy. The U.S. Government retains for itself, and others acting on its behalf, a paid-up, nonexclusive, irrevocable worldwide license in said article to reproduce, prepare derivative works, distribute copies to the public, and perform publicly and display publicly, by or on behalf of the Government.

References

1. Pankiewicz, K.W. Novel nicotinamide adenine dinucleotide analogues as potential anticancer agents: quest for specific inhibition of inosine monophosphate dehydrogenase. *Pharmacol. Ther.*, **1997**, *76*, 89-100.
2. Andrei, G.; De Clercq, E. Molecular approaches for the treatment of hemorrhagic fever virus infection. *Antiviral Res.*, **1993**, *22*, 45-75.
3. Halloran, P.F. Molecular mechanisms of new immunosuppressants *Clin. Transplant,* **1996**, *10*, 118-123.
4. Carr S.F., Papp E., Wu J.C., Natsumeda Y. Characterization of human type I and type II IMP dehydrogenases. *J Biol Chem.*, **1993**, 268, 27286-90.
5. Zhang R., Evans G., Rotella F.J., Westbrook E.M., Beno D., Huberman E., Joachimiak A., Collart F.R. Characteristics and crystal structure of bacterial inosine-5'-monophosphate dehydrogenase. *Biochemistry* **1999**; 38, 4691-700.
6. Hupe D.J., Azzolina B.A., Behrens N.D. IMP dehydrogenase from the intracellular parasitic protozoan Eimeria tenella and its inhibition by mycophenolic acid. *J. Biol. Chem.* **1986**, 261, 8363-8369.
7. Kerr K.M., Cahoon M., Bosco D.A., Hedstrom L. Monovalent cation activation in Escherichia coli inosine 5'-monophosphate dehydrogenase. *Arch Biochem Biophys* **2000**, 375, 131-137.
8. Sintchak, M.D., Fleming, M.A., Futer, O.; Raybuck, S.A.; Chambers, S.P.; Caron, P.R., Murcko, M.A., Wilson, K.P. Structure and mechanism of inosine monophosphate dehydrogenase in complex with the immunosuppressant mycophenolic acid. *Cell* **1996**, 85, 921-930.
9. Whitby, F.G., Luecke, H.; Khun, P., Somoza, J.R., Huete-Perez, J.A., Phillips, J.D., Hill, C.P., Fletterick, R.J., Wang, C.C. Crystal structure of Tritrichomonas foetus inosine-5'-monophosphate dehydrogenase and the enzyme-product complex *Biochemistry* **1997**, 36, 10666-10674.
10. McMillan F.M., Cahoon M., White A., Hedstrom L., Petsko G.A., Ringe D. Crystal structure at 2.4 A resolution of Borrelia burgdorferi inosine 5'-monophosphate dehydrogenase: evidence of a substrate-induced hinged-lid motion by loop 6. *Biochemistry* **2000**, 39, 4533-42.
11. Nijkamp, H.J.J. and De Hann, P.G. Genetic and Biochemical studies of the guanosine 5'-monophosphate pathway in Escherica coli. *Biochim Biophys Acta* **1967**, 145:31-40.
12. Fürste, J.P., Pansegrau, W., Frank, R., Blöcker, Scholz, P., Bagdasarian, M., and Ianka, E. Molecular cloning of he plasmid RP4 primase region in a multi-host-range tacP expression vector. *Gene* **1986**, 48:119-131.

Chapter 15

Mizoribine: Experimental and Clinical Experience

Hiroaki Ishikawa, Masahiko Tsuchiya, and Hiromichi Itoh

Asahi Kasei Corporation, 9-1 Kanda Mitoshirocho, Chiyoda-ku, Tokyo
101-9481, Japan

An immunosuppressant demonstrated to be a potent IMPDH
inhibitor, mizoribine (MZR) was discovered in 1971 by
scientists of Asahi Kasei Corporation in Ohito, Japan. Studies
conducted in Japan have confirmed its clinical effectiveness
and relative safety for the prevention of rejection in renal
transplantation, and for the treatment of lupus nephritis,
rheumatoid arthritis, and nephrotic syndrome. MZR (trade
name: Bredinin) was put on the market in 1984, and has been
widely used for the aforesaid indications for 17 years,
contributing to overall patient care. MZR offers outstanding
clinical benefits in that after nearly two decades of use MZR
has exhibited a low incidence of severe adverse drug reactions
and no enhancement of oncogenicity.

Historical Perspective

Eupenicillium brefeldianum M-2166, the ascomycetes of which were
harvested from the soil of Hatijo Island, Tokyo, Japan, in 1971, produces
mizoribine (MZR). This compound, a nucleoside of the imidazole class, was
found to have weak antimicrobial activity against *Candida albicans,* but it
proved ineffective against experimental candidiasis (*1*).

MZR inhibits the *de novo* biosynthesis of purines. Unlike azathioprine (AZT), the compound is not taken up by nucleic acids in the cell. Instead, after phosphorylation, mizoribine-5'-monophosphate (MZR-5'P) inhibits GMP synthesis by the antagonistic blocking of IMPDH (Ki = 10^{-8} M) and GMP-synthetase (Ki = 10^{-5} M) in the pathway from IMP to GMP of the purine synthesis system. The drug was found to inhibit both humoral and cellular immunity by selectively inhibiting the proliferation of lymphocytes, leading to its development as an immunosuppressive agent. The clinical efficacy of MZR as an immunosuppressant in renal transplantation was studied in various Japanese institutes from 1978 to 1982. MZR was approved by the Japanese Ministry of Health, Labor and Welfare in 1984 as a drug indicated for "the prevention of rejection in renal transplantation" (2). Recently, it has been most commonly used in combination with other immunosuppressants, such as cyclosporin (CyA) or tacrolimus, and corticosteroids, in transplantation.

The differentiating characteristics of MZR, in contrast to AZT, are that it has been shown in animal experiments to lack oncogenicity, and has been shown clinically to be associated with a low incidence of severe adverse drug reactions, e.g., myelosuppression and hepatotoxicity. These findings indicate the usefulness of the drug for long-term immunosuppressive therapy. Therefore, several clinical trials for the treatment of autoimmune diseases were carried out, and the clinical usefulness of this drug became obvious (3-6). In addition to the above-mentioned approval for "the prevention of rejection in renal transplantation," MZR has been approved in Japan for the treatment of "lupus nephritis" (1990), "rheumatoid arthritis" (1992), and "primary nephrotic syndrome" (1995). In these diseases, MZR has often been used in combination with corticosteroids and/or anti-inflammatory drugs.

Mechanism of Action

MZR (Figure 1) has a very specific mechanism of action on the antiproliferation of lymphocytes, which does not interfere with purine synthesis in other cell lines. Purine synthesis is accomplished by two separate pathways: the *de novo* pathway and the salvage pathway (Figure 2). In the *de novo* pathway, the ribose phosphate portion of purine nucleotides is derived from 5-phosphoribosyl 1-pyrophosphate (PRPP), which is synthesized from ATP and ribose 5-phosphate. Lymphocytes are primarily dependent on the *de novo* pathway (7-9). In the salvage pathway, purine bases, sugars, and other products are basically recycled. Most cells, including polymorphonuclear leucocytes and neurons, are able to utilize the salvage pathway. The specific effect of MZR on the antiproliferation of lymphocytes arises because it acts only against the *de novo* pathway, not the salvage pathway in purine biosynthesis.

Using the mouse lymphoma cell line L5178Y, which is very sensitive to MZR, Sakaguchi et al. (*10, 11*) found that MZR strongly inhibited DNA as well

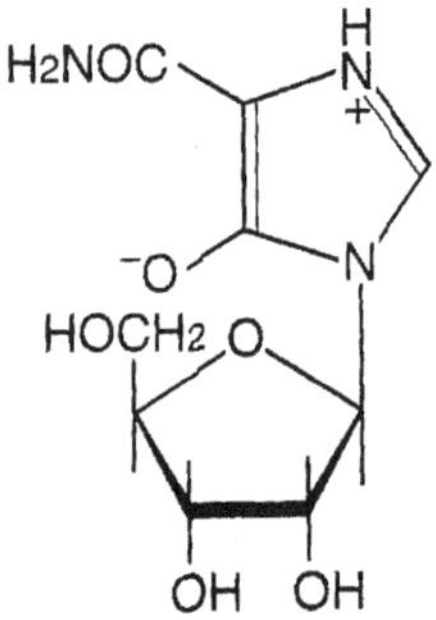

4-carbamoyl-1-β-D-ribofuranosylimidazolium-5-olate

Figure 1. Structure of mizoribine.

as RNA synthesis, but not protein synthesis. AZT, which also exerts its immunosuppressive effect through antimetabolism, is known to be incorporated into nucleic acid in place of thioguanosine 5'-triphosphate (*12*). However, MZR was not found to be incorporated into DNA or RNA in a study using [^{14}C]MZR, but did inhibit the synthesis of nucleic acid specifically (*13*). MZR almost completely suppresses the growth of L5178Y cells at a concentration of 10^{-5} M. The addition of 2×10^{-4} M GMP to this culture system liberates the cells from growth inhibition by MZR (Figure 3). However, other purine nucleotides or pyrimidine nucleotides do not reverse the effect of MZR (*13*). From these results, it can be concluded that MZR inhibits the synthesis of GMP from inosine 5'-monophosphate (IMP) in the purine metabolism pathway, without being incorporated into DNA or RNA.

Koyama et al. (*14*) confirmed that MZR is metabolized into an active form, MZR 5'P, by adenosine kinase (AK) in carcinoma cells (Figure 4). They used various drug resistant cells that were obtained by treating mouse mammary carcinoma cells (FM3A) with N-methyl-N'-nitro-N-nitrosoguanidine. These MZR-resistant (MZR^r) mutants were 15- to 19-fold less sensitive than wild-type cells. Like wild-type cells, MZR^r mutants were capable of incorporating radioactivity from ring-labeled adenosine into the acid-insoluble macromolecular fraction. However, hypoxanthine-guanine phosphoribosyltransferase deficient (HGPRT^{r-}) mutants derived from the MZR^r cells did not incorporate the radioactivity at all or incorporated it at a markedly reduced rate. There are two different pathways through which exogenous adenosine enters the purine nucleotide pool. In one pathway, it is phosphorylated by AK and then metabolized into adenosine 5'-monophosphate (AMP); in the other, it becomes

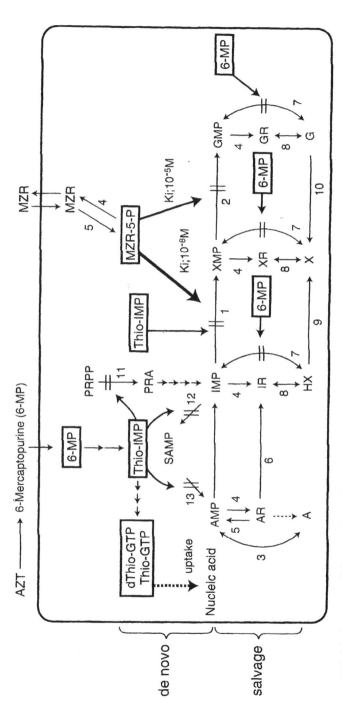

Figure 2. Action mechanisms of MZR and AZT.

1=IMP dehydrogenase (EC 1.2.1.14), 2=GMP synthetase (EC 6.3.5.2), 3=adenine phosphoribosyltransferase, APRT (EC 2.4.2.7), 4=5'-nucleotidase (EC 3.1.3.5), 5=adenosine kinase, AK (EC 2.7.1.20), 6=adenosine deaminase (EC 3.5.4.4), 7=hypoxanthine-guanine phosphoribosyltransferase, HGPRT (EC 2.4.2.8), 8=purine nucleoside phosphorylase (EC 2.4.2.1), 9=xanthine oxidase (EC 1.2.3.2), 10=guanine deaminase (EC 3.5.4.3), 11=phosphoribosylpyrophosphate amidotransferase (EC 2.4.2.14), 12=adenylosuccinate synthetase (EC 6.3.4.4), 13=adenylosuccinate lyase (EC 4.3.2.2).

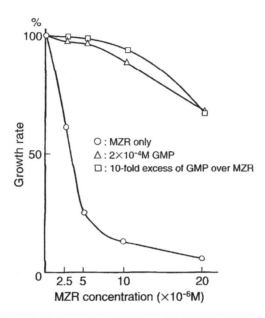

Figure 3. Effect of GMP on the inhibition of L5178Y cells by MZR. Various concentrations were added to L5178Y cell cultures, and the cells were incubated for 40 h. Cell numbers were determined with a microcell counter.

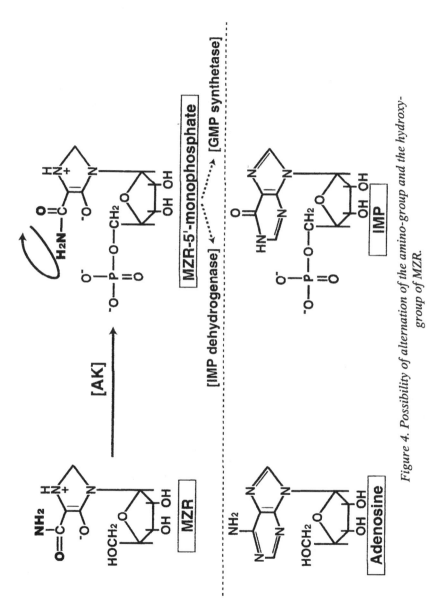

Figure 4. Possibility of alternation of the amino-group and the hydroxy-group of MZR.

inosine through the action of adenosine deaminase, which is then converted to hypoxanthine (by purine nucleoside phosphorylase) and is metabolized into IMP (by HGPRT). The two pathways of adenosine metabolism described above are blocked. Enzyme assays using cell-free extracts revealed that the MZR^r mutants had less than 3% of the AK activity found in wild-type cells. From these results, it was clear that MZR suppressed cell growth with the help of AK. This strongly suggested that MZR exerts its suppressive effect on cell growth only after it is metabolized to MZR-5'P by adenosine kinase.

Kuzumi et al. (15) investigated the inhibitory effects of MZR and MZR-5'P using cell-free extracts from rat liver on IMPDH and Walker sarcoma cells on GMP synthetase. MZR inhibited neither enzyme, while MZR-5'P inhibited both, the Ki values being 10^{-8} M against IMPDH and 10^{-5} M against GMP synthetase. These results demonstrate that the suppressive effect of MZR on cell growth is due to MZR-5'P and not to MZR itself, and that MZR-5'P primarily inhibits IMPDH, and secondarily inhibits GMP synthetase. Thus, MZR-5'P inhibits these two enzymes which act on two sequential steps in the GMP synthesis process. It seems that MZR-5'P almost completely inhibits guanine nucleotide synthesis. Furthermore, quantitative changes in the purine nucleotides in MZR-treated cells have been studied to confirm the enzyme-inhibiting effect of MZR-5'P. L5178Y cells, in which de novo synthesis of purine nucleotides were arrested with aminopterin, were incubated with ^{14}C-labeled hypoxanthine, in the presence or absence of MZR. Then, purine nucleotides were isolated, and radioactivity was measured in each of the nucleotides. The amount of GMP-containing guanine nucleotide decreased considerably after incubation in the presence of MZR, compared to incubation in the absence of MZR.

MZR is metabolized to MZR-5'P by AK, so it is a mimic for adenosine. MZR-5'P inhibits both IMPDH and GMP synthetase, meaning MZR-5'P has a chemical structure analogous to IMP and GMP. These results show that the phosphorylation of MZR may cause the amino-group and hydroxy-group to alternate between their relative positions, in effect, to spin (Figure 4).

Turka et al. (16, 17) studied the effect of MZR on human peripheral blood cells stimulated with anti-CD3 monoclonal antibody or pharmacological mitogens. MZR (1-50 µg/ml) inhibited T cell proliferation by 10-100% in a dose dependent fashion for all stimuli tested. In addition, MZR causes a dose dependent decrease in GTP pools, and the addition of guanosine both prevents this GTP depletion and reverses the antiproliferative effects at all but the highest doses of MZR. They utilized cell cycle analysis to further characterize the point at which MZR inhibited T cell proliferation. Cells were stimulated with anti-CD3 monoclonal antibody, harvested after 48 hours and stained with propidium iodide before flow cytometric analysis. As seen in Figure 5 (A and C), cultures containing MZR showed significantly fewer cells in the S, G2 and M phases of

Purified T cells were stimulated with anti-CD3 MAb, harvested after 48 h, stained with propidium iodide, and DNA content was analyzed by flow cytometry. Cells were stimulated in the absence (A and B) or presence (C and D) of 2μg/ml of MZR. The cultures shown in B and D were supplemented with 50μM guanosine plus 100μM 8-aminoguanosine to replete intracellular GTP (B and D). Each panel displays fluorescence in arbitrary units on a linear scale. Equal numbers of cells were analyzed in each graph. The percentage of cells in S+G2+M phases are : (A)–33% ; (C) (MZR)–21% ; (B) (GTP repletion)–45% ; and (D) (MZR plus GTP repletion) 36%.

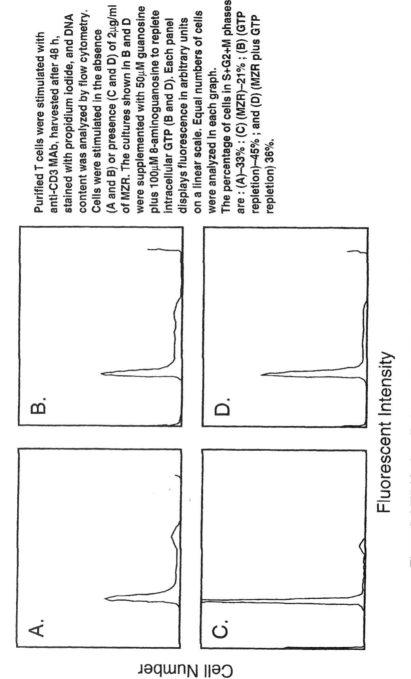

Fluorescent Intensity

Cell Number

Figure 5. MZR blocks cells from exiting G1 and entering S phase.

the cell cycle. This observation indicates that MZR blocks the movement of cells from the G1 to the S phase (Figure 5). Furthermore, they showed that MZR does not affect steady-state mRNA levels of c-myc, IL-2, C-myb, histone and cdc2 kinase, or the expression of IL-2 receptors, which are observed in the early stage of T cell activation.

In Vitro Effects

Growth Inhibitory Effects for Various Cells

Mizuno et al. investigated the growth inhibitory effect of MZR on some cultured cells. They showed that MZR has a strong inhibitory effect against lymphoma cell line L5178Y and L-cells, with IC_{50} values of >100 (Figure 6) (1).

Effect on Lymphocytes Stimulated by Mitogens or Allogenic cells

Kamata et al. (18) studied the effect of MZR on lymphocytes from beagle dogs. Dose-dependent inhibition by MZR was observed in the blastogenic response of lymphocytes induced by concanavalin A, phytohemagglutinin, or pokeweed mitogen and mixed lymphocyte reaction (MLR). Significant inhibition was evident at a concentration of 2.5 mcg of MZR/ml (Figure 7).

Furthermore, Ichikawa et al. (19) investigated the effect of MZR on the proliferation of human lymphocytes. They showed that MZR suppresses the blastogenic response of human lymphocytes that were stimulated by the above mentioned three mitogens and MLR. The mitogen responses and MLR were significantly suppressed at a concentration of 10 mcg MZR/ml, and the 50% inhibition dose was between 1.0 and 10 mcg of MZR/ml for the three mitogens and MLR (Figures 8A and 8B).

In Vivo Effects

Transplantation

A trial of MZR to test its immunosuppressive effectiveness on canine kidney allograft rejection, and to determine the adverse reactions to the drug, was carried out by Uchida et al. (20). They used beagle dogs originating from two different farms in the United States. In order to avoid histocompatibility problems, the kidney of a dog from one farm was always transplanted to a dog from the other farm. MZR had a remarkable effect on canine kidney allograft

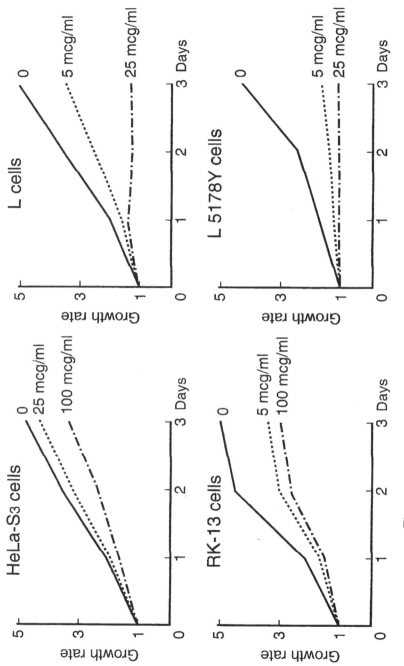

Figure 6. Growth inhibitory effect of MZR on various cells.

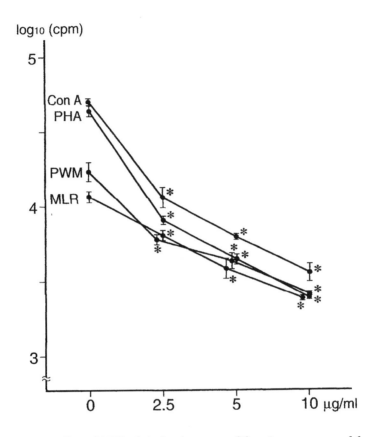

Figure 7. Effect of MZR given in vitro on proliferative responses of dog PBLs stimulated by mitogens or allogenic cells. MZR was added to culture fluid at 0 to 10 µg/ml, and its effect on the above responses was studied. Each bar represents mean ± SE of six measurements in two dogs. (∗) P < 0.01 versus MZR 0.

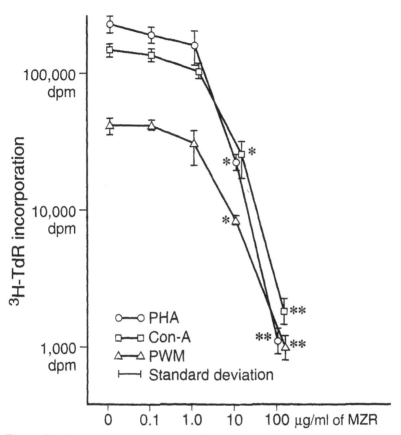

Figure 8A. Suppression of response of human PBL to mitogens by various doses of MZR. Statistical significance compared with control: (∗) P < 0.05; (∗∗) P < 0.01.

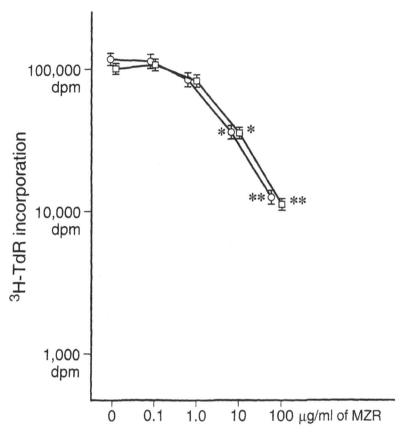

Figure 8B. Suppression of MLR by various doses of MZR. (○) responder-HLA-A2, Aw31, B7, Bw51, DR1, and DRw9; stimulator-HLA-A11, Aw24, B27, Bw54, DR4, and DRw9. (□) responder-HLA-A2, A-, Bw48, Bw54, DR4, and DR6; stimulator-HLA-A11, Aw24, B27, Bw54, DR4, DRw9. Statistical significance compared with control: (*) P < 0.05; (**) P < 0.01.

survival. MZR had a dose-dependent effect on survival, which reached a peak and plateau after a mean survival period of 34.7 ± 4.9 (SE) days at 5.0 mg/kg/day MZR (Figure 9). Its effectiveness in the suppression of the immune response is dependent on the dosage. Atrophy of the spleen and the mesenteric lymph nodes is common in treated dogs. Septic complications are rare, with no reduction in the number of peripheral white blood cells. Erosive changes of the mucosa of the entire bowel are the major adverse reactions of the drug treatment, resulting in a relatively short mean survival period. Abnormal liver function has not been observed.

Cyclosporin, which suppresses the T cell-mediated immune response, is a major immunosuppressive drug used in organ transplantation. However, it does not affect T cell-independent antibody production (21). Therefore, Suzuki (22) and Amemiya (23) carried out experimental heterotopic heart-lung transplantation in rats and allogenic renal transplantation in dogs to investigate the synergistic effects of MZR and CyA. In rat heart and partial-lung transplantation, adult male WKAH rats were used as recipients of grafts obtained from ACI rats. The mean survival period of 6.5 ± 1.2 days with MZR at 2 mg/kg/day and 8.2 ± 2.5 days with CyA at 2 mg/kg/day was significantly prolonged to 46.8 ± 4.6 days or longer with MZR + CyA in combination at the same doses (Table I) (22). In dog renal transplantation, kidneys from mongrel dogs were transplanted to beagle dogs. The animals were divided into 4 groups of 5 animals each: a non-treated group, a CyA group (10 mg/kg/day), an MZR group (2.5 mg/kg/day), and a CyA + MZR group (10 mg CyA + 2.5 mg MZR/kg/day). As shown in Table II, the mean survival period was 20.2 ± 7.1 days in the CyA group, 10.6 ± 1.5 days in the MZR group, and 97.0 ± 60.8 days in the CyA + MZR group (23). Gregory et al. tried to determine if immunosuppression with a combination of low-dose MZR/CyA could prolong the survival of canine renal allograft recipients compared with either drug alone. Their study showed results similar to those obtained by Amemiya et al. for canine renal transplantation (24).

Lupus Nephropathy of New Zealand Black/White F1 Hybrid Mice

Kamata et al. (25) investigated the therapeutic effect of MZR on the experimental lupus nephropathy of New Zealand black/white F1 mice. Treatment with 20 mg MZR/kg every 2 days, starting at 14 weeks of age, significantly prolonged the survival time of the mice. The mean life span was 54.3 ± 4.2 weeks, compared with the untreated controls, which had a mean survival time of 38.1 ± 2.9 weeks (P < 0.01). In addition, MZR suppressed the elevation of serum anti-DNA antibody titers, the spontaneous development of splenomegaly, and the histological development and progression of glomerulonephritis observed in

308

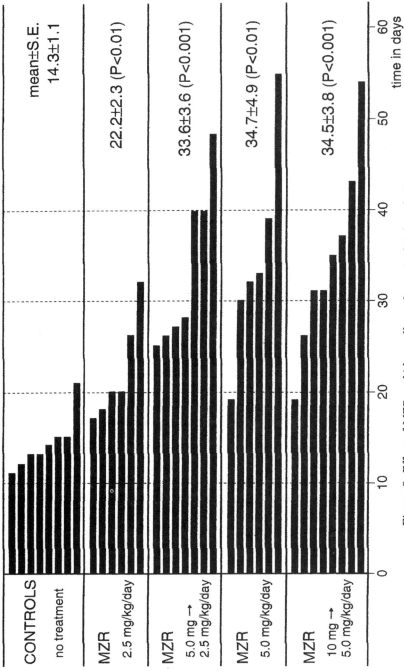

Figure 9. Effect of MZR on kidney allograft survival in beagle dogs.

untreated animals. The authors deduced that MZR lead to lower anti-DNA antibody production in vivo by directly suppressing the primary hyper-functioning B cells. This work showed that the anti-trinitrophenyl (TNP) antibody production in vivo by spleen lymphocytes of DBA/2 mice stimulated by TNP-Brucella abortus, a T cell-independent antigen, was effectively blocked by MZR.

Table I. Graft Survival in Different Experimental Groups of Heart and Partial-Lung Transplantation in Rat

	Graft Survival (days)	Mean ± SD (days)	Significance[a]
· No treatment (control)	7, 7, 9, 9, 10, 11	8.8 ± 1.6	-
· CyA 5 mg/kg/day	12, 15, 17, 23	16.8 ± 1.6	$P < 0.001$
· CyA 2 mg/kg/day	5, 6, 7, 10, 10, 11	8.2 ± 2.5	NS[b]
· MZR 4 mg/kg/day	5, 6, 6, 7, 7, 8, 9	6.9 ± 1.3	NS
· MZR 2 mg/kg/day	5, 5, 7, 7, 8, 9	6.5 ± 1.2	NS
· CyA 2 mg/kg/day + MZR 2 mg/kg/day	40, 44, >50, >50, >50	>46.8 ± 4.6	$P < 0.001$

[a] Mean comparison conducted between control group and experimental group.
[b] Not significant.

Primary Nephritis

Kobayashi et al. (26) performed an experiment to ascertain whether the progression of crescentic Masugi nephritis in rabbits could be prevented by treatment with MZR. Crescentic glomerulonephritis can be induced reproducibly by intramuscular preimmunization on day 1, followed by intravenous injections of nephrotoxic duck globulin on days 3 and 5. Thirty rabbits were divided into three groups, including controls (group 1). Two groups of 10 nephritic rabbits were each treated with 10 mg MZR/kg either after or before the development of proteinuria (groups 2 and 3). In group 3, the onset of proteinuria was significantly delayed and the duration of survival was significantly prolonged, compared with controls. Serum antibody titers after day 8, and creatinine levels after day 10, as well as the initial amounts of proteinuria, were also significantly lower during treatment in group 3 than in the control group. Histologically, the prominent diffuse intra- and extra-capillary proliferation, with monocyte accumulations, observed in the control group, were markedly diminished in group 3. These results suggest that early treatment in crescentic

Table II. Markedly Prolonged Survival with CyA + MZR in Canine Renal Allograft Recipients

Group	n	CyA[a]	MZR[a]	Final BUN	Final S-Cr	Histological Score of Rx[b]	Recipient Survival (days)	Mean Recipient Survival (days)	P value
Untreated	5	0	0	208, 211, 437, 303, 329	9.9, 11.9, 15.2, 10.6, 11.2	4, 4, 3, 4, 4	9, 9, 10, 16, 16	12.0 ± 3.7	
CyA	5	10	0	394, 288, 203, 348, 139	15.4, 16.7, 6.9, 11.1, 6.6	3, 4, 4, 3, 4	13, 17, 19, 20, 32	20.2 ± 7.1	<0.5 vs untreated group
MZR	5	0	2.5	185, 134, 217, 327, 310	5.5, 6.1, 7.4, 7.6, 13.6	3, 4, 4, 3	9, 9, 11, 12, 12	10.6 ± 1.5	NS[c] vs untreated group
CyA + MZR	5	10	2.5	195, 230, 208, 135, 66	10.1, 7.9, 7.5, 5.8, 3.2	3, 4, 4, 4	43, 57, 83, 105, 197	97.0 ± 60.8	<0.01 vs untreated group, CyA group, and MZR group

[a] mg/kg/day.
[b] Rejection (Rx).
[c] Not significant.

glomerulonephritis with MZR will suppress the production of humoral antibody and prevent the progression of glomerular lesions.

Shimizu et al. (27) undertook immunopathological studies to determine whether the glomerular injuries in ddY mice, a model for IgA nephropathy, are influenced by the administration of MZR. The ddY mice were treated with a low (0.05 mg/ml) or a high (0.1 mg/ml) dose of MZR, which was diluted with drinking water, for 35 weeks. Flow cytometric analysis showed that there was a marked decrease in the number of B cells and IgA-bearing B cells. The deposition of IgA in the glomerular mesangial areas and capillary walls of the high dose MZR-treated mice was markedly decreased compared with that of the control mice. The glomerular mesangial expansion in the high-dose MZR-treated mice was milder than that found in the control mice. In 45-week old ddY mice, the average number of intra-glomerular cells in the high-dose and low-dose MZR-mice was slightly lower than that in the control mice. The levels of urinary protein excretion in the high-dose MZR-treated mice were also lower than those in the low-dose MZR-treated mice or control mice. It appears that treatment with MZR might influence the proliferation of B cells, especially IgA-bearing B cells, and improve the glomerular IgA deposition and glomerular expansion in early-stage IgA nephropathy of ddY mice.

Experimental Arthritis

We studied the effects of MZR on adjuvant and collagen arthritis in rats. In adjuvant arthritis, twice daily doses of 2.5 and 5.0 mg MZR/kg showed a marked inhibition of the development of secondary lesions of arthritis compared with the corresponding single daily doses of 5 and 10 mg MZR/kg. Moreover, MZR treatment improved the bone lesions of the hind legs (28, 29).

Animal Toxicology

Animal toxicity studies demonstrated that the toxicological effects of MZR were (1) reduced body weight gain, which was reversible after cessation of administration; (2) mild toxic effects on the digestive tract, bone marrow, lymph nodes, and reproductive organs (reversible after cessation of administration), but no deaths in multiple dose tests (20 mg/kg in rats and 60 mg/kg in dogs) and no remarkable histopathological changes in the liver or other main organs (Table III summarizes the parameters and relevant findings of a comparative subacute toxicity study of MZR and azathioprine in dogs) (30-34) and (3) teratogenicity and chromosome aberrations at dose levels close to the proposed therapeutic dose, but no mutagenic or oncogenic activity (35-37).

Table III. Comparative Subacute Toxicity of MZR and AZT in Dogs

Study Method

		MZR group	AZT group
Species		beagle dogs (male; age: 7-9 months)	beagle dogs (male; age: 7-9 months)
Groups (mg/kg/d; p.o.)		5, 10, 20	5, 10
No. of animals		4 per group	4 per group
Administration period		62 days	62 days

Study Results

		MZR group	AZT group
Cases of death		None	2 animals at 5 mg/kg (at day 43 and 57) 3 animals at 10 mg/kg (at day 22, 32, and 46)
General signs		No particular anomalies	Reduced food intake, diarrhea, melena and stomatitis (1–2 weeks before death)
Body weight		No particular anomalies	Marked decrease as the animals were nearing death
Laboratory tests	WBC	No particular anomalies	Significant decrease at 5 mg/kg from day 28; Significant decrease at 10 mg/kg from day 7 ($P < 0.05$)
	RBC	No particular anomalies	No particular anomalies
	GOT	No particular anomalies	Significant rise at 10 mg/kg from day 14 ($P < 0.05$)
	GPT	No particular anomalies	Significant rise at 10 mg/kg from day 7 ($P < 0.05$)
	ALP	No particular anomalies	Significant rise at 10 mg/kg from day 7 ($P < 0.05$)
	ChE	No particular anomalies	Tendency towards drop at 10 mg/kg from day 42
	BUN	No particular anomalies	No particular anomalies

Histo-pathology	Liver	Sporadic edematous swelling of cells in 2 animals each of the 10 and 20 mg/kg groups	Cell atrophy, nucleus degeneration and lipid droplet degeneration were noted in the dead animals and the survivors of the 10 mg/kg group, while sporadic edematous swelling of cells occurred in 1 survivor of the 5 mg/kg dose group
	Kidneys	No particular anomalies	Glomerula degeneration, atrophy and flattening of tubular epithelium were noted in the dead animals and the survivors of the 10 mg/kg dose group, while stagnation in 1 animal of the 10 mg/kg dose group and necrotic foci in 1 animal of the 5 mg/kg group were sporadically noted
	Bone marrow	Slight decrease in myelocytes in 1 animal of the 20 mg/kg group	Marked decrease in myelocytes in the dead animals
	Mesent. lymph nodes	No particular anomalies	Severe atrophy of lymph follicles in the dead animals
	Thymus	No particular anomalies	Marked atrophy in the dead animals

NOTE: Only changes due to drug administration are listed.

SOURCE: Data are from Reference 32.

The toxicological findings are consistent consequences of the inhibitory action of MZR on nucleic acid synthesis.

Pregnant and nursing women should avoid using MZR because of its teratogenic activity (35-37).

Absorption, Distribution, Metabolism, and Excretion

The absorption, distribution, metabolism, and excretion of MZR were examined by giving ^{14}C-MZR to rats. After oral administration (38), MZR is rapidly absorbed and blood concentrations of MZR peak 1.5 hours after the drug has been administered, then decrease steeply. MZR is almost completely eliminated within 24 hours. In whole-body autoradiography, radioactivity was found to be high in the stomach, small intestine, liver, kidney, spleen, and thymus at one hour after the administration of MZR. Within 24 hours, 85% of the administered dose was excreted in urine and 1.0% in bile. More than 99% of the radioactivity in plasma one hour after dosing was unchanged ^{14}C-MZR by inverse isotope dilution analysis. Furthermore, the 85% of MZR excreted in urine within 24 hours after administration was unchanged.

In six kidney-transplanted patients with good renal function, whose serum creatinine levels were under 2.7 mg/dl, the serum concentrations of MZR peaked at about 2.3 g/ml two hours after oral administration of 100 mg MZR, then gradually decreased with a $T_{1/2}$ value of 2.2 hours. About 82% of the oral dose of MZR was excreted in the urine of these transplanted patients six hours after administration (39). In patients with renal dysfunction, the serum concentrations of MZR remained high even at 24 hours after administration. In these cases, MZR can be removed by hemodialysis. The elimination rate of MZR from serum is closely related to renal function (Figure 10) (40).

Clinical Efficacy

Renal Transplantation

The clinical efficacy of MZR as an immunosuppressant in renal transplantation was studied in various Japanese institutes from 1978 to 1982, the period when immunosuppression was performed mainly with AZT and corticosteroids without immunophilin-binding drugs such as cyclosporin (CyA) or tacrolimus. During that period, 200 to 300 renal transplants were performed in Japan each year. The immunosuppressive effect in 57 cases of triple-drug therapy (MZR + AZT + corticosteroid) was compared with that obtained in 72 historical controls treated with AZT + corticosteroid in combination. The graft survival rate for the MZR treated group, 89.6%, was significantly higher than that for the non-MZR treated group (74.6%).

Immunosuppression with 3 different MZR-containing regimens, as shown in Table IV, was studied in a total of 226 renal transplant recipients (2). Two drugs (MZR plus a corticosteroid; 60 patients) or three drugs (MZR, AZT, and a corticosteroid; 166 patients) were administered concomitantly, mostly once a day, with the initial and maintenance doses of MZR being 100-200 and 75-100 mg/day, respectively.

Table IV. Patient Characteristics

Group	Start of MZR (after transplant)	Living Donor Transplants	Cadaveric Donor Transplants	Total	Reason for use of MZR
A	0 days	66	14	80	1)
B	2 days – 1 year	75	9	84	2), 3), 4)
C	1 year or more	57	5	62	2), 3), 4)
Total		198	28	226	

1) To prevent acute or delayed rejection response by scheduled administration immediately following transplant, including difficult transplant cases caused by immunosuppressive therapy using mainly azathioprine and adrencortical hormones that resulted in hepatic disorders and leukopenia prior to transplant.

2) To prevent acute or delayed rejection response due to use of previously administered drugs.

3) Use of azathioprine and other drugs was difficult, or dose reduction became inevitable due to hepatic disorders and leukopenia caused by the use of previously administered drugs.

4) Occurrence of hepatic disorders and leukopenia due to use of previously administered drugs was anticipated.

In patients treated with MZR because the conventional immunosuppressive regimens could not be tolerated due to adverse effects, the changes in their adverse effects were followed up. Table V shows the improvements in leukopenia and hepatic dysfunction compared with the laboratory data prior to MZR treatment. Leukopenia was observed before MZR treatment in 37 of the 146 patients in groups B and C. It improved with treatment in 28 of these cases, and tended to improve in almost all the remaining cases, with no patient

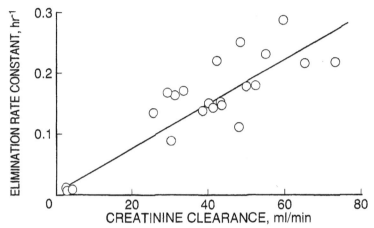

Figure 10. Relationship between the elimination rate constant of MZR and endogenous creatinine clearance. Open circles represent 19 renal transplant patients and 3 anuri patients. The line plotted from the 22 data points is described by the formula y = 0.0036x + 0.0052, r = 0.854.

Table V. Improvement of Leukopenia and Impaired Hepatic Function (Groups B and C) in Patients Treated with MZR because of Adverse Effects due to Conventional Therapy

	No. of Patients	Improvement			No Change	Aggra- vation	Improvement Rate (%)
		Marked	Moderate	Slight			
Leukopenia	37	2	26	8	1	0	75.7 (28/37)
Hepatic dysfunction	88	12	36	20	14	6	54.5 (48/88)

NOTE: Leukopenia was observed before MZR treatment in 37 of the 146 cases of groups B and C. It improved with treatment in 28 cases, and tended to improve in almost all other cases, but was not aggravated in any of the cases. Impaired hepatic function was observed in 88 of the 146 cases. It improved markedly or at least moderately in 48 patients. A slight improvement was noted in a further 20 cases. Only in 6 cases was aggravation observed.

experiencing aggravation. Impaired hepatic function was observed in 88 of these 146 patients. It improved markedly or at least moderately in 48 patients, a slight improvement was noted in a further 20, and aggravation was observed in only 6 patients.

As a part of the same retrospective analysis, data for group A patients (scheduled to receive MZR immediately following transplantation) were compared with those for patients who had received standard therapy, i.e., corticosteroids and azathioprine, from 1978-1980. The efficacy of MZR was evaluated in 61 renal allograft recipients on an immunosuppressive regimen including MZR, as compared with the findings in 84 similar recipients on conventional therapy. As shown in Figure 11, the patient survival rate at one year was over 90% for both groups of recipients. However, the graft survival rate for the MZR-treated group (89.6%) was significantly higher than that (74.6%) for the conventionally treated group ($P < 0.05$). Furthermore, the incidence of acute rejection was lower in the MZR-treated group (61.4% vs 73.4%), as was the occurrence of side effects (37.7% vs 46.4%).

From 1988 through 1999, the Transplantation Study Group of Tokyo Women's Medical University designed and executed a prospective, randomized, long-term study to evaluate the immunosuppressive effect of MZR in renal transplantation, by comparing MZR and AZT under cyclosporin-corticosteroid-based immunosuppression. A total of 116 patients were enrolled between January 1988 and April 1989, and equally divided into two groups of 58 patients each. There was no significant difference between the two groups in terms of recipient sex, donor sex, donor source (such as living and cadaveric donor) HLA-AB, DR mismatches, and ABO-compatibility. Patients received MZR 4-5 mg/kg per day or AZT 2-1 mg/kg per day for nine years after transplantation. Oral administration of immunosuppressive drugs was started two days before transplantation. Primary endpoints of this study were 9-year patient survival rate, 9-year graft survival rate, and discontinuance rate.

The 9-year patient survival rates of the MZR group and AZT group were 88% and 83%, respectively. Similarly, the 9-year graft survival rates of the MZR group and AZT group were 58% and 52%, respectively (Figure 12). There was no significant difference between the two groups in terms of graft and patient survival rates. On the other hand, sixteen AZT group patients (27.6%) were forced to discontinue AZT administration and change to MZR due to adverse effects, consisting of myelosuppression in 11 patients and liver dysfunction in 5 patients. Since no MZR related adverse effect occurred, no patients discontinued MZR in the MZR group (Table VI). According to these results, MZR has almost the same immunosuppressive effect compared to AZT. However, since MZR showed much fewer adverse effects, no patients treated with MZR converted to AZT, whereas 27.6% of the patients treated with AZT were forced to change to MZR because of adverse effects (41).

318

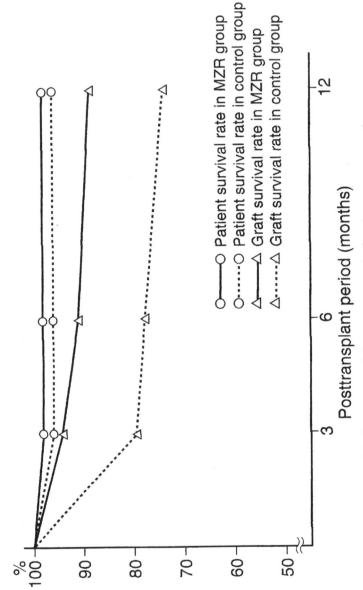

Figure 11. Comparison of the one-year survival and acceptance rate between the MZR group and the conventional therapy group (living donor kidneys).

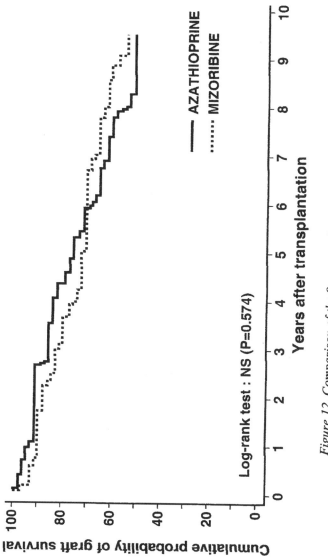

Figure 12. Comparison of the 9-year graft survival rate between MZR group and azathioprine group.

Table VI. Discontinuance Rates in Long-term Randomized Trial Comparing AZT and MZR

	AZT Group	MZR Group
Completed patients	42 (73.4%)	58 (100%)
Discontinued patients	16[a] (27.6%) [11: myelosuppression 5: liver dysfunction]	0

[a] All cases were changed to MZR treatment.

Lupus Nephritis

From 1986 through 1988, Homma et al. (3) initiated a 24-week prospective, randomized, single-blind trial to evaluate the efficacy and safety of MZR in patients with lupus nephritis that was refractory to treatment. Forty-nine patients, who had been treated with corticosteroids, in 11 institutes, were enrolled, and assigned randomly to receive either MZR (150 mg/day) or placebo. The results of the study of 23 patients in each group were evaluated both by the physicians in charge and the study committee. There were no significant differences between the patients' backgrounds, except for male predominance in the MZR-treated group. The global improvement rate was significantly higher in the MZR group (26.1%) than the placebo group (4.3%). Laboratory findings were evaluated according to criteria prepared prior to the study. The serum creatinine level and creatinine clearance of the MZR group improved significantly, and the urinary protein, BUN, and serum total protein levels, and other clinical symptoms, such as eruption, tended to improve in the MZR group. There was no significant difference between the incidences of side effects in the MZR (16.0%) and placebo (16.7%) groups. Side effects were reported by four patients in the MZR group—one each of rash, leukopenia, infection, and gastrointestinal distress—but were not serious and required no special treatment.

Rheumatoid Arthritis

From 1987 through 1990, Shiokawa et al. (4, 5) conducted two prospective, randomized, double-blind, controlled trials to confirm the efficacy and safety of MZR in patients with rheumatoid arthritis.

The first was a 16-week trial using non-active placebo as the control drug. To be eligible for this trial, patients had to meet the 1958 American Rheumatism

Association classification criteria for rheumatoid arthritis and be between the ages of 16 and 65. They were required to have at least 3 swollen and 6 tender joints, and their disease duration had to have been at least 6 months. Two hundred three patients who were refractory to treatment with nonsteroidal anti-inflammatory drugs (NSAIDs) and low-dose corticosteroids participated in this study. They were required to discontinue treatment with disease-modifying antirheumatic drugs (DMARDs) before entering the study with a washout period of at least 12 weeks. They were allowed to remain on their NSAID and/or oral corticosteroid regimens, providing that the daily steroid dose was ≤5 mg prednisolone or equivalent. The patients were assigned randomly to receive either MZR (300 mg/day) or placebo, and efficacy was assessed in a total of 178 patients (89 each in the MZR and placebo groups) who were screened for study eligibility. The global improvement rate evaluated by the physicians in charge was significantly higher in the MZR group (27.7%) than the placebo group (12.4%). In the MZR group, the swollen and tender joint counts, duration of morning stiffness, pain score, and modified Health Assessment Questionnaire (m-HAQ) showed amelioration. Furthermore, the C-reactive protein (CRP) and erythrocyte sedimentation rate (ESR) elevations were suppressed. The placebo group showed no such amelioration or suppression. Further immunological profiling was performed in this study: the rheumatoid factor positivity and CD4/CD8 value of the MZR group decreased slightly. The overall safety rating revealed the incidences of the adverse reactions in the MZR and placebo groups were 32.7% and 23.5%, respectively, and the difference was attributable to the differing incidences of gastro-intestinal abnormalities and hypersensitivity. No serious adverse reactions occurred in either group.

The second study was a 24-week trial to compare MZR with an established DMARD, lobenzarit disodium (LZR), which was used in the late 1980s in Japan. The study involved 72 patients in the MZR group and 77 in the LZR group, and there were no significant differences in the patients' backgrounds. The global improvement rates, evaluated by the physicians in charge, were 26.8% for the MZR group and 12.3% for the LZR group, and the respective incidences of side effects were 31.9% and 48.7%. No serious side effects were observed in either group. According to the physicians in charge, the utility of MZR was superior to that of LZR.

The radiographic films of 41 of the patients participating in this study were evaluated in a random order by two readers, and the radiographic scores were calculated using the Larsen method. Only the LZR group (21 patients) showed significant increases in mean erosion damage scores. These results show that MZR retarded radiographic progression of joint destruction in patients with rheumatoid arthritis.

Table VII. Adverse Reactions to MZR

	Renal Transplantation	Lupus Nephritis	Rheumatoid Arthritis	Nephrotic Syndrome	Total
No. of surveyed cases	916	275	3478	240	4909
No. of adverse reaction cases	186	33	462	38	719
No. of adverse reaction episodes	229	47	658	49	983
Incidence of adverse reaction cases	20.31%	12.00%	13.28%	15.83%	14.65%
Incidence of Adverse Reactions (%)					
Blood					
Leukopenia	6.99	2.55	0.20	0.83	1.63
Thrombocytopenia	0.98	0.73	0.14		0.33
Anaemia	0.44	0.36	0.72	0.42	0.63
Infection					
Pneumonia	0.55		0.03	0.42	0.14
Mycosis pulmonary	0.55				0.10
Herpes zoster	0.76	1.45	0.17		0.35
Other viral infection	0.76		0.03		0.16
Liver					
Hepatic function abnormal	4.15	1.09	1.18	2.92	1.81
Hypersensitivity					
Rash		1.09	1.64	2.08	1.32
Prurigo			1.09		0.77

Metabolic					
Uric acid increased	1.64	1.45	0.43	2.50	0.81
Gastrointestinal					
Celialgia		0.36	2.21	1.67	1.67
Anorexia	0.98	1.09	1.12	0.83	1.08
Vomiting	0.55	1.09	1.18	0.42	1.02
Nausea	0.11	1.09	0.23		0.24
Diarrhea	0.22	0.73	0.63	0.42	0.55
Stomatitis	0.76	0.73	0.66	0.42	0.67
Skin					
Epilation	1.09	1.09	0.29	1.67	0.55

NOTE: Listed adverse reactions occurred in 0.5% of cases or greater for at least one indication of MZR.

Nephrotic Syndrome

From 1989 through 1991, Koshikawa et al. (6) performed a 24-week, prospective, randomized, double-blind, placebo-controlled, comparative trial to assess the efficacy of MZR in patients with steroid-resistant nephrotic syndrome. Efficacy was assessed in a total of 158 patients (80 in the MZR group and 78 in the placebo group). The global improvement rate, evaluated by the physicians in charge, was significantly higher for the MZR (33.8%) group than the placebo (14.1%) group. In particular, the difference between the improvement rates of the two groups was marked when the subgroup taking corticosteroids at a daily dose of less than 20 mg as prednisolone at baseline (30.0% vs 5.3%) was evaluated. Laboratory studies revealed an average 25.2% reduction in the urinary protein level of the MZR group compared with 10.0% for the placebo group. The incidences of side effects in the MZR (13.6%) and placebo (11.9%) groups did not differ significantly.

Adverse Reactions

Various kinds of clinical trials and a postmarketing surveillance study involved a total of 4906 cases receiving MZR therapy for kidney transplantation and three disease patients. The principal adverse reactions associated with the use of MZR are leukopenia, abnormal hepatic function, rash, uric acid increased, gelialgia, arenexia, and vomiting. Adverse reactions that occurred of a rate of 0.5% or greater in the patient population for at least one indication of MZR are presented in Table VII.

Immunosuppressive patients are at increased risk for cancers. Especially in transplantation, of the many thousands of renal transplant patients at risk, the incidence of cancers has ranged from 2% to 16%, with an average of 6% in western countries (42).

Among 916 renal transplantation patients (226 from clinical trials, 690 from the postmarketing surveillance study) receiving MZR, cancer was found in four patients, corresponding to an incidence rate of 0.44% (43). A questionnaire survey conducted in a Japan renal transplant group reported 204 cancer cases out of 9010 renal transplant cases: an incidence rate of 2.26% (Table VIII) (44).

Conclusions

MZR is a nucleoside of the imidazole class, isolated from a culture medium of an ascomycete. After phosphorylation by adenosine kinase, MZR-5'-monophosphate inhibits GMP synthesis by the antagonistic blocking of IMPDH ($Ki = 10^{-8}$ M) and GMP-synthetase ($Ki = 10^{-5}$ M). MZR has been found to inhibit both humoral and cellular immunity. MZR has been shown in animal experiments to lack oncogenicity and have fewer toxic effects than azathioprine. In an experimental transplantation model, combination therapy with MZR plus a calcineurin inhibitor, such as cyclosporin or tacrolimus, manifested a synergistic

Table VIII. Incidence of Cancer after Renal Transplantation with MZR

Survey	Incidence	Author
MZR Study: until approval and PMS[a] (1978–1990)	0.44% (4/916)	
Questionnaire survey (1970–1994) in Japan (covering all immunosuppressive drug therapies)	2.26% (204/9010)	M. Imanishi[b] (Kink Univ.)
Major transplantation centers in western countries	2–16% average 6%	I. Penn[c, d] (Univ. of Cincinnati)

[a] Postmarketing surveillance (PMS).

[b] Data are from Reference 44.

[c] Data are from Reference 42.

[d] Curr Probl Cancer **1982,** 6, 1-64.

immunosuppressive effect, prolonging graft survival. MZR has been shown clinically to be associated with a low incidence of severe adverse reactions. MZR has been approved in Japan for the prevention of rejection in renal transplantation, and for the treatment of lupus nephritis, rheumatoid arthritis, and the nephrotic syndrome, and has been widely used for the aforementioned indications, contributing to overall patient care.

Abbreviations

MZR	mizoribine
IMPDH	inosine monophosphate dehydrogenase
AZT	azathioprine
MZR-5'P	mizoribine-5'-monophosphate
AK	adenosine kinase
HGPRT	hypoxanthine-guanine phosphoribosyltransferase
GMP	guanosine 5'-monophosphate
AMP	adenosine 5'-monophosphate
CyA	cyclosporin
DMARD	disease-modifying antirheumatic drug
NSAID	nonsteroidal anti-inflammatory drug
LZR	lobenzarit disodium

References

1. Mizuno, K.; Tsujino, M.; Takada, M.; Hayashi, M.; Atsumi, K.; Asano, K.; Matuda, T. *J. Antibiot.* **1974**, *27*, 775-782.
2. Inou, T. *Japan J. Transplant.* (Japanese) **1982**, *17 (Supple)*, 547-561.
3. Homma, M.; Akizuki, M.; Yokohari, R.; Hashimoto, H.; Kashiwazaki, S.; Kondo, H.; Irimajiri, S. *J. Clin. Therap. Med.* (Japanese) **1989**, *5*, 795-824.
4. Shiokawa, Y.; Honma, M.; Shichikawa, K.; Miyamoto, T.; Hirose, S.; Nobunaga, T.; Mizushima, Y.; Sugawara, S.; Warabi, H.; Kondo, H.; Ogawa, N. *J. Clin. Exper. Med. (Igaku no Ayumi: Japanese)* **1991**, *156*, 811-831.
5. Shiokawa, Y.; Honma, M.; Shichikawa, K.; Miyamoto, T.; Hirose, S.; Nobunaga, T.; Mizushima, Y.; Sugawara, S.; Warabi, H.; Kondo, H.; Ogawa, N. *J. J. Inf.* (Japanese) **1991**, *11*, 375-396.
6. Koshikawa, S.; Sato, M.; Narita, K.; Sakai, O.; Nakajima, M. *Kidney and Dialysis* (Japanese) **1993**, *34*, 631-650.
7. Allison, A. C.; Hovi, T.; Watts, R. W. E.; Webster, A. D. B. *Lancet* **1975**, *3*, 1179-1183.
8. Allison, A. C.; Hovi, T.; Watts, R. W. E.; Webster, A. D. B. *Ciba Found Symp.* **1975**, *48*, 207-223.
9. Hovi, T.; Allison, A. C.; Raivio, K. O.; Vaheri, A. *Ciba Found Symp.* **1977**, *48*, 225-248.
10. Sakaguchi, K.; Tsujino, M.; Yoshizawa, M.; Mizuno, K.; Hayano, K. *Cancer Res.* **1975**, *35*, 1643-1648.
11. Sakaguchi, K.; Tsujino, M.; Mizuno, K.; Hayano, K.; Ishida, N. *J. Antibiot.* **1975**, *28*, 798-803.
12. LePage, G. A.; Whitecar, J. P. Jr. *Cancer Res.* **1971**, *31*, 1627-1631.
13. Sakaguchi, K.; Tsujino, M.; Hayashi, M.; Kawai, K.; Mizuno, K.; Hayano, K. *J. Antibiot.* **1976**, *29*, 1320-1327.
14. Koyama, H.; Tsuji, M. *Biochem. Pharmacol.* **1983**, *32*, 3547-3553.
15. Kusumi, T.; Tsuda, M.; Katsunuma, T.; Yamamura, M. *Cell Biochem. Funct.* **1988**, *7*, 201-204.
16. Turka, L. A.; Dayton, J.; Sinclair, G.; Thompson, C. B.; Mitchell, B. S. *J. Clin. Invest.* **1991**, *87*, 940-948.
17. Dayton, J. S.; Turka, L. A.; Thompson, C. B.; Mitchell, B. S. *Mol. Pharmacol.* **1992**, *41*, 671-676.
18. Kamata, K.; Okubo, M.; Ishigamori, E.; Masaki, Y.; Uchida, H.; Watanabe, K.; Kashiwagi, N. *Transplantation* **1983**, *35*, 144-149.
19. Ichikawa, Y.; Ihara, H.; Takahara, S.; Takada, K.; Shrestha, G. R.; Ishibashi, M.; Arima, M.; Sagawa, S.; Sonoda, T. *Transplantation* **1984**, *38*, 262-267.
20. Uchida, H.; Yokota, K.; Akiyama, N.; Masaki, Y.; Aso, K.; Okubo, M.; Okudaira, M.; Kato, M.; Kashiwagi, N. *Transplant. Proc.* **1979**, *11*, 865-870.
21. Kunkel, A. ; Klaus, G. G. B. *J. Immunol.* **1980**, *125*, 2526-2531.

22. Suzuki, S.; Hijioka, T.; Sakakibara, I.; Amemiya, H. *Transplantation* **1987,** *43,* 743-744.
23. Amemiya, H.; Suzuki, S.; Niiya, S.; Watanabe, H.; Kotake, T. *Transplantation* **1988,** *46,* 768-771.
24. Gregory, C. R.; Gourley, I. M.; Cain, G. R. *Transplant. Proc.* **1988,** *20,* 223-225.
25. Kamata, K.; Okubo, M.; Uchiyama, T.; Masaki, Y.; Kobayashi, Y.; Tanaka, T. *Clin. Immunol. Immunopathol.* **1984,** *33,* 31-38.
26. Kobayashi, Y.; Shigematsu, H.; Masaki, Y. *Virchows Arch. [Cell Pathol.]* **1983,** *43,* 103-119.
27. Shimizu, M.; Shou, I.; Tsuge, T.; Abe, M.; Tomino, Y. *Nephron* **1998,** *79,* 67-72.
28. Tsujino, M; Ishikawa, H.; Mizuno, K. *J. J. Inf.* (Japanese) **1990,** *10,* 63-66.
29. Ishikawa, H.; Shibata, K. *J. J. Inf.* (Japanese) **1991,** *11,* 507-511.
30. Saito, M.; Sano, M.; Narama, I. *Pharmacometrics* (Japanese) **1983,** *26,* 293-306.
31. Nagata, T.; Nagata, T.; Aramaki, Y.; Enomoto, M.; Satou, K.; Isaka, H. *Pharmacometrics* (Japanese) **1983,** *26,* 327-346.
32. Matsumoto, K.; Morino, T.; Hara, T.; Yano, J.; Tsujino, M.; Mizuno, K.; Yamamoto, H. *Japan J. Transplant.* (Japanese) **1982,** *17 (Supple),* 603-614.
33. Saito, M.; Sano, M.; Narama, I. *Pharmacometrics* (Japanese) **1983,** *26,* 307-325.
34. Nagata, T.; Nagata, T.; Aramaki, Y.; Enomoto, M.; Satou, K.; Enomoto, M. *Pharmacometrics* (Japanese) **1983,** *26,* 327-346.
35. Kubota, H.; Sasaki, M.; Suda, M.; Kobayashi, Y.; Hayano, K. *Pharmacometrics* (Japanese) **1983,** *26,* 377-387.
36. Sasaki, M.; Kubota, H.; Suda, M.; Kobayashi, Y.; Hayano, K. *Pharmacometrics* (Japanese) **1983,** *26,* 409-414.
37. Okamoto, K.; Kobayashi, Y.; Yoshida, K.; Nozaki, Y.; Kawai, Y.; Kawano, H.; Mayumi, T.; Hama, T. *Cong. Anom.* (Tokyo) **1978,** *18,* 227-233.
38. Murase, J.; Mizuno, K.; Kawai, K.; Nisiumi, S.; Kobayashi, Y.; Hayashi, M.; Morino, T.; Suzuki, T.; Baba, S. *Pharmacometrics* (Japanese) **1978,** *15,* 829-835.
39. Kusaba, R.; Otubo, O.; Inou, T.; Tsujino, M.; Ishikawa, H. *Japan J. Transplant.* (Japanese) **1982,** *17 (Supple),* 585-593.
40. Takada, K.; Muranishi, S.; Asada, S.; Takahara, S.; Nagano, S.; Fukunishi, T.; Ichikawa, Y.; Sonoda, T. *Japan J. Transplant.* (Japanese) **1982,** *17 (Supple),* 595-601.
41. Tanabe, K.; Tokumoto, T.; Ishikawa, N.; Kanematu, A.; Oshima, T.; Harano, M.; Inui, M.; Yagisawa, T.; Nakajima, I.; Fuchinoue, S.; Takahashi, K.; Toma, H. *Transplant. Proc.* **1999,** *31,* 2877-2879.
42. Penn, I.; Brunson, M.E. *Transplant. Proc.* **1988,** *20,* 885-892.
43. Asahi Kasei Corporation; internal report, 1998.
44. Imanishi, M.; Kunikata, S.; Akiyama, T.; Kurita, T. *Japan J. Transplant.* **1996,** *31,* 100-107.

Indexes

Author Index

Subject Index

I

Immunosuppressant agents
  IMPDH inhibitors, 5*f*, 48, 51–52,
    141, 143, 212–213
  mechanism, IMPDH cofactor
    analogs, 247
  mizoribine, 248–249*f*, 294–295
  mycophenolate mofetil, 159, 185,
    248–249*f*
  See also Mizoribine, Mycophenolate
    mofetil, Transplantation rejection
IMP
  and guanylate metabolism, 235*t*
  as substrate for AMP, 1, 2*f*
  catalysis via IMPDH, 1, 2*f*
  complexes with bacterial IMPDH, 14
  de novo guanine nucleotide
    synthesis, 200*f*
  hydride transfer, 140
  metabolic enzymes, 21, 22*f*
  role in purine metabolism, 185
  See also IMPDH reaction, IMP
    binding site
IMP binding site
  active site cleft, 153–156
  IMP ligand, 147*f*
  IMPDH, 192
  IMPDH cleft, 151*f*
  IMPDH conformational change, 252
  IMPDH inhibitors, 158–159
  IMPDH type II inhibition, 12–13
  inhibitors, 142*f*
  ribavirin, 37
  sequential IMPDH mechanism, 283
  therapeutic relevance, 285*t*
  See also Active site, NAD binding
    site
IMPDH gene
  5′ region in *P. carinii*, 118–121*f*
  relationship to p53 gene, 59–60
  gene knockout, 91–92
  gene, 94*f*
  mutant mice, breeding strategy, 97*f*
  mutants, 107–108

mutations and lymphocyte activation,
  99
IMPDH inhibition
  anticancer and antiviral activity, 1
  biological effects, 143, 213–214
  compared with other NAD–
    dependent enzymes, 252–253
  effect on guanine pools, 51, 186
  effect on interleukin 2 secretion, 105*f*
  effect on serum immunoglobulin
    levels, 106*f*
  effect on splenocyte proliferation,
    101–102*f*
  E–XMP*–NAD complex, 252–253
  molecular target regions, 285*t*
  MPA, 53–54
  new strategies for drug development,
    161–166
  species specific, 13–14
  specificity and IMPDH reaction, 283
  type I vs. type II, 277–279*f*
  type II specific, 12–13
  See also Drug design, IMPDH
    inhibitors, IMPDH reaction,
    IMPDH type II, NAD analogs,
    Species-specific IMPDH
IMPDH inhibitors
  and guanylate function, 232
  anti-HIV activity, 213
  circumvention by salvage pathway,
    50
  clinical limitations, 161
  COMPARE analysis, 233*t*–234
  general categories, 248
  human IMPDH, 5–12
  identification by screening method,
    282, 292
  immunosuppressant, 294
  IMP and NAD binding sites, 142*f*,
    157–159
  inhibitor screen, 284–286, 292
  isoform targeted drug design, 49,
    144, 161–163, 201, 214, 252
  molecular target regions, 285*t*
  NAD analogs, 202*t*

342